# Photodetectors Handbook

# Photodetectors Handbook

Edited by **Kate Brown**

CLANRYE
INTERNATIONAL

New Jersey

Published by Clanrye International,
55 Van Reypen Street,
Jersey City, NJ 07306, USA
www.clanryeinternational.com

**Photodetectors Handbook**
Edited by Kate Brown

International Standard Book Number: 978-1-63240-407-7 (Hardback)

Printed in the United States of America.

# Contents

# Preface

Few recent developments in enhancement of photodetectors and photodetection systems for particular applications are covered in this all inclusive book. It consists of theoretical aspects and simulations. It also provides information regarding the optimization of photodetection systems for transfers of time, measurement of vibrations, imaging, particle size analysis, magnetic field, particle energy, and polarization of light. The aim of this book is to serve as a valuable source of reference for researchers, students, and engineers engaged in the field of photonics and advanced technologies.

Significant researches are present in this book. Intensive efforts have been employed by authors to make this book an outstanding discourse. This book contains the enlightening chapters which have been written on the basis of significant researches done by the experts.

Finally, I would also like to thank all the members involved in this book for being a team and meeting all the deadlines for the submission of their respective works. I would also like to thank my friends and family for being supportive in my efforts.

**Editor**

# Part 1

# Theoretical Modelling and Simulations

# Avalanche Process in Semiconductor Photo Detectors in the Context of the Feedback Theory

Vasily Kushpil

*Nuclear Physics Institute of Academy Science of the Czech Republic*
*Czech Republic*

## 1. Introduction

The feedback theory is a quite advantageous tool in the process of description of complex systems' behaviour or the procedure of complex processes. We can characterize the complexity of the system from its structural properties when there is a large amount of bounds between many elements of system. We can also characterize the complexity of the system as its functional complexity when the current state of system is defined as a result of many self-consistent states originating during the evolution of the system. It can be said in most general words that the feedback theory (FBT) describes the behaviour of a system or a process when the current status is defined as a result of achieving the self-consistency of the main system parameters. The feedback theory is most fruitfully applied to the description of those complex systems for which, due to their statistical character it is impossible to construct the full physical-mathematical model with a simple solution, but for which it is easy to select statistical parameters of the system and define their connections with the system properties that are of our interest. This approach allows applying FBT to describe the evolution of complex systems in chemistry, physics, biology, sociology and economics [1].

In this paper, we discuss the avalanche process in semiconductor avalanche photo detectors (APD) proceeding from three different points of view on this process. Then, we sum up the conclusions made in relation to the avalanche process in connection with avalanche photo detectors. We present APD as a converter of photo radiation to electric current with further current amplification. As a result, we can interpret the APD operation from various points that will allow us, finally, both to make several interesting conclusions on their practical application and analyze a possibility to choose optimal parameters in APD manufacturing.

To describe efficiently the processes in the frames of the feedback theory, it is necessary to fetch out the main physical parameters that determine the process procedure and establish quantitative ratios between the character of the process procedure and the change of these parameters. Further, we discuss the main characteristic parameter – the multiplication factor of the avalanche photo detectors M as an applied biasing voltage function (of the material parameters and topology) and attempt to describe the conditions for the achievement of self-consistency with the choice of the feedback coefficient function. It should be mentioned that at present FBT is most widely used in the control theory [2], i.e. in the description of the behaviour of the systems for which it is necessary to achieve the desired character of behaviour. It can be usually attained by the choice of the corresponding system feedback function if the system transfer function without feedback is known. From this point of view, the application of FBT should allow us to approach the designing of avalanche photo detectors from the taken characteristics. We will obtain an opportunity to build the necessary APD transfer function and define the necessary feedback function with it. We have to learn to connect the topology and technology with the parameters of the established Feedback Function (FBF) and then, proceeding from the given FBF, we will be able to obtain topological and technological parameters. The practical application of the described physical-technological algorithm is not discussed in this paper. Our task is to show how it is possible to come from different APD models to their generalized description using FBT.

## 2. Feedback theory basics

Let us regard the main notions of the feedback theory in radio technique. Hereinafter, we will have to describe an avalanche photodiode as an amplifying device with feedback and present it in the form of the corresponding equivalent scheme.

In radio technique the behaviour of a system with feedback is described by the function of a special type – the system function (in a complex form, as the transfer function $H(s)$). Let us discuss a simple scheme of the signal amplifier shown in Fig.2.1 to whose input a signal is given that depends on time $x(t)$, and the signal $y(t)$ is received in the output.

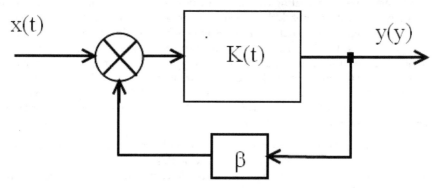

Fig. 2.1. Feedback in amplifier.

The system coefficient of gain function is defined as a ration $K(t)=y(t)/x(t)$. If a part if the signal from the system output is transmitted to the input it is supposed that the feedback is active in the system and the feedback coefficient is defined as $\beta(t)=((x'(t)-x(t))/y(t))$, where $x'(t)$ – is the signal in the system input in the action of the feedback in the system. At $\beta(t)>0$ the feedback is positive; at $\beta(t)<0$ the feedback is negative; at $\beta(t)=0$ there is no feedback in the system. The system gain is written in the system with acting feedback in the form (2.1)

$$K=K_o/(1-\beta*K_o) \tag{2.1}$$

where $K_o$ is gain of the system without the feedback. Negative feedback (NFB) is most often used in electronic systems as in this case the system stability increases. Besides, providing

$K_0 \gg 1$ the gain is defined mainly with the feedback coefficient $K \approx 1/\beta$ that allows a desired coefficient of gain of the system be set by the choice of only one parameter $\beta$.

It should be stressed that the transfer function (1.1) in radio technology is given in frequency representation [3]. Speaking more accurately, it is regarded as a result of Laplace transformations on the complex plane. In this case the transfer function is equivalent to the frequency characteristic of the chain if active sources are absent.

$$H(s)=Y(s)/X(s)=L[y(t)]/L[x(t)] \tag{2.2}$$

The frequency representation for radio technical amplifying systems is more informative and suitable for their designing and study. We will discuss here only the time representation of the system coefficient of gain (amplifying) for avalanche detectors. The reason is the suitability and simplicity of this representation for the description of the avalanche photo detectors' operation and sufficiently simple procedure of measuring of the APD pulse characteristic. The time representation of the system gain is easily obtained if the APD pulse characteristic is known. In this stage, we do not consider the transfer function for APD, not to make the description and comprehension of the idea to describe the process of avalanche multiplication in the FBT frames too complicated.

## 3. Probability model of the avalanche multiplication

Let us consider an idealized model of avalanche multiplication of electrons in a homogeneous semiconductor with the following suppositions: electron multiplication takes place in a limited region with the length $L_a$ with constant electron field voltage $E_a$=const in this region. The probability of multiplication along the $L_f$ electron free flight is P<1. See Figure 3.1.

If there are $N_0$ electrons at the beginning of multiplication in the output of this region there will be $N \gg N_0$ electrons, as a result of multiplication. It can be expressed quantitatively in the form (3.1).

The multiplication factor is defined as $M=N/N_0$

$$N=N_0*(1+P+P^2+P^3+...+P^n) \text{ where } n=L_a/L_f \tag{3.1}$$

For n>>1 this expression can be written in the form

$$N=N_0(1-P)^{-1} \tag{3.2}$$

Formally, let us re-write (2.1) in the form

$$N_x=N_0*(1+(P-\text{ß})+(P-\text{ß})^2+(P-\text{ß})^3+...+(P-\text{ß})^n)=N_0*(1-(P-\text{ß}))^{-1} \tag{3.3}$$

In this expression ß<1 and it describes a probability of electron multiplication suppression along its free flight. From expressions (3.1) and (3.3) we get

$$M=(N_x/N_0)=M_0/(1+\text{ß}*M_0), \text{ where } \qquad M_0=(1-P)^{-1} \tag{3.4}$$

Therefore, we see that with this simple model the physical sense of the feedback coefficient is obvious. It describes a probability of electron multiplication suppression in its free flight.

It is to imagine a situation when to account for the multiplication probability of not only electrons but also holes, we upgrade the expression (3.3) so as $P=P_e+P_h$ and $ß= ß_e+ß_h$, while $P_e$, $P_h$ are multiplicity probabilities for electrons and holes in their free flight length. In this case $N_x=N_e+N_h$.

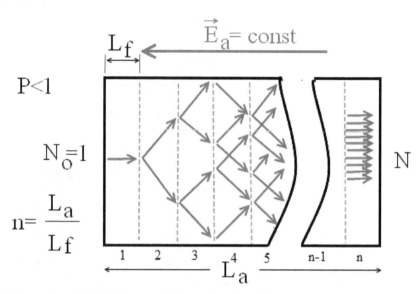

Fig. 3.1. Illustration of Probability model for avalanche process.

This simple model presupposes independence of multiplication processes of electrons and holes. It is not easy to implement it in practice.

$$N_x=N_e+N_h.=N_{0e}*(1-(P_e-ß_e))^{-1}+ N_{0h}*(1-(P_h-ß_h))^{-1} \qquad (3.5)$$

We would like to note that it is the first illustration of moving from a certain definite (in this case, simple probability) model for avalanche photo-detectors to further interpretation of this model in the frames of FBT.

We shall discuss in more detail some not very obvious statements of the suggested model. First, we suppose that all electron multiplication events are independent as a whole. The second supposition comes from the sum row notion in the form of the algebraic expression (3.2). This expression has uncertainty at $P=1$. It leads to a natural restriction $P<1$ and shows that the multiplication probability can be as close to the entity but cannot be equal to it. Both statements are not significant for the main conclusion – the result of the mathematical description of the avalanche generation process can be presented in the form of a formula for the coefficient of gain of the amplifier with feedback after some identical transformations, and, consequently, we can apply it for the analysis of the operation of FBT avalanche photo-detectors. To illustrate some advantages of the interpretation, we consider (3.4) in the case $M_0 \gg 1$. Then $M \sim 1/ß$ and, consequently, the multiplication factor is definitely determined by the feedback coefficient $ß$. In other words, the avalanche suppression processes in APD define the multiplication factor. As such requirements correspond to the operation principle by convention Geiger Mode Avalanche Photo Diode

(GAPD) [4], the probability interpretation of multiplication coefficient has quite a definite meaning. Namely, as GAPD operates in the region where the field intensity is critical and P~1, the avalanche suppressing should be defined with a certain internal process whose probability should be practically constant to provide for low noise of the detector. The cell structure of GAPD is the simplest and most natural way to develop them, as the cell dimensions and the intensity of the electric field in it define the maximal amplifying. The smaller is the dimension of the cell the better are noise properties of GAPD due to electron fluctuations decrease in it. Further, we will continue the discussion of this subject in the phase of more accurate mathematical calculations.

## 4. Physical model of the avalanche multiplication

Let us discuss a classical one-dimensional system of continuity equations for P-N transition shown in Fig. 4.1 [5]

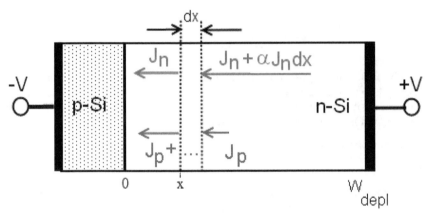

Fig. 4.1. Avalanche multiplications of electrons in high electrical field region of P-N junction.

$$\frac{dJ_n}{dx} = \frac{dJ_p}{dx} = \alpha_n \cdot J_n + \alpha_p \cdot J_p \qquad (4.1)$$

at boundary conditions

$$J_n(0) = J_{no} \qquad J_p(0) = J - J_{po}$$

$$J_n(Z) = J - J_{po} \qquad J_p(W_{depl}) = J_{po}$$

$$J(x) = J_n(x) + J_p(x) = const$$

The solution of this system will be interesting for us in the form of multiplication factor Mn=Jn/Jno for the case of purely electronic injection.

$$M_n = \frac{1}{1 - \int \left( \alpha_n \cdot \exp\left( -\int \left( \alpha_n - \alpha_p \right) dx \right) \right) dx'} \qquad (4.2)$$

if $M_{no}$ is written in the form

$$M_{no} = \frac{1}{1 - \int\left(\alpha_n \cdot \exp\left(-\int\left(\alpha_n\right)dx\right)\right)dx'}$$

(4.3)

It is possible to represent (4.2) in the following form

$$M_n = \frac{M_{no}}{1 + \beta_n \cdot M_{no}}$$

(4.4)

In this expression the feedback coefficient is in the form

$$\beta_n = \int \alpha_n \cdot e^{-A(x)} \cdot \left(1 - e^{-B(x)}dx\right)$$

(4.5)

where

$$A(x) = \int \alpha_n dx \quad \text{and} \quad B(x) = \int \alpha_p dx$$

We can easily obtain the expression for the avalanche multiplication factor in the case of purely "holes" injection, simply changing n to p in formula (4.4) and vice versa. This formal transformation, as it may seem, of a well-known expression makes it possible for us to regard electron and hole multiplication from the FBT point of view.

Let us study some limit cases for formula (4.5).

Firstly, the trivial case $e^{B(x)} = 1$ means that there is no feedback or $B(x) = \int \alpha_p dx = 0$ - there is no multiplication of the hole component. On the other hand, the presence of the holes multiplication process is a factor that means the presence of positive feedback. This fact easily shows the case $e^{B(x)} \gg 1$ when we get from formula (4.5)

$$\beta_n = -\int \alpha_n \cdot e^{-(A(x) - B(x))}dx$$

(4.6)

In the obtained expression, the minus mark indicates the feedback type – namely positive feedback, and the ratio of two components of the exponent shows the FB depth.

Concerning the restrictions on the transfer from representation (4.2) to representation (4.4), they are connected with the necessity to follow the implementation of requirement

$$\int(\alpha_p \exp(-\int \alpha_n)dx)dx' \neq 1$$

(4.7)

For FB coefficient the following condition should be implemented

$$\beta_n \cdot M_n > 1 \qquad \text{or}$$

$$0 \leq \int \alpha_n \cdot e^{-(A(x) - B(x))}dx < \frac{1}{M_{no}}$$

Let us consider the character of possible behaviour of the feedback coefficient in the region of avalanche multiplication. We differentiate (4.6) in coordinate. As a result we get three variants of behaviour.

Case $A(x)>B(x)$ $d\beta n/dx$ – the rate of hole multiplication in the direction of differentiation gradually decreases.

Case $A(x)=B(x)$ $d\beta n/dx$ – the rate of hole multiplication in the direction of differentiation is constant.

Case $A(x)<B(x)$ $d\beta n/dx$ – the rate of hole multiplication in the direction of differentiation gradually increases.

Thus, having done a formally identical transformation of the initial equation, we get an opportunity to mark three characteristic cases for positive FB. The method discussed above will be further applied and allow us to make a clear and generalizing classification of various APD types.

There two reasons that show that there is no need to analyze deeper the obtained expressions. Firstly, our main aim is an acquaintance with a possibility itself to interpret an avalanche process on various levels of abstraction and for various models. Secondly, further studies will have sense if we use a specific APD topology. There exists a simpler model of FBT for avalanche detectors to make general description of some common properties of APD. This model allows one to use a more illustrative and simpler approach, from the point of practical application of APD.

## 5. Miller's formula

It is well-known in APD applications that in practice the use of the physical model to describe a specific APD will be too complex and far from informative. That is why various empiric or semi-empiric models are used. Miller suggested the first model of this type in 1955 [6]. It turns out that the new approach based on the identical transformation of the main formula of this model allows one to obtain sufficiently interesting results.

In the first variant of formula (5.1) the influence on avalanche amplification was not considered of the APD internal base and the contact resistance (the equivalent APD scheme is in Fig. 5.1a).

$$M_{no} = \frac{1}{1-\left(\dfrac{V}{V_{br}}\right)^n} \tag{5.1}$$

Later, the formula was modified and presented in the form (5.2) for equivalent APD scheme shown in Fig. 5.1в.

$$M_n = \frac{1}{1-\left(\dfrac{V-i\cdot R_{fb}}{V_{br}}\right)^n} \tag{5.2}$$

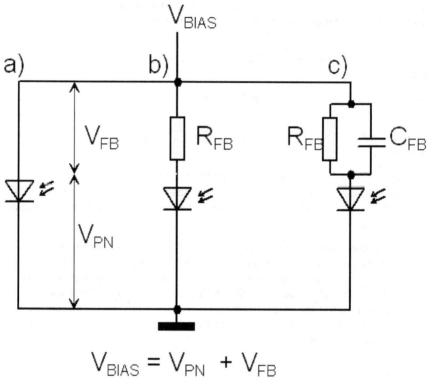

$$V_{BIAS} = V_{PN} + V_{FB}$$

Fig. 5.1. The evolution of equivalent circuit in Miller's formula.

It turns out that the Miller formula can be transformed to the form of expression (2.0). Let us do transformation for formula (5.3) (the equivalent APD scheme is suggested in [7] and shown in Fig. 5.1c).

$$M_n = \frac{1}{1 - \left(\dfrac{V - V_{fb}}{V_{br}}\right)^n} \tag{5.3}$$

In the case under consideration the transformation has sense as the model exactly describes the APD operation with an internal FB. In (5.4) we show the feedback coefficient in the general form to indicate that the feedback value is defined by the change of the voltage $V_{fb}$. On the other hand, in this expression a significant FB difference is clearly seen if the FB resistance consists of purely active component and if it contains the reactive component [8].

$$M(V) = \frac{M_o(V)}{1 - \beta \cdot M_o(V)} \tag{5.4}$$

here $\beta = \dfrac{V_{fb}}{V_{br}} = \dfrac{i \cdot R_{fb}}{V_{br}}$

In the first case external chains of APD connection define the rate of the avalanche multiplication suppression; in the second case, the influence of the reactive component on the rate of the avalanche multiplication suppression is strong. Due to this, the suggested model allows one to describe accurately the operation of APD with internal local negative FB. The FB locality implies that the process of the avalanche suppressing is much quicker than the changes in the parameters of external chains of APD power supply. Thus, we can choose parameters $C_{fb}$ and $R_{fb}$ in such a way that we can accurately describe the APD behaviour with negative local FB. A step further is more complicated as it is necessary to connect the obtained distributed parameters with topological and technological parameters of the APD under study.

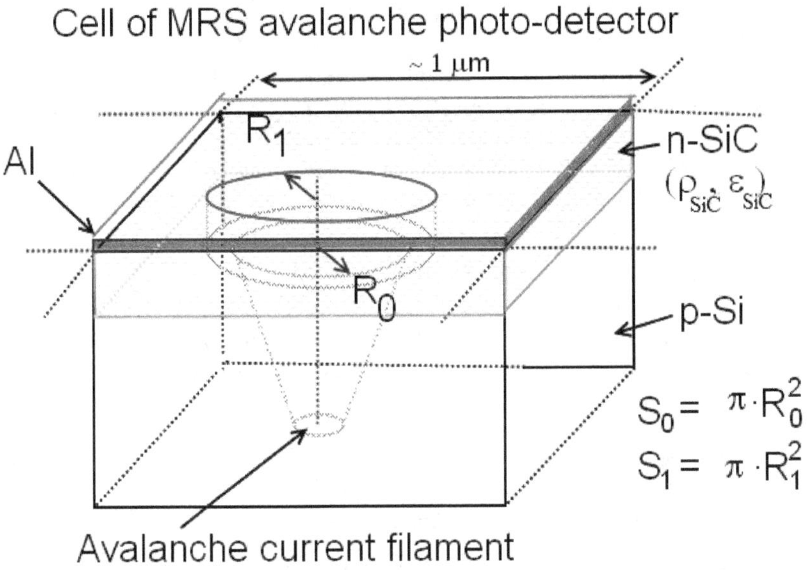

## Cell of MRS avalanche photo-detector

Fig. 5.2. Metal - Resistive Layer - Silicon (MRS APD) cell. The diameter of current filament must be change on boundary n-SiC – p-Si. ($\rho_{SiC}/ \rho_{p\text{-}Si}$)~10

In the discussed example (Figure 5.2) we can proceed from the fact that we have already known the parameters of the resistive layer (resistivity) $\rho_{SiC}$ and area N$^+$ region $S_N$. The value $C_{fb}$ is more significant for the description of the dynamic mode, while for the static one $R_{fb}$ is more important, so we can expect that $R_{fb} = \rho_{SiC} \cdot \dfrac{d_{SiC}}{S_0}$ where $S_0$ determines the cross section area of the avalanche in stationary mode (avalanche current is constant).

To describe the dynamic mode, the value $C_{fb}$ is more significant, while for the static mode $R_{fb}$ is more significant. We can expect here that $R_{fb} = \rho_{SiC} \cdot \dfrac{d_{SiC}}{S_1}$ where $S_1$ determines the dimension of the spreading area of the filament of the current on the boundary $N_{SiC}$-$P_{Si}$, proceeding from the term of achieving the current stationary when M(t)=const, and

$$C_{fb} = \frac{\varepsilon_0 \cdot \varepsilon_{SiC} \cdot S_1}{d_{SiC}}$$

where $S_1$ defines the maximum dimension of the avalanche current from which the avalanche suppression starts.

We can define a characteristic constant $\tau_{fb}$ (5.5) that we will call the time of feedback of the process of avalanche multiplication suppression. This value is determined by the APD technology and topology. It should be noted that the constant does not depend on the resistive layer thickness, but does depend on the spreading area cross-section and the thickness of the avalanche current area cross-section.

$$\tau_{fb} = \varepsilon_0 \cdot \varepsilon_{SiC} \cdot \rho_{SiC} \cdot \frac{S_1}{S_0} = \tau_{SiC} \cdot \frac{S_1}{S_0} \qquad (5.5)$$

It seems obvious that for local negative FB it is necessary to implement the condition $\tau_{fb} \geq \tau_{av}$, namely, the time of FB suppression must be a little longer than the time of the avalanche generation $\tau_{av}$. If we consider the model of the abrupt P-N junction and presuppose that the avalanche region $W_{av} \sim 0.1$ of the thickness of the depletion region $W_{depl}$ and the velocity of the electron movement in the multiplication region $V_{sat}$ we get an evaluation in the form (5.6).

$$\varepsilon_0 \cdot \varepsilon_{SiC} \cdot \rho_{SiC} \cdot \frac{S_1}{S_0} \geq \frac{W_{av}}{V_{sat}} \qquad (5.6)$$

As $V_{br}$ is depend on the silicon parameters [9], and $V_{sat}$=const, we get the evaluation (for the n-SiC/p-Si structure) of the minimal concentration of acceptors in p-Si in the form (5.7).

$$W_{av} \leq V_{sat} \cdot \tau_{SiC} \frac{S_1^2}{S_0^2} \qquad (5.7)$$

$\tau_{SiC} = \varepsilon_0 \cdot \varepsilon_{SiC} \cdot \rho_{SiC}$ — Maxwell time for n-SiC.

This implies (for simplicity we suppose that $S_1=S_0$) that at $\rho_{SiC}$=156 [ohm/cm] the acceptors concentration in silicon must be lower $N_p < 2*10^{15}$ 1/cm³. But at the same time the concentration cannot be too low. Otherwise FB will not have time to suppress locally the avalanche production process. If we take into account that an agreement of this kind is necessary for each new value $\rho_{SiC}$ and, besides, that in the general case $S_1 \neq S_0$ depends on the impurity concentration and conditions on the border n-SiC/p-Si, it becomes clear why the suggested FB mechanism was implemented by only one research group [10].

The presented derivations are one more example of FBT application for the description of APD operation. In the Chapter 6 that considers examples of application of the models based on FBT ideas we discuss the model of the n-SiC/p-Si structure in more detail.

Let us regard the general analysis of formula (5.4)[11]. We mark that in the first approximation we thought that n=1. If we take the general case n>1, the feedback coefficient must be presented in the form (5.8)

$$\beta = \frac{V^n - (V - V_{br})^n}{V_{br}^n} \tag{5.8}$$

We study the character of behaviour of the FB coefficient depending on the voltage changes V.

If we define the avalanche gain efficiency coefficient in the form

$G_{ef} = \frac{dM_0}{dV} - M_0^2 \cdot \frac{d\beta}{dV}$ the derivative on V from (4.4) will be in the form (5.9)

$$\frac{dM}{dV} = \frac{G_{ef}}{(1 - \beta \cdot M)^2} \tag{5.9}$$

If we define the FB efficiency parameter in the form (5.10) it is possible to outline three characteristic cases of FB demonstration in APD in Fig. 5.3.

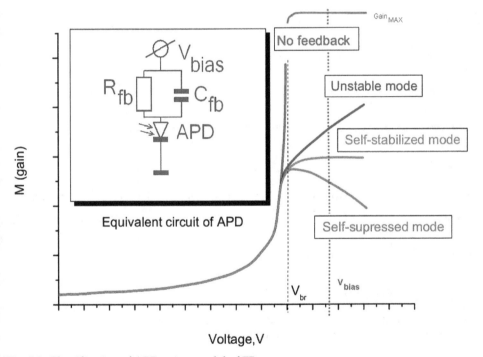

Fig. 5.3. Classification of APD using model of FB.

$$K_{ef} = \frac{\dfrac{d\beta}{dV}}{\dfrac{1}{M_0} \cdot \dfrac{dM_0}{dV}} \tag{5.10}$$

$K_{ef} < 1$ – with the growth of voltage the feedback increases slower than the avalanche gain that corresponds to the APD operation in the unstable mode;

$K_{ef} = 1$ – with the growth of voltage the feedback increases as quickly as the avalanche gain that corresponds to the APD operation in the self stabilization mode;

$K_{ef} > 1$ – with the growth of voltage the feedback increases quicker than the avalanche gain that corresponds to APD operation in the self-suppressing mode.

A classification of this type allows us to define the character of the FB behavior. For some particular cases $\frac{d\beta}{dV} \approx \frac{1}{M_0} \cdot \frac{dM_0}{dV}$ for example we can state:

$\frac{d\beta}{dV} > 0$ FB coefficient monotonously grows with the voltage increase; thus, theoretically, there must exist voltage when, if higher than its value, relative changes of the avalanche gain $\frac{1}{M_0} \cdot \frac{dM_0}{dV}$ are lower than the FB introduced suppressing. Initially, the condition $\frac{d\beta}{dV} < \frac{1}{M_0} \cdot \frac{dM_0}{dV}$ is implemented.

$\frac{d\beta}{dV} = 0$ The FB coefficient is constant with the voltage growth; therefore, theoretically there must be voltage for which relative changes in avalanche gain $\frac{1}{M_0} \cdot \frac{dM_0}{dV}$ correspond to FB introduced suppressing. Initially, the condition $\frac{d\beta}{dV} < \frac{1}{M_0} \cdot \frac{dM_0}{dV}$ is implemented.

$\frac{d\beta}{dV} < 0$ The FB coefficient monotonously decreases with the voltage growth; thus, theoretically there must exist voltage when, if high than its value, relative changes of the avalanche gain $\frac{1}{M_0} \cdot \frac{dM_0}{dV}$ become higher than the FB introduced suppressing. Initially, the condition $\frac{d\beta}{dV} > \frac{1}{M_0} \cdot \frac{dM_0}{dV}$ is implemented.

All the above stated, despite the fact that only cases of quite particular character were considered, can further assist in sorting out correctly the main mechanism of FB for each specific case. And this will be the first and basic step how to connect experimental data with the APD physics, topology and technology.

## 6. Examples of FBT application for APD

### 6.1 Logistic model of front of avalanche in GAPD

In this chapter we consider model that describes the process of avalanche generating in GAPD. The peculiarity of the model is in the fact that we will not solve the Fundamental Equation System for semiconductors and will not actually use the Poisson equation. The dynamics of an avalanche process in GAPD is determined by two important characteristics of this detector. The first is the difference of the bias voltage and breakdown voltage and the second is the dimension and structure of the GAPD cell. Let us discuss a GAPD cell as it is shown in Fig. 6.1.

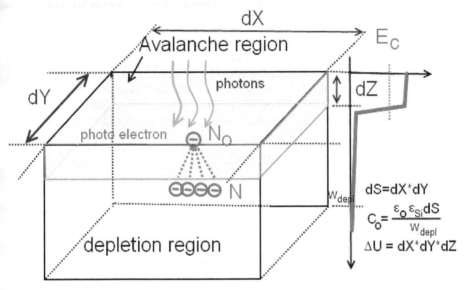

Fig. 6.1.1. Structure of GAPD for model.

The electric field in the cell is close or more to the critical one; therefore, the probability of the avalanche generation of a secondary carrier P~1. Moreover, there exists a certain threshold value of photo-generated electrons $N_o$ necessary for the start of the avalanche process. Thus, to suppress the avalanche in cell from thermal noises for the thermo-generated carriers $N_T$ the $N_T < N_o$ inequation should be implemented. We can evaluate $N_T$ supposing that the noise generates only by the detector dark current. This supposition will allow us evaluate theoretically the dependence of the size of cell on the concentration of the doping applied for the APD material. It is obvious that the higher is the doping concentration the lower is the maximal cell volume $\Delta \bar{U}$ at the same cell area. At backward bias voltage $V_{bias}$, the dark current $J_d$ and equals to $N_T$ will determine the concentration of minor carriers generated per time unit in the depletion region. For APD with a long base, presupposing that the dark current is defined by the volume component from [12] it follows $J_d \approx \dfrac{e \cdot N_i \cdot \Delta U}{2 \cdot \tau_{bulk}}$ Let us consider to be definite a cell of the S area in the form of a abrupt P-N junction on silicon of the N type with the donor concentration $N_d$ and depletion region size $W_{depl}(N_d, V_{br})$ at break-down voltage $V_{br}$.

$$N_T \approx \frac{N_i \cdot \Delta U}{2 \cdot \tau_{bulk} \cdot U_{sat}} \tag{6.1.1}$$

Consequently $N_0 > \dfrac{N_i \cdot \Delta U}{2 \cdot \tau_{bulk} \cdot U_{sat}}$ or $\Delta U = S \cdot W_{depl}(N_d, V_{br}) < \dfrac{2 \cdot \tau_{bulk} \cdot U_{sat}}{N_i}$

It should be noted that, if the threshold value $N_0$ is large, the avalanche probability is small and the front time increases. Big lifetime of the minor charge carriers in SCR (Space-Charge Region) makes it possible to produce GAPD cells of a larger area. After we have evaluated the maximal cell area let us consider the following model.

Let the maximal number of electrons in the cell available for multiplication be equal to N. At the moment of time t after the start of the avalanche multiplication process there are already x electrons; therefore, the number of electrons potentially available for multiplication will be (N-x). If the length of the avalanche gain region is $L_{av}$ and the velocity of electron in the electric field $U_{sat}$=const, (the time of front of avalanche in the order $T_{av} \sim L_{av}/U_{sat}$). The differential equation for the described situation is given in the form (6.1.2).

$$\frac{dx}{dt} = k \cdot x \cdot (N - x) \tag{6.1.2}$$

with the initial requirement $x(t = 0) = N_0$

The solution of this equation is well known as a logistic equation [13].

$$x(t) = \frac{N}{(1 - a \cdot \exp(-k \cdot N \cdot t))} \tag{6.1.3}$$

where $a = \dfrac{N - N_0}{N_0}$

The constant $1/k$ must have time dimensionality, so we input a certain constant of the time of the avalanche front and define it on the basis of the dimension analysis.
$\tau_0 = \dfrac{1}{k} = T_{av} \cdot \dfrac{N_0}{N} = \dfrac{L_{av}}{V_{sat}} \cdot \dfrac{N_0}{N}$ (at $N_0$=1 $\tau_0$=$T_{av}/N$ has a simple physical interpretation – it is the time of one multiplication act, and if the threshold number $N_0$ grows the time of one multiplication act becomes longer). Now we write (5.1.3) in the form

$$x(t) = \frac{N}{(1 - a \cdot \exp(-\frac{t}{\tau_0}))} \tag{6.1.4}$$

To transfer to the interpretation of the solution from the FBT point of view, we introduce new variables. $M(t) = x(t)/N_0$ - the avalanche gain coefficient at the moment of time t, $M_0$=$N/N_0$ - maximal avalanche gain in GAPD.

We write (6.1.4) in the form

$$M(t) = \frac{M_0}{(1 + \beta(t) \cdot M_0)} \tag{6.1.5}$$

where $\beta(t) = (1 - \dfrac{1}{M_0}) \cdot \exp(-\dfrac{t}{\tau_0})$

The model given above describes the exponential change of the FB factor from the maximal value $\beta_{max} = (1 - \dfrac{1}{M_0})$ to zero that corresponds to the maximal current through GAPD.

In Fig. 6.1.2 we show the results of calculations of the avalanche front line for different values of $N_0$. The lower are the threshold values the faster augment the avalanche front. It corresponds to the idea that higher probability of the charge carrier ionization in APD corresponds to the fast front of the avalanche process.

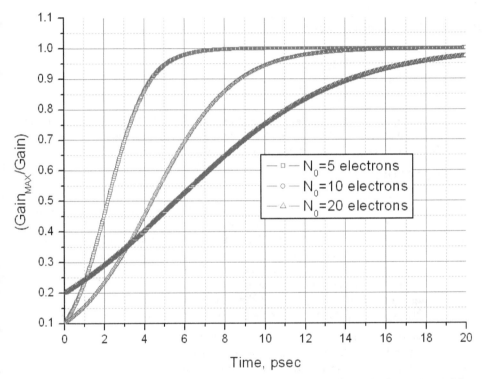

Fig. 6.1.2. Simulation of raising current in avalanche versus time by using Logistic model

## 6.2 GAPD Analysis. FBT interpretation

Paper [11] suggests a model where the feedback factor was given in the form $\beta = \dfrac{V_{fb}}{V_{br}}$.

Proceeding from the supposition that in GAPD the voltage should change by the value $V - V_{br}$ with the avalanche suppressing due to the voltage drop on the amplifying cell to suppress the avalanche, we think it is the very similar FB voltage

$V - V_{br} = V_{fb}$. Let the FB factor be equal to $\beta = V_{fb}/V_{br}$. We consider the charge gain in the GAPD cell whose capacity is $C_0$. The cell discharge charge is $Q_0 = C_0 * V_{fb}$. If the threshold number of electrons $N_0$, the avalanche gain coefficient of the charge $M_x \sim Q_0/e * N_0 = C_0 * V_{fb}/e * N_0$. On the other hand, we can define the voltage of a separate gain at $V=V_{br}$ when $\beta\sim1$, so the choice $\beta=\dfrac{V_{fb}}{V_{br}}$ will be justified if $M_0$ is chosen in the corresponding way. From the expression (6.2.1) we obtain the value $M_0$.

$$M = \frac{C_0 \cdot V_{fb}}{e \cdot N} = \frac{M_0}{1 + \dfrac{V_{fb}}{V_{br}} \cdot M_0} \qquad (6.2.1)$$

The result of the solution is given by way of the charge gain coefficient.

$$M_0 = \frac{C_0 \cdot V_{fb}}{e \cdot N_0 - C_0 \cdot V_{fb} \cdot \dfrac{V_{fb}}{V_{br}}} = \frac{Q_0}{Q_i + \dfrac{V_{fb}}{V_{br}} \cdot Q_0} \qquad (6.2.2)$$

In this expression $Q_0$ is an intensified charge from the cell and $Q_i$ is the initial charge that switches the avalanche in GAPD. Thus, we have managed to present the FB factor in the desired form and, at the same time, keep the general kind of the expression for the avalanche gain in the form $M = \dfrac{M_0}{1 + \beta \cdot M_0}$.

Let us discuss now the feedback model when the avalanche restrictions happen due to limitation of the charge available for the avalanche multiplication in the cell. As the main parameter is the type and concentration of GAPD material doping, we will try to connect our abstract model with silicon parameters. The maximal charge available for multiplication in the cell is defined by the cell and SCR dimensions and equals to $Q_{max}=C_0*V_{fb}\sim N_a^{1/2}$ (we suppose that the basic material is silicon of the N type with donor concentration $N_d$). The possible charge of the avalanche gain $Q_{av}$ is defined by the electric field intensity in the gain area, $Q_{av}=e*N_0*M\sim N_a^m$ where $m>1.2$ than the length of the avalanche region.

As the change of the field intensity per length unit in low-ohm material is higher than in high-ohm material, the rate of the avalanche gain change in the low-ohm material in the SCR avalanche part is higher and the avalanche generates in smaller space. It leads to its greater instability and imposes strict requirements to the time of FB establishment, as it was indicated above (formula (5.7)). In this case the FB mechanism due to the limitation of the charge available for gain in the cell is more preferable.

There should exist conditions when FB due to charge quantity limitation in the cell is possible and this condition will be $Q_{av}\geq Q_{max}$. In carrying out this inequation, the maximal charge is defined by the cell geometry and not by electric field intensity in it. Meeting this requirement will allow one to develop GAPD with high field intensity without fear of destroying the cell with high currents due to local instability of the avalanche process.

Let us study a simple model shown in Fig. 6.1 the abrupt P-N junction on the N-type silicon. If $N_d>>N_a$ the dimension $W_{depl}$ of SCR at the bias voltage equal to breakdown voltage is known and the maximal charge accumulated in the cell at the excess of breakdown voltage by $\Delta V=V-V_{br}$ is defined from equation (6.2.3). Let us suppose that the region of the avalanche multiplication is $W_{av} \sim 0.1 * W_{depl}(V_{br})$, and $R_o$ – is the dimension of the region of avalanche current filament. The maximal cell charge is:

$$Q_{max} = C_0(N_d, V_{br}) \cdot \Delta V \approx \frac{\varepsilon_0 \cdot \varepsilon_{Si} \cdot S}{W_{depl}(V_{br})} \cdot \Delta V = \frac{\varepsilon_0 \cdot \varepsilon_{Si} \cdot \pi \cdot R_0^2}{\sqrt{\dfrac{2 \cdot \varepsilon_0 \cdot \varepsilon_{Si} \cdot V_{br}}{e \cdot N_d}}} \cdot \Delta V \qquad (6.2.3)$$

To calculate the maximally possible charge of the avalanche gain let us suppose that the electric field in the region of the avalanche gain $W_{av}$ is constant E=const and the ionization coefficient $\alpha(E)$ is known. We think that an avalanche appears as a micro-plasma current filament of a constant and independent of material thickness and $S_0/S_1 = 1$ (see chapter 6). Then, the maximally achievable charge of the avalanche multiplication can be calculated from equation (6.2.4).

$$Q_{av} \approx e \cdot N_0 \cdot \alpha(E) \cdot S_1 \cdot W_{av} = e \cdot N_0 \cdot \alpha \cdot (-\frac{E_c}{E}) \cdot S_1 \cdot W_{av} \qquad (6.2.4)$$

In the expression $\alpha_0$, $E_c$ the model parameters and e- is the electron charge. Let us suppose that the field intensity E is equal to the maximal for the given PN junction at voltage $V = V_{br}$.

$$E = E_{max} = \frac{e \cdot N_d \cdot W_d(N_d, V_{br})}{\varepsilon_0 \cdot \varepsilon_{Si}}$$

As a result, we obtain expression (6.2.5)

$$Q_{av} = e \cdot N_0 \cdot \alpha \cdot (-\frac{\varepsilon_0 \cdot \varepsilon_{Si} \cdot E_c}{e \cdot N_d \cdot W_d(N_d, V_{br})}) \cdot S_1 \cdot W_{av}(N_d, V_{br}) \qquad (6.2.5)$$

where $W_{av}(N_d, V_{br}) \approx 0.1 \cdot \sqrt{\dfrac{2 \cdot \varepsilon_0 \cdot \varepsilon_{Si} \cdot V_{br}}{e \cdot N_d}}$

In Fig. (6.2.1) and (6.2.2) calculation results are presented in the given formulas (6.2.3) and (6.2.5) for $N_0 = 1$, $R_0 = 2,3,5,7,10,15$ um, $S_1 = 1$мкм$^2$ and $\Delta V = 0.1, 0.25, 0.5, 1.0, 2.0, 3.0, 5.0$ V. Although the dimension of the current filament is not taken into account in calculations of the dependence that shows the maximal charge change of the avalanche gain $Q_{av}(N_d, W_{av}(N_d, V_{br}))$ achievable in multiplying, while it is easy to make the corresponding correction, the main thing is to connect it rightly with real GAPD topology and technology. The observed decrease of the charge at high $N_d$ concentrations is explained by the effect of decrease $W_{av}(N_d, V_{br})$.

In Fig. (6.2.1) the red line shows how the maximal charge $Q_{av}(N_d, W_{av}(N_d, V_{br}))$ achievable in multiplying changes if the donor concentration in the base silicon changes in the limits from $10^{12}$ cm$^{-3}$ to $10^{16}$ cm$^{-3}$, and the straight lines show how the available charge in the cell changes $Q_{max}(N_d, W_{av}(N_d, V_{br}, R_o))$ at various values of the avalanche region radius $R_o$ and $\Delta V = 1$. The region where the feedback mechanism operates to restrict the avalanche at the expense of the limitation of the charge available for gain in the cell is defined by the inequation $Q_{av} \geq Q_{max}$.

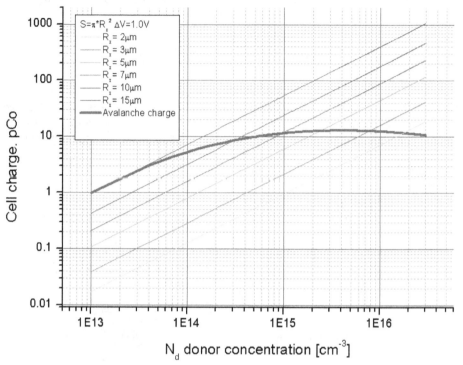

Fig. 6.2.1. The selection optimal size of cell for GAPD. The concentration of donors in cell must be selected near cross point of red line of maximum avalanche charge and the lines of different radius of cell.

In Fig. (6.2.2) the red line shows how the maximal charge $Q_{av}(N_d, W_{av}(N_d, V_{br}))$ changes if it is attained in multiplying and the donor concentration in the base silicon changes in the limits from $10^{12}$ cm$^{-3}$ to $10^{16}$ cm$^{-3}$, while the straight lines show how the charge available in the cell changes $Q_{max}(N_d, W_{av}(N_d, V_{br}, R_0))$ at various voltages $\Delta V$ if $R_0 = 5$. The region of application of the feedback mechanism to restrict the avalanche at the expense of the limitation of the charge available for gain in the cell is defined by the inequation $Q_{av} \geq Q_{max}$.

Let us discuss an example of application of the obtained dependences. If we choose silicon with $N_d = 10^{16}$ cm$^{-3}$ at the expected $\Delta V = 1$ the size $R_0$ of the charge region available for multiplication should be less than 3 μm (for example, 2 μm) and the mode is implemented when the maximal avalanche gain is restricted by the cell geometry and not by the intensity of the cell electric field. At $\Delta V = 2$ the size $R_0$ of the charge region available for multiplication should be less than 2 μm. In another Figure we see that if we choose silicon with $N_d = 10^{16}$ cm$^{-3}$ at the chosen $R_0 = 5$ the excess voltage $\Delta V$ for the charge multiplication should be less than 0.5 (for example, 0.25); if $R_0 = 2$ we can choose $\Delta V = 2$ but not larger. The discussed examples show that the implementation of the restriction mode of the charge in the cell is quite complicated due to the complex interconnections of many parameters. It is necessary to calculate the GAPD construction from the point that the emerged avalanche will discharge the cell to the full extent and switch off due to the carriers' deficit, with no unstable unlimited by anything avalanche process. In practice, the discussed above mechanisms of avalanche quenching are implemented simultaneously, both in the form of decreasing the voltage in the cell and achieving depletion of free multiplication carriers in the avalanche region. To answer this question it is necessary to conduct more elaborate studies.

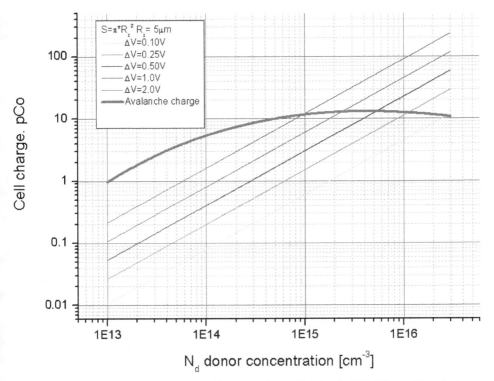

Fig. 6.2.2. The selection optimal over voltage $\Delta V = (V_{bias} - V_{br})$ for GAPD. The red line is the maximal avalanche charge can be the limit for over voltage for given GAPD with fixed size and concentration donor in cell.

### 6.3 APD equivalent scheme – SPICE model

A model to shape an avalanche front in APD on the basis of n-SiC/p-Si P-N junction is suggested for application in paper [11]. The form of the FB factor (its time dependence during the front shaping) was obtained on the basis of experimental data, with an initial supposition that the n-SiC layer properties determine the characteristic time of FB establishment.

The full equivalent schemes of MRS APD and Micro Channel Avalanche Photo Diode (MCAPD) are shown in Fig. 6.3.1. We are interested in one of its elements – a nonlinear source of current I(t) (APD) controlled by current ($R_{fb}$, $D_{fb}$, $C_{fb}$). The source amplifies the photocurrent $I_0(t)$ M times and by a certain time dependence of the FB factor $\beta(t)$ it correctly describes the shaping of the APD front.

Let us consider the method of definition of time dependence of the FB factor to be applied in the further suggested model. The method can be applied to linear APD, but not to GAPD.

Let the APD under study be under the effect of a rectangular light pulse that generates current $I_0(t)$ in APD. An intensified current pulse I(t) can be given in the form

$$I(t) = I_0(t) \cdot \frac{M_0}{1 + \beta(t) \cdot M_0} \tag{6.3.1}$$

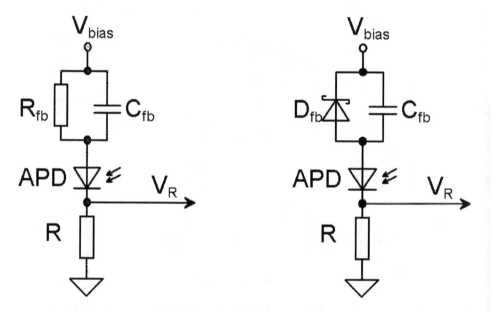

Fig. 6.3.1. Two different model were investigated for analysis by SPICE. APD - is source of current that controlled by voltage. $R_{fb}$ - linear element of FB can be used for MRS APD [14], the $D_{fb}$ is not linear element of FB and can be used for MC APD [15].

We suppose that $M_0$=const does not depend on time $M_0 \neq M_0(t)$ and only $\beta = \beta(t)$.

Then it is easy to obtain the dependence for the FB factor from (5.3.1)

$$\frac{I_0(t)}{I(t)} = \beta(t) + \frac{1}{M_0} \tag{6.3.2}$$

As we suppose that $M_0^{-1}$=const any time dependence of the FB factor is obviously reflected by equation (6.3.2).

For test of method were  defined of FB factors for APD of two types described in papers [14] and [15].

The dependences that correspond to the definition are presented in Table 6.1.1 as fragments of the SPICE code and the simulation result and its comparison with experimental data is given in Fig. 6.3.2 and 6.3.3.

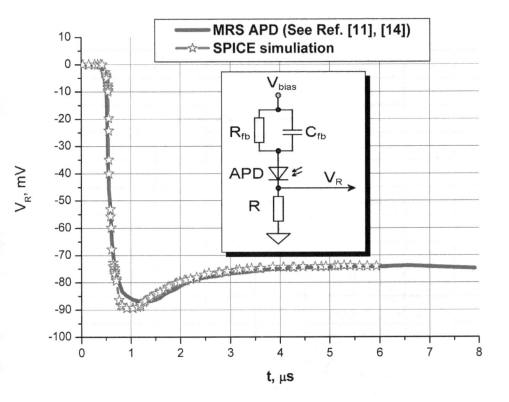

Fig. 6.3.2. Simulation result for MRS APD [11], [14].

| Model MRS | $R_{fb}$=1K | $C_{fb}$=1200pF | R=1K | |
|---|---|---|---|---|
| Model MP APD | $C_{fb}$=5pF | $D_{fb}$(CJO)=10pF | $D_{fb}$(BV)=100mV | $D_{fb}$(IBV)=20nA |

Table 6.1. (Parameters for simuliation)

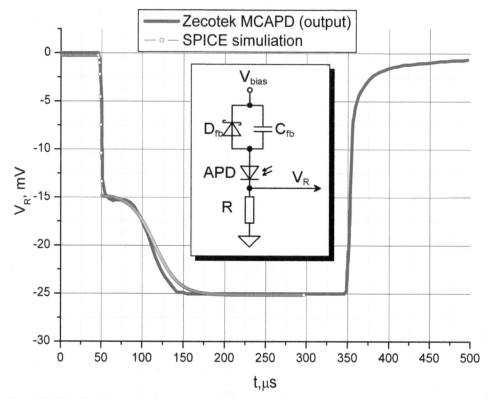

Fig. 6.3.3. Simulation result for MCAPD [15].

**APD SPICE model:**

*.FUNC DPWR(D) {I(D)\*V(D)}*
*.FUNC BPWR(Q) {IC(Q)\*VCE(Q)+IB(Q)\*VBE(Q)}*
*.FUNC FPWR(M) {ID(M)\*VDS(M)}*
*.FUNC HOTD(D,MAX) {IF((V(D)\*I(D)>MAX),1,0)}*
*.FUNC HOTB(Q,MAX) {IF((VCE(Q)\*IC(Q)+IB(Q)\*VBE(Q)>MAX),1,0)}*
*.FUNC HOTF(M,MAX) {IF((VDS(M)\*ID(M)>MAX),1,0)}*
*.PARAM LOW3MIN={IMPORT(LOW3MIN.OUT,LOW3THRES)}*
*.PARAM HIGH3MAX={IMPORT(HIGH3MAX.OUT,HIGH3THRES)}*
*.PARAM LOWLVDS={IMPORT(LOWLVDS.OUT,LOWLIMIT)}*
*.PARAM HILVDS={IMPORT(HILVDS.OUT,HILIMIT)}*
*.PARAM LIMTLVDS={IMPORT(LIMTLVDS.OUT,LVDSLIMITS)}*
*.FUNC SKINAC(DCRES,RESISTIVITY,RELPERM,RADIUS) {((PI\*RADIUS\*RADIUS)/*
*((PI\*RADIUS\*RADIUS)-PI\*(RADIUS-*
*SKINDEPTHAC(RESISTIVITY,RELPERM))\*\*2))\*DCRES}*
*.FUNC SKINDEPTHAC(RESISTIVITY,RELPERM)*
*{503.3\*(SQRT(RESISTIVITY/(RELPERM\*F)))}*
*.FUNC SKINTR(DCRES,RESISTIVITY,RELPERM,RADIUS,FREQ) {((PI\*RADIUS\*RADIUS)/*

*((PI\*RADIUS\*RADIUS)-PI\*(RADIUS-*
*SKINDEPTHTR(RESISTIVITY,RELPERM,FREQ))\*\*2))\*DCRES}*
*.FUNC SKINDEPTHTR(RESISTIVITY,RELPERM,FREQ)*
*{503.3\*(SQRT(RESISTIVITY/(RELPERM\*FREQ))))}*
*BG1 0 PINA I = {V(IN)\*POWER((40/(40-(V(0)-V(PINA)))),1)*

## 7. Conclusion

The FBT approach allows to hide information about real physics processes in different types of APD. We can exchange a physical model describing a complex behavior of the system by a simple universal model with predictable results. The basic steps are in definition of main statistical parameters and their connections with external conditions of APD. Practically in all cases it is possible to implement a formal transfer from the physical model of an avalanche generation to the description in FBT. The formal separation of processes into those of avalanche multiplication and avalanche suppression or restriction allows a simpler understanding of the physical origin of the phenomena. Based on FBT notions, it is possible to construct simple enough model to describe processes of avalanche gain. This model can be connected with important parameters of APD under study and allows to determine the link of some technological parameters with APD properties.

Application of the FBT approach to construct the SPICE model allows one to facilitate the procedure of definition of the equivalent APD circuit main parameters and correctly describe the peculiarities of the front signal shaping.

## 8. References

[1] F.L. Lewis, (1993). Applied Optimal Control and Estimation, Prentice-Hall.
[2] John Doyle, Bruce Francis, Allen Tannenbaum, (1990). Feedback Control Theory, Macmillan Publishing Co.
[3] William McC. Siebert, (1986). Circuits, signals, and systems, Published by MIT Press, McGraw-Hill in Cambridge, Mass, New York .
[4] K. Jradi, et.al., (2011). Computer-aided design (CAD) model for silicon avalanche Geiger mode systems design: Application to high sensitivity imaging systems, Nuclear Instruments and Methods in Physics Research Section A. Vol. 626-627, pp. 77-81
[5] S. M. Sze, (1981). Physics of Semiconductor Devices, New York: Wiley.
[6] S. L. Miller, (1955). Avalanche breakdown in germanium, Phys. Rev. 99, pp. 1234-1241
[7] D. Bisello, et al., (1995). Metal-Resistive layer-Silicon (MRS) avalanche detectors with negative feedback, Nuclear Instruments and Methods in Physics Research Section A, Vol. 360, Issue 1-2, pp.83-86
[8] F. Zappa, A. Lacaita, C. Samori, (1997). Characterization and Modeling of Metal-Resistance-Semiconductor Photodetectors, IEEE Trans. On Nucl. Sc., Vol. 44, No. 3
[9] McKay, K. (1954). Avalanche Breakdown in Silicon, Physical Review 94:pp. 877-880
[10] Z. Sadygov et al., (1991). Proc. SPIE. Optical Memory and Neural Networks, pp. 158-168
[11] V. V. Kushpil, (2009). Application of simple negative feedback model for avalanche photodetectors investigation, Nuclear Instruments and Methods in Physics Research Section A, Vol. 610, Issue 1, pp.204-206
[12] (2003). Encyclopedia of Optical Engineering, pp. 128-130

[13] Gershenfeld, Neil A. (1999).The Nature of Mathematical Modeling. Cambridge, UK: Cambridge University Press.

[14] A.V Akindinov, et al., (1997). New results on MRS APD, Nuclear Instruments and Methods in Physics Research Section A, Vol. 387, pp.231-234

[15] Z. Sadygov et al., (2010). Microchannel avalanche photodiode with wide linearity range, arXiv.org>physics>ArXiv:1001.3050

# 2

# Mathematical Modeling of Multi-Element Infrared Focal Plane Arrays Based on the System 'Photodiode – Direct-Injection Readout Circuit'

I.I. Lee and V.G. Polovinkin
*A.V. Rzhanov Institute of Semiconductor Physics Siberian Branch of*
*Russian Academy of Sciences*
*Russia*

## 1. Introduction

The direct-injection readout circuit was proposed in 1973 by A. Steckl and T. Koehler (Steckl & Koehler, 1973). With such readout circuits, first hybrid multi-element InfraRed Focal Plane Arrays (IR FPAs) were implemented (Steckl, 1976; Iwasa, 1977). The direct-injection readout circuits are being used in the majority of hybrid multi-element Long-Wave InfraRed (LWIR) FPAs, which hold more-than-70% a share in the world market of thermography systems (Rogalski, 2000).

A diagram of FET-based readout circuit is shown in Fig.1a, the most popular design involving a charge-coupled device, Fig.1b (Longo, 1978; Felix,1980; Takigawa, 1980; Rogalski, 2000). On application of a dc voltage $U_G$ to the input gate, a certain voltage, defined by the surface potential under the input gate of the photodetector channel, sets across the photodiode, and the current generated in the IR detector is integrated at the storage capacitor $C_{int}$. On applying of a transfer pulse Ft (Fig.1a), the charge accumulated in the storage capacitor transfer to the column read bus.

The equivalent circuit of the direct injection readout circuit is shown in Fig. 2.

Normally, the current through the photodiode is in the range of $10^{-7} \div 10^{-10}$A, the input FET being therefore operated in subthreshold mode. The transconductance of the input FET channel is:

$$g_{in} = \frac{\partial I_{in}}{\partial V_G} = \frac{qI_{in}}{N^*kT} \tag{1}$$

Here, q is the electron charge, k is the Boltzmann constant, $N^* = (C_{OX}+C_D^*+C_{SS}^*)/C_{OX}$ ($C_{OX}$ is the specific capacitance of gate dielectric, $C_{SS}^*=qN_{ss}^*$ is the specific capacitance of fast surface states, $C_D^*$ is the specific capacitance of the depletion region, T is temperature, and the asterisk * indicates that the parameter value is taken for the conditions with surface Fermi level equal to 3/2 $\varphi_{FB}$.

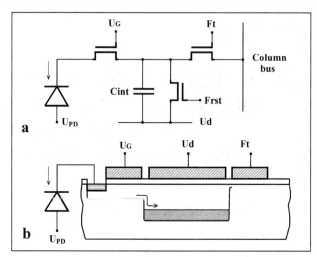

Fig. 1. Readout circuit with charge direct injection.

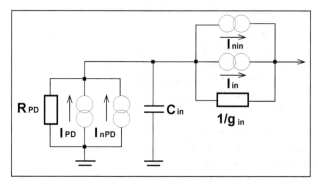

Fig. 2. The equivalent circuit of the photodetector channel of a hybrid IR FPA based on the system 'photodiode – direct-injection readout circuits'. Here, $I_{PD}$ is the current that flows through the photodiode, $I_{nPD}$ is the noise current that flows through the photodiode, $R_{PD}$ is the dynamic resistance of the photodiode, $I_{in}$ is the current integrated under the storage gate of the readout circuit, $I_{nin}$ is the noise current integrated under the storage gate, and $C_{in}$ is the capacitance of the input direct injection.

From the equivalent circuit, all basic relations for the readout circuit of interest can be inferred. In particular, for injection efficiency $\eta_I$ we obtain (Longo, 1978; Felix,1980):

$$\eta_I = \frac{g_{in}}{g_{PD} + g_{in} + j\omega C_{in}} \qquad (2)$$

where $g_{PD} = 1/R_{PD}$, $R_{PD}$ is the dynamic resistance of the photodiode.

Analysis of transfer characteristics of the system 'photodiode – direct-injection readout circuit' has revealed several drawbacks inherent to such readout circuits (Longo, 1978; Felix,1980; Takigawa, 1980; Rogalski, 2000):

- an IR FPA responsivity value comparable with that of IR photodiodes operated in photovoltaic regime can be reached only on the condition that the injection efficiency approaches unity. This requirement entails rigid requirements on the dynamic resistance of photodiodes, these requirements often being hard to meet, especially in LWIR FPAs;
- since the bias voltage across the photodiode is defined by the surface potential of the input-gate modulated channel, for multi-element IR FPAs the variations of transfer characteristics (fixed-pattern noise) and the excess noise level proved to be largely defined by the spread of input-FET threshold voltages $\Delta V_{th}$ (Longo, 1978; Gopal, 1996).

Several circuit designs were proposed to eliminate the above drawbacks. For instance, R. Bluzer and R. Stehlik (Bluzer & Stehlik,1978) proposed to use, with the aim of reducing the input impedance of readouts, the so-called buffered direct injection, implemented through introduction of an inverting amplifier in between the input direct injection and input FET gate.

J. Longo et al. (Longo, 1978) proposed a design of direct injection readout circuit capable of diminishing the spread of photodiode bias voltages to the typical difference of threshold voltages of two closely spaced FETs. Nonetheless, being capable of improving certain characteristics of direct injection readouts, the additional components (the amplifier normally involves 4 to 6 FETs) take off a substantial fraction of surface area from the photodetector cell, doing simultaneously the direct injection readout circuits out of their main advantageous features, low energy consumption and a large storage capacity, high in comparison with readout circuits of other types.

The advances in silicon technology and in the technology of IR photodiodes based on narrow bandgap semiconductors have allowed a substantial increase of photodiode dynamic resistance and a reduced variation of photodiode bias voltages across FPA from the values ~ 20-50 mV, typical of the technology level of the 1970s, to ~ 1-10 mV. For this reason, as a rule, the simplest design of direct injection readout circuit shown in Fig.1 is mainly used.

An analysis of the system 'photodiode – direct injection readout circuit' given in (Steckl & Koehler 1973; Steckl, 1976; Iwasa, 1977; Rogalski, 2000; Longo, 1978; Felix,1980; Takigawa, 1980; Gopal, 1996) was carried out versus photodiode biasing, although the bias voltage across the photodiode is defined by the surface potential in the input-FET channel and can be adjusted through the proper choice of the potential $U_G$ applied to the input-FET gate. That is why such estimates do not allow the prediction of responsivity of multi-element IR FPAs and evaluation of fixed-pattern noise level and Noise-Equivalent Temperature Difference (NETD) figures of thermography systems based on such IR FPAs. In other words, such estimates do not permit numerical modeling of multi-element IR FPAs.

In the present article, we propose a mathematical model of the system 'photodiode – direct injection readout circuit', a computer program, and an analysis procedure for thermography systems based on multi-element IR FPAs; these model and analysis procedure were developed to solve the above-indicated problems. We describe a procedure for revealing the effects due to the spread of electrophysical parameters of photodiodes and Si readouts based on plotting calculated fixed-pattern noise, detectivity, and NETD histograms of thermography systems based on multi-element IR FPAs. It is significant that all dependences were obtained as dependent on the input-gate voltage of readouts; this enables a comparison with the experiment and formulation of requirements to electrophysical and design parameters of photodiodes and Si readouts necessary for implementation of design characteristics.

## 2. A mathematical model for the system 'IR photodiode - direct-injection readout circuit'

Basic assumptions adopted in the proposed mathematical model of the system 'IR photodiode – direct injection readout circuit' were first formulated in (Kunakbaeva, Lee & Cherepov, 1993) and further developed in (Kunakbaeva & Lee, 1996; Karnaushenko;  Lee et al., 2010; Gumenjuk-Sichevska, Karnaushenko, Lee & Polovinkin, 2011). In calculating the transfer characteristics of the system, for the input FET we employ the model of long-channel transistor in weak inversion (Overstraeten, 1975). The noise charge Q(t) integrated under the storage gate of the direct injection readout circuit is calculated as a McDonalds function expressed in terms of spectral current density $S_i(\omega)$ (Buckingham, 1983):

$$\overline{Q^2}(t) = \frac{1}{\pi}\int_0^\infty \frac{S_i(\omega)}{\omega^2}(1-\cos\omega t)d\omega \tag{3}$$

The spectral density of the noise current in the input-gate modulated channel is given by

$$S_i(\omega) = 4kTg_{in}\alpha_1\left|1-\eta_I\right|^2 + \frac{2\pi B_{in}I_{in}^2}{\omega}\left|1-\eta_I\right|^2 + \left(2qI_{PD} + \frac{4kT\alpha_2}{R_{PD}}\right)\left|\eta_I\right|^2 + \frac{2\pi B_{PD}}{\omega}\left|\eta_I\right|^2 \tag{4}$$

The first and second terms in (4) stand for the thermal noise of the input FET and for the $1/f$ noise induced by this FET (here, $\alpha_1$ and $\alpha_2$ are coefficients)  and the third and forth terms stand for the thermal noise and $1/f$ noise of the photodiode, here, $B_{in}= 2\pi KI_{in}^2/WL(C_{OX}+C_D^*)^2$, K is a coefficient (D'Souza, 2002), and $B_{PD} = \alpha_3\left(I_{PD}-\eta_K I_P\right)^2$, $\alpha_3$ is a coefficient (Tobin, 1980). Substituting the first term of (4) into (3), after integration we obtain the following analytical expression for $Q_1$ - the thermal noise of the gate-modulated channel expressed as the amount of noise electrons:

$$Q_1^2 = \frac{2kTC_{in}g_{in}R_{PD}\alpha_1}{q^2(1+g_{in}R_{PD})}\left[1-\exp\left(-T_{in}\frac{1+g_{in}R_{PD}}{R_{PD}C_{in}}\right)\right] +$$
$$+ \frac{2kTg_{in}\alpha_1}{q^2(1+g_{in}R_{PD})^2}\left[T_{in} - \frac{R_{PD}C_{in}}{1+g_{in}R_{PD}}\left(1-\exp\left(-T_{in}\frac{1+g_{in}R_{PD}}{R_{PD}C_{in}}\right)\right)\right] \tag{5}$$

where $T_{in}$ is the integration time.

In estimating the $1/f$ noise of the input FET $Q_2$, also expressed as the amount of noise electrons, while performing integration, we have to take the fact into account that the readout regime involves a high-pass filter with transfer characteristic $\omega^2/(\omega_0^2 + \omega^2)$, where $\omega_0 = \pi/T_{in}$.

$$Q_2^2 = \frac{2KI_{in}^2}{qWL(C_{OX}+C_D)^2} \int_0^\infty \frac{1+(C_{in}R_{PD}\omega)^2}{(1+g_{in}R_{PD})^2+(C_{in}R_{PD}\omega)^2} \frac{1}{\omega(\omega_0^2+\omega^2)} \sin^2\frac{\omega T_{in}}{2} d\omega \qquad (6)$$

We substitute the third term of (4) into (3) and obtain, after integration, the following analytical expression for $Q_3$ - the noise component due to photodiode current noise expressed as the amount of noise electrons:

$$Q_3^2 = \frac{(g_{in}R_{PD})^2}{q^2(1+g_{in}R_{PD})^2}\left(qI_{PD}+\frac{2kT\alpha_2}{R_{PD}}\right)\left(T_{in}-\frac{R_{PD}C_{in}}{1+g_{in}R_{PD}}\left(1-\exp\left(-T_{in}\frac{1+g_{in}R_{PD}}{R_{PD}C_{in}}\right)\right)\right) \qquad (7)$$

The forth term in (4) stands to allow for the 1/f noise of the photodiode:

$$Q_4^2 = 2\alpha_3^2(I_{PD}-\eta_K I_P)^2 \frac{(g_{in}R_{PD})^2}{q^2}\cdot$$
$$\cdot\int_0^\infty \frac{1}{(1+g_{in}R_{PD})^2+(C_{in}R_{PD}\omega)^2}\frac{1}{\omega(\omega_0^2+\omega^2)}\sin^2\frac{\omega T_{in}}{2}d\omega \qquad (8)$$

The detectivity $D^*$ of the photodetector channel in which photosignals from the IR photodiode are read into the direct injection input is given by (Rogalski, 2000):

$$D^* = \frac{\lambda}{\hbar c}\frac{(A_{PD}T_{in}/2)^{1/2}\eta_I\eta_K}{(Q_1^2+Q_2^2+Q_3^2+Q_4^2+Q_{other}^2)^{1/2}} \qquad (9)$$

where $A_{PD}$ is the photodiode area and $Q_{other}$ stands for all other noise components such as, for instance, the measuring channel induced noise.

## 3. Examples of model calculations

### 3.1 Modeling procedure for multi-element IR FPAs with direct injection readout circuits

The modeling procedure for the system 'IR photodiode – direct injection readout circuits' can be described using the simple IR photodiode model:

$$I_{PD} = \eta_K I_P + I_0\left[1-\exp(-\beta V_{PD})\right]+\frac{V_{PD}}{R_S} \qquad (10)$$

Here, $\eta_K$ is photodiode quantum efficiency, $I_0$ is the photodiode saturation current, $R_S$ is the photodiode shunt resistance, and $I_P$ is the current induced by background radiation.

The calculation starts with specifying the values of electrophysical and design parameters of readout circuits, photoelectrical parameters of photodiodes, and operating conditions of IR FPA, which all are listed in Table. 1.

At a given input gate voltage $V_G$, the voltage across the photodiode can be determined from the condition $I_{in} = I_{PD}$. We identify the points of intersection of the current-voltage characteristic of the photodiode and the transfer characteristics of the input FET to determine the photodiode bias voltage.

| Designation | Parameter | Numerical value |
|---|---|---|
| $\mu$ | Mobility of minorities in the inversion channel of FET | 500 cm$^2$ V$^{-1}$·c$^{-1}$ |
| $N_D$ | Donor concentration in the substrate | 7·10$^{14}$ cm$^{-3}$ |
| $C_{OX}$ | Specific capacitance of gate dielectric | 1.24·10$^{-7}$ F·cm$^{-2}$ |
| $V_{FB}$ | Flat-band voltage | 0 V |
| $N_{ss}^{*-}$ | Surface-state density | 1·10$^9$ cm$^{-2}$·eV$^{-1}$ |
| W, L | Length and width of input-gate modulated channel | 30 μm, 3 μm |
| $C_{in}$ | Input capacitance of FPA cell | 0.5 pF |
| $\alpha_1$ | Numerical coefficients | 2 |
| $\alpha_2$ | Numerical coefficients | 2 |
| $\alpha_3$ | Numerical coefficients | 10$^{-3}$ |
| K | Numerical coefficients | 1.5·10$^{-24}$ F$^2$·cm$^{-2}$ |
| $A_{PD}$ | Photodiode area | 9·10$^{-5}$ cm$^2$ |
| $\eta_K$ | Photodiode quantum efficiency | 0.8 |

Table 1. Design and electrophysical parameters of readout circuits

Figure 3 shows the curves of currents $I_{in}$ and $I_{PD}$ for the photodiode model (10). With the parameters of photodiode, input FET, and radiation environment adopted in Fig. 3, the voltage drop across the photodiode is zero at $V_{G0}$ = 1.198 V. Given the voltage $V_G$, the quantities $\eta_L$, $I_{in}$, and D* can be calculated. In the next cycle, a new value is assigned to $V_G$, and all the characteristics are to be recalculated. In this way, we obtain the main performance characteristics of the system 'photodiode-direct injection readout' as a function of the gate voltage at the input gate. The dependences in Fig. 3 can be used in a joint analysis of the effect due to noise characteristics of photodiodes, measured versus photodiode bias voltage, and the responsivity of photodetector channels based on the system 'photodiode – direct injection readout circuits'.

Figures 4a, b, c, and d shows the calculated curves of photodiode bias voltage $V_{PD}(V_G)$, injection efficiency $\eta_I(V_G)$, input current $I_{in}(V_G)$, and detectivity D*$(V_G)$, respectively.

At $V_{PD}$ = 0 V, the product $g_{in}R_{PD}$ for curves 1-5 in Fig. 4 equals respectively 794, 11.1, 11.1, 3.4, and 1.4. As it is seen from Figs. 4a and 4b, the dependences $V_{PD}(V_G)$ and $\eta_I(V_G)$ small informative in evaluating the uniformity level of transfer characteristics and responsivities of photodetector channels, more helpful here being the dependences $I_{in}(V_G)$ and D*$(V_G)$ (see Figs 4c and 4d). As it is seen from Fig. 4, curves 1, at $g_{in}R_{PD}$ >100 the photodiode-direct injection readout system can be considered "ideal". With increasing the gate voltage $V_G$, as the photodiode bias voltage approaches zero, the injection efficiency $\eta_I$ tends to unity, the current integrated in the readout circuit becomes roughly equal to $I_{PD}$, $I_{in} \approx I_{PD}$, and the detectivity reaches D* $\approx$ 2.87·10$^{11}$ cm·Hz$^{1/2}$·Wt$^{-1}$, this value being comparable with the theoretical limit of D* for a photodiode with $\eta_K$=0.8 operating in photovoltaic regime in BLIP mode. With further increase of $V_G$, for the "ideal" system the performance characteristics of the photodetector channel based on the system 'photodiode – direct injection readout circuit' (namely, $\eta_I$, $I_{in}$, and D*) become almost independent of both $V_G$ and electrophysical and design parameters of readout circuits. In the latter case, performance characteristics of IR FPAs based on direct injection readout circuits can be evaluated using standard simplifications (Longo, 1978; Felix,1980). In "non-ideal" systems, in which the relation $g_{in}R_{PD} \gg 1$ is fulfilled not too strict (see curves 2-5 in Figs. 4c and 4d), curves $I_{in}$ ($V_G$) and $D^*$($V_G$) show a different behavior. The current $I_{in}$ integrated in the readout circuit increases with increasing $V_G$. The dependences $D^*$($V_G$) exhibit a pronounced maximum. Being considered as a function of photodiode electrophysical parameters and background illumination current, the maximum detectivity is attained at a voltage $V_G$ at which the photodiode gets driven by 5-30 mV in reverse direction (curve 4 in Fig. 4); this detectivity rather weakly depends on the value of $g_{in}R_{PD}$. For instance, as the value of $g_{in}R_{PD}$ decreases from 794 to 3.4 (curves 1 and 4 in Fig. 4) the detectivity $D^*$ falls in value from $2.87 \cdot 10^{11}$ cm $\cdot$Hz$^{1/2}$$\cdot$Wt$^{-1}$ to $1.96 \cdot 10^{11}$ cm $\cdot$ Hz$^{1/2}$$\cdot$Wt$^{-1}$, i.e. within a factor of 1.5. Note that the values of $g_{in}R_{PD}$ for curves 2 and 3 are identical; nonetheless, the dependences $I_{in}$ ($V_G$) (Fig. 4c) and $D^*$ ($V_G$) (Fig.4d) for those cases differ substantially.

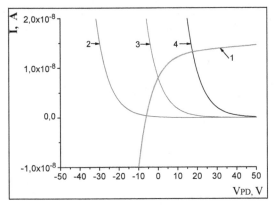

Fig. 3. Current-voltage characteristics of photodiode (curve 1) and input FET in weak inversion (curves 2-4) for three values of $V_G$, 1.17, 1.198, and 1.22 V. $I_0 = 5 \cdot 10^{-9}$ A, $R_S = 3 \cdot 10^{7}$ Ohm, $I_P = 1 \cdot 10^{-8}$ A, and $\eta_K = 0.8$.

The factors causing these differences can be clarified if one considers the calculated dependences of Fig. 5; in this figure, in addition to curves $D^*$ ($V_G$), also dependences $Q_1(V_G)$, $Q_2(V_G)$, $Q_3(V_G)$, and $Q_4(V_G)$ are shown. In the "ideal" system (see Fig.5 a), with increasing the voltage $V_G$, when the injection efficiency approaches unity, the total noise charge $Q_5$ becomes defined just by the photodiode noise current $Q_3$, having almost the same magnitude as the noise induced by background radiation fluctuations (9) since $I_{in} \approx I_P$ and $\eta_I$ $\approx 1$. In "non-ideal" systems (see calculated dependences in Fig. 5 b-e), the charge $Q_5$ grows in value with increasing $V_G$. The increase of the noise and the related reduction of $D^*$ is primarily defined by the growth of $Q_3$ and $Q_4$. For the dependences shown in Fig. 5 b-e the difference between the curves $D^*$($V_G$) is primarily defined by the growth of the 1/f noise of photodiode (component $Q_4$). The difference of the dependences $Q_4(V_G)$ is due to the higher current $I_{in}$ and a lower value of $\eta_I$ for curves 4. With identical values of $g_{in}R_{PD}$ at zero voltage drop across the photodiode (compare Fig. 5b and Fig. 5c), a better IR FPA responsivity can be reached with diodes exhibiting a larger dynamic resistance on their biasing in reverse direction by 10-30 mV. For state-of-the-art level of silicon technology ($C_{ox} \sim (0.5\text{-}1.2) \cdot 10^{-7}$ F cm$^{-2}$, $N^* \approx 1$), the noise induced by the input FET (cp. curves 1 and 2) is only substantial at voltages $V_G < V_{G0}$ even if the product $g_{in}R_{PD}$ has a value close to unity, and this noise does not limit the IR FPA detectivity. Knowing of the values of individual noise components enables goal-directed optimization of electrophysical and design parameters of photodiodes and readout circuits.

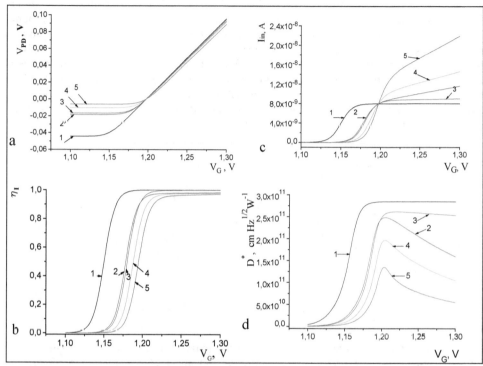

Fig. 4. Calculated dependences of performance characteristics of a photodiode-direct injection system on input-gate voltage. a – photodiode bias voltage, b – injection efficiency $\eta_I$, c – current $I_{in}$ integrated in the readout circuit, d - detectivity D*. Integration time $T_{in}$ =$5 \cdot 10^{-4}$ s, $I_P$ =$1 \cdot 10^{-8}$ A, $\eta_K$ =0.8, for curve 1 - $I_0$ =$1 \cdot 10^{-11}$ A, $R_P$=$1 \cdot 10^{11}$ Ohm; for curve 2 - $I_0$ =$5 \cdot 10^{-10}$ A, $R_P$ =$3 \cdot 10^7$ Ohm; for curve 3 - $I_0$ =$7 \cdot 10^{-10}$ A, $R_P$ =$3 \cdot 10^8$ Ohm; for curve 4 - $I_0$ = $2 \cdot 10^{-9}$ A, $R_P$ = $2 \cdot 10^7$ Ohm; for curve 5 - $I_0$ = $5 \cdot 10^{-9}$ A, $R_P$ =$1 \cdot 10^7$ Ohm.

Important figures of merit of multi-element IR FPAs are the fixed-pattern noise and the responsivity uniformity of photodetector channels. The spread of photoelectric parameters is defined by the variation of electrophysical parameters of readouts and by the variation of photoelectrical parameters of photodiodes. For direct injection readouts operated in weak inversion regime, the main effect is due to input-FET threshold variations rather than due to non-uniformity of geometric dimensions of input-FET gates.

Histograms of photodiode bias voltages and injection efficiencies calculated with allowance for FET threshold variations are shown in Fig. 6. In performing the calculations, it was assumed that the spread of threshold voltages obeys a normal distribution law, and a total of 1000 realizations were considered. As it is seen from Fig. 6, the spread of FET threshold voltages leads to a spread of photodiode bias voltages (Fig. 6 a), and also to a spread of injection efficiencies $\eta_I$ and, as a consequence, to variation of transfer characteristics of the 'photodiode-direct injection readout' system.

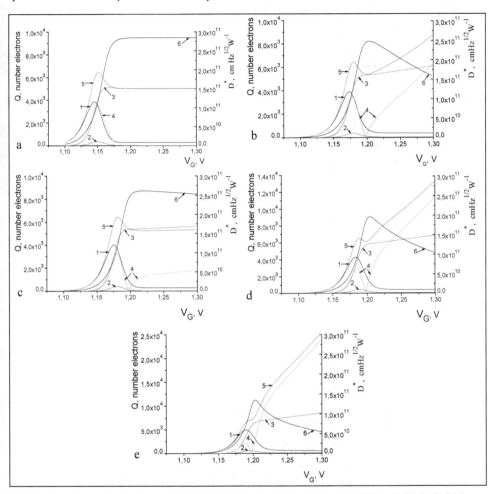

Fig. 5. Calculated dependences of $D^*(V_G)$ (curve 6, right axis) and $Q_1(V_G)$, $Q_2(V_G)$, $Q_3(V_G)$, $Q_4(V_G)$, and $Q_5(V_G)$ (curves 1-5, left axis). $Q_5 = (Q_1^2 + Q_2^2 + Q_3^2 + Q_4^2)^{1/2}$. The values of photodiode electrophysical parameters adopted in the calculations are the same as in Fig. 4 for curves 1-5, respectively.

Figure 7 shows the calculated curve $I_{in}$ $(V_G)$ that illustrate the influence of the dispersion of threshold voltages with $\sigma(V_{th}) = 10$ mV on the characteristics of a multi-element IR FPA (the values of photoelectrical parameters for IR photodiodes are the same as those adopted in Fig. 4 for curves 2). It is seen from Fig. 7 a-b that for a "non-ideal" system the dispersion of threshold voltages brings about a fixed-pattern noise and a spread of currents integrated in the readouts, $I_{in}$ $(V_G)$. For multi-element IR FPAs, with allowance for the spread of threshold voltages, at a voltage $V_G$ =1.205 V (Fig.5b), for which the detectivity of a single channel $D^*$

attains a maximum, the spread of currents $I_{in}$ integrated in the readouts falls in the range from $4 \cdot 10^{-9}$ to $1.15 \cdot 10^{-8}$ A (Fig. 7a), whereas the current integrated in the "ideal" system is $I_{in}$ = $0.8 \cdot 10^{-8}$ A.

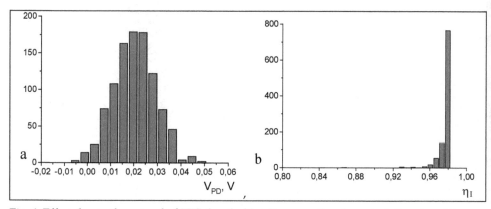

Fig. 6. Effect due to the spread of FET threshold voltages on the characteristics of multi-element IR FPAs; a – histogram of photodiode voltages, b – histogram of injection efficiencies $\eta_I$ for $I_0 = 5 \cdot 10^{-10}$ A, $R_P = 3 \cdot 10^7$ Ohm, $I_P = 1 \cdot 10^{-8}$A, $\eta_K = 0.8$, and $V_G = 1.225$ V. The standard deviation of input-FET threshold voltages is $\sigma(V_{th}) = 10$ mV.

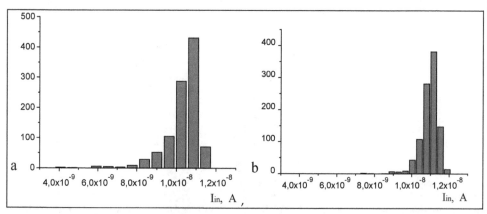

Fig. 7. Calculated histograms $I_{in}(V_G)$ for a normal distribution of input-FET threshold voltages with $\sigma(V_{th}) = 10$ mV. For Figs. 7a and 7b the voltages $V_G$ are respectively 1.205 Vand 1.225 V. The calculations were performed for photodiodes with parameter values the same as those for curves 2 in Fig. 4.

Figure 8 show histograms $D^*(V_G)$ calculated with allowance for the spread of threshold voltages. In the calculations of detectivity, the accumulation time was defined by the charge capacity of the readout circuit and by the maximal (over 1000 realizations) current $I_{in}(V_G)$.

In multi-element IR FPAs, in view of non-uniformity of threshold voltages, at $V_G = 1.205$ V about 1% of photodetector channels have a detectivity $D^*$ lower than $1 \cdot 10^{11}$ cm $\cdot Hz^{1/2} \cdot Wt^{-1}$ (see Fig. 8b), whereas for a single channel we have $D^* = 2.48 \cdot 10^{11}$ cm $\cdot Hz^{1/2} \cdot Wt^{-1}$. In the case of multi-element IR FPAs because of the spread of threshold voltages, there arises a necessity to adjust the gate voltage (for the values of electrophysical parameters of photodiodes adopted in the calculations, Fig. 8b); the optimum value here is $V_G =1.225$ V. The increase of $V_G$ results in a substantial reduction of fixed-pattern noise (see Fig.7b). In the latter situation, the minimal detectivity is not lower than $1.85 \cdot 10^{11}$ cm $\cdot Hz^{1/2} \cdot Wt^{-1}$, and more than 98% pixels have a detectivity $D^*$ greater than $2.0 \cdot 10^{11}$ cm $\cdot Hz^{1/2} \cdot Wt^{-1}$. With further increase of $V_G$, "dark" parasitic photodiode current components grow in value and the detectivity of the majority of photodetector channels decreases (Fig. 8c).

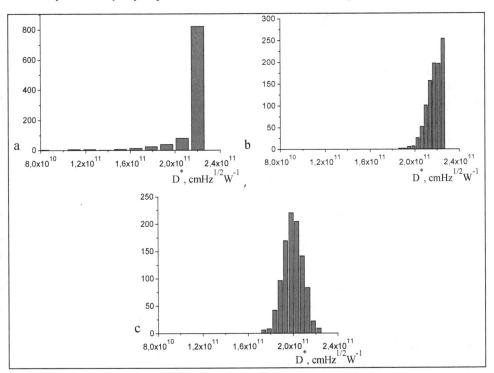

Fig. 8. Histograms $D^*(V_G)$ calculated for FET threshold voltages normally distributed with $\sigma(V_{th}) = 10$ mV for $V_G = 1.201, 1.225,$ and $1.249$ V (respectively Figs. 8a, 8b, and 8.6c). The calculations were performed for photodiodes with parameter values adopted for curves 2 in Fig. 4.

The effect due to dispersion of threshold voltages in fabricated readout circuits is illustrated by a comparison of calculated histograms $I_{in}(V_G)$ and $D^*(V_G)$ in Fig. 8 with histograms calculated for $\sigma(V_{th}) = 2$ mV, Fig. 9. As it is seen from the calculated dependences in Fig. 9, in the latter case the detectivity of the multi-element IR FPA is at the level of about 90% of $D^*$ of the single photodetector channel.

Along with non-uniformity of input-FET threshold voltages, the photoelectric parameters of photodiodes, $\eta_K$, $R_{PD}$, and $I_0$, also inevitably display some scatter of values. Figure 10 shows histograms of $I_{in}$ (Fig. 10a) and $D^*$ (Fig.10b) calculated on the assumption that the distributions of $R_{PD}$ and $I_0$ obey normal distribution laws.

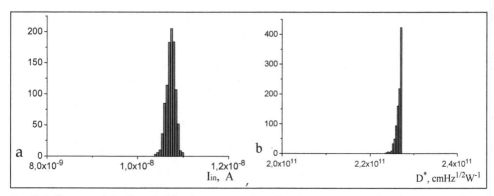

Fig. 9. Influence of the threshold voltage spread on the characteristics of multi element IR FPAs; a – histogram of $I_{in}$, b – histogram of D*. The standard deviation of input-FET threshold voltages is $\sigma(V_{th})$ = 2 mV, $I_0$=5 ·10⁻¹⁰ A, $R_{PD}$ =3 ·10⁷ Ohm, $I_P$ =1 ·10⁻⁸ A, $\eta_K$ =0.8, $V_G$ =1.213 V.

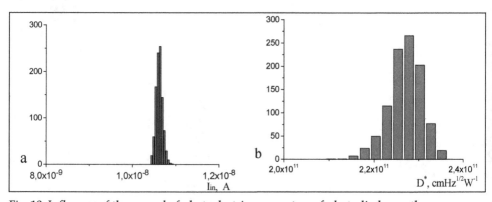

Fig. 10. Influence of the spread of photoelectric parameters of photodiodes on the performance characteristics of multi-element IR FPAs; a – histogram of $I_{in}$, b – histogram of D*. Parameter values adopted in the calculations: $\sigma(R_{PD})$ = 3 ·10⁶ Ohm, $\sigma(I_0)$ = 1 ·10⁻¹⁰ A, $I_0$=5 ·10⁻¹⁰ A, $R_{PD}$ =3 ·10⁷ Ohm, $I_P$ =1 ·10⁻⁸ A, $\eta_K$ =0.8, $V_G$ =1.225 V.

Plotting histograms of performance characteristics of multi-element IR FPAs, namely, currents integrated in the readout circuits and detectivity) versus the input-gate voltage $V_G$ allows one to carry out numerical experiments and compare their results with experimental data. Such histograms can be considered generalized characteristics of multi-element IR FPAs indicative of their quality.

### 3.2 Modeling procedure for multi-element IR FPAs with direct injection readout circuits using the experimental-current voltage characteristics of photodiodes

The developed procedure enables performing an analysis of the system 'photodiode –direct-injection readout' using the experimental current-voltage characteristics of photodiodes. Figure 11 shows a family of "dark" and "light" current-voltage characteristics of one hundred photodiodes measured under room-temperature background conditions. The photodiodes were fabricated on the basis of a $Hg_{1-x}Cd_xTe$ variband heteroepitaxial structure. The stoichiometric composition of the photosensitive layer was x=0.225 (Vasilyev, 2010).

In performing the calculations, the experimental current-voltage characteristics were approximated with the expressions:

$$I_{PD}(V_{PD}) = C_0 + C_1 \exp(C_2 V_{PD}) + C_3 \exp(C_4 V_{PD}) + C_5 V_{PD} \tag{11}$$

where the values of coefficients $C_0$-$C_5$ were chosen individually for each photodiode.

The characteristics of multi-element IR FPA were calculated by the procedure, described in section 3.1, that allows calculation of all photodetector-channel characteristics, $\eta_I(V_G)$, $I_{in}(V_G)$, and $D^*(V_G)$.

Figure 12 shows the histograms $D^*(V_G)$ and $I_{in}(V_G)$ calculated for the standard deviation value of threshold voltages of the readouts $\sigma(V_{th})$ =10 mV. An analysis of dependences calculated with lower values of $\sigma(V_{th})$ suggests that the use of silicon technology permitting values $\sigma(V_{th})$ < 2-3 mV will allow a substantial (by 20-30%) reduction of the fixed-pattern noise level. The changes in the calculated detectivity histograms proved to be less substantial since the spread of detectivity values is primarily defined by the variations of photodiode quantum efficiency (Vasilyev, 2010).

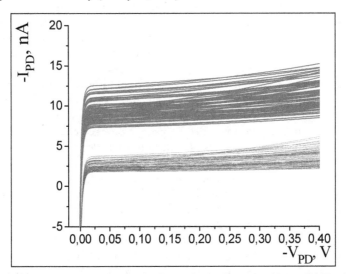

Fig. 11. A family of experimental current-voltage characteristics of $Hg_{1-x}Cd_xTe$ photodiodes; "dark" characteristics are shown with blue lines, and characteristics measured under room-temperature radiation background with red lines are shown with red lines.

It should be noted that the numerical values of coefficients $\alpha_1$, $\alpha_2$, $\alpha_3$, and K were borrowed from literature sources. Experimentally measured values of these coefficients or more elaborated model for photodiode noises can easily be incorporated into the computer problem, allowing an improved accuracy in predicting the characteristics of IR FPAs.

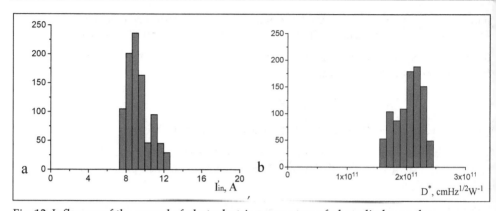

Fig. 12. Influence of the spread of photoelectric parameters of photodiodes on the performance characteristics of multi-element IR FPAs; a – histogram of $I_{in}$, b – histogram of $D^*$. Parameter values adopted in the calculations: $\sigma(R_{PD}) = 3 \cdot 10^6$ Ohm, $\sigma(I_0) = 1 \cdot 10^{-10}$ A, $I_0 = 5 \cdot 10^{-10}$ A, $R_{PD} = 3 \cdot 10^7$ Ohm, $I_P = 1 \cdot 10^{-8}$ A, $\eta_K = 0.8$, $V_G = 1.225$ V.

### 3.3 Analysis of LWIR thermography systems based on $Hg_{1-x}Cd_xTe$ photodiodes

The simple photodiode model (10) proved to be insufficient for adequately predicting the current-voltage characteristics of $Hg_{1-x}Cd_xTe$ photodiodes in the spectral region 8 to 14 µm. In modeling current-voltage characteristics of $Hg_{1-x}Cd_xTe$ photodiodes, all main mechanisms of charge transport in $p$-$n$ junctions must be taken into account, including the background radiation current $\eta_K I_P$, the diffusion current, the thermal generation/recombination current, the tunneling current through localized levels within the bandgap of the depleted region of the $p$-$n$ junction, the interband tunneling current (Anderson, 1981), the Shockley-Reed-Hall current through traps in quasi-neutral $n$ and $p$ regions, the Auger current, and the radiative generation/recombination current in the $p$-$n$ junction and in the quasi-neutral regions (Anderson, 1981; Rogalski, 2000). All the currents listed above are independent currents to be taken into account additively, except for the thermal generation-recombination and trap-assisted tunneling through traps of the Shockley-Reed-Hall type in the depleted region of the $p$-$n$ junction, since the rates of the latter processes is determined by the trap occupation. The two latter mechanisms were modeled in the approximation of balance equations for charge carriers at trap levels (Anderson, 1981; Gumenjuk-Sichevska, 1999; Sizov, 2006; Yoshino, 1999). We assumed the presence of localized donor-type centers in the bandgap with energy $E_t = 0.6$-$0.7 E_g$ over the valence-band edge (Krishnamurthy, 2006). In calculating the band-gap energy of the material as a function of stoichiometric composition and temperature, we used the expression (Rogalski, 2000).

$$E(x,T) = \left[ -0.302 + x\left[ 1.93 + x(-0.81 + 0.832x) \right] + 5.32 \cdot 10^{-4} \cdot (1 - 2x)\frac{-1822.0 + T^3}{255.2 + T^2} \right] \quad (12)$$

The main electrophysical parameters of $Hg_{1-x}Cd_xTe$ photodiodes adopted in the calculation of their current-voltage characteristics are listed in Table 2.

By way of example, Figure 13 shows the current-voltage characteristics of $Hg_{1-x}Cd_xTe$ photodiodes for different long-wave spectral-response cutoffs, and Figure 14 shows the calculated dependences of $R_{PD}$ $A_{PD}$ on the cutoff wavelength of the spectral response of the photodiodes.

As it was shown in (Gumenjuk-Sichevska, 1999; Sizov, 2006), the photodiode model that was used in the present calculations permits calculation of the current-voltage characteristics as a function of the stoichiometric composition of $Hg_{1-x}Cd_xTe$, temperature, and electrophysical parameters of substrate material, the obtained dependences being consistent with those reported in the literature (Rogalski, 2000; Yoshino,1999).

Figure 15 shows the calculated dependences of the maximum values of $D^*(V_G)$ of a photodetector channel based on the system 'HgCdTe photodiode – direct injection readout circuit' on the cutoff wavelength $\lambda_2$. In calculating the background radiation flux reaching the photodiodes in the spectral region from $\lambda_1$ to $\lambda_2$ ($\lambda_1 = 4$ μm and $\lambda_2$ is the cutoff wavelength), for the blackbody temperature a value 300 K was adopted; the aperture angle was assumed defined by the relative aperture of the optical system $F/f = 0.5$, and for the transmission of the optical system, a value 0.9 was adopted.

| Designation | Parameter | Numerical value | |
|---|---|---|---|
| | | Case 1 | Case 2 |
| $N_a$, $N_d$ | Acceptor and donor concentrations | $1.0 \cdot 10^{22}$ m$^{-3}$, $2.0 \cdot 10^{21}$ m$^{-3}$ | $2.0 \cdot 10^{22}$ m$^{-3}$, $2.0 \cdot 10^{21}$ m$^{-3}$ |
| $N_t$, $N_{tV}$ | Concentration of traps in the p-n junction and in the quasi-neutral regions | $6.0 \cdot 10^{21}$ m$^{-3}$, $2.0 \cdot 10^{21}$ m$^{-3}$ | $3.0 \cdot 10^{21}$ m$^{-3}$, $1.0 \cdot 10^{21}$ m$^{-3}$ |
| $\tau_n$, $\tau_p$, $\tau_{nV}$, $\tau_{pV}$ | Electron and hole lifetimes in the p-n junction and in the quasi-neutral regions | $0.2 \cdot 10^{-6}$ s, $0.2 \cdot 10^{-6}$ s, $6.0 \cdot 10^{-6}$ s, $6.0 \cdot 10^{-6}$ s | $0.2 \cdot 10^{-6}$ s, $0.2 \cdot 10^{-6}$ s, $1.0 \cdot 10^{-7}$ s, $1.0 \cdot 10^{-7}$ s |
| $E_t$ | Energy position of trap level | $E_t = 0.7$ $E_g$ eV | |
| $P$ | Interband matrix element | $8.3 \cdot 10^{-10}$ eV·m | |
| $W_c^2$ | Squared matrix element for the tunneling of charge carriers from the trap level into the band | $3 \cdot 10^{-67}$ J$^2$·m$^3$ | |
| $\Delta$ | Spin – orbital interaction constant | 0.96 eV | |
| $\varepsilon_r$ | Static dielectric permittivity | 17.5 | |
| $A_{PD}$ | Photodiode area | $9 \cdot 10^{-5}$ cm$^2$ | |

Table 2. Electrophysical parameters of $Hg_{1-x}Cd_xTe$ photodiodes.

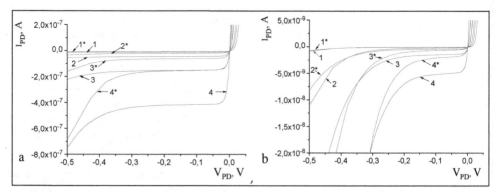

Fig. 13. Calculated current voltage characteristics of $Hg_{1-x}Cd_xTe$ photodiodes at temperatures 77 K (a) and 60 K (b). For curves 1 – 4 (1* - 4*) the cutoff wavelength is respectively $\lambda_2$ = 11, 12, 13, and 14 µm. The electrophysical parameters of photodiodes are listed in Table 2; case 1 – curves 1-4, case 2 - curves 1* - 4*.

The detectivity of the photodetector channel at temperature T=77 K for photodiodes with electrophysical-parameter values indicated as case 1 in Table 2 (curve 1 in Fig. 15) at $\lambda_2$ > 10 µm becomes lower than Background Limited Performance (BLIP) detectivity (curve 3). On cooling the photodiodes to temperature 60 K, the detectivity of photodetector channel approaches BLIP detectivity at wavelengths below ~ 13 µm (curve 3). For photodiodes with electrophysical-parameter values indicated as case 2 in Table 2.2, due to a larger dynamic resistance, the long-wave spectral-response cutoff at which the FPA detectivity attains its maximum value shifts towards longer wavelengths (see curves 2 and 4 in Fig. 15).

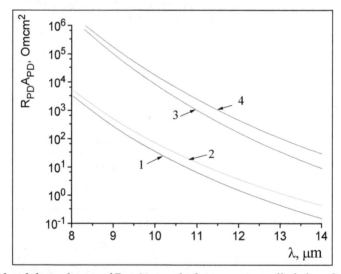

Fig. 14. Calculated dependences of $R_{PD} A_{PD}$ on the long-wave cutoff of photodiodes; curves 1 and 2 refer to temperature T=77 K, and curves 3 and 4, to temperature T=60 K. Electrophysical-parameter values: curves 1 and 3 – case 1, curves 2 and 4 – case 2.

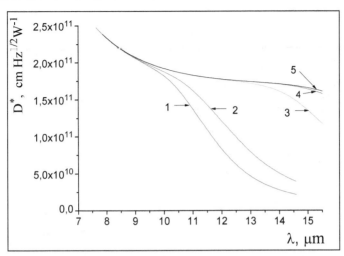

Fig. 15. Calculated curves $D^*(\lambda_2)$ for a photodetector channel based on the system 'Hg$_{1-x}$Cd$_x$Te photodiode-direct injection readout'; 1, 2 – photodiode temperature 77 K, 3, 4 – photodiode temperature 60 K. Curves 1, 3 and 2, 4 were calculated for parameter values indicated in Table 1 as cases 1 and 2, respectively, for direct injection charge capacity $Q_{in}=5 \cdot 10^7$ electrons. Curve 5 – calculated dependences for BLIP detectivity.

Figure 16 shows the curves $D^*(V_G)$ and the dependences $Q_1(V_G)$, $Q_2(V_G)$, $Q_3(V_G)$, $Q_4(V_G)$ and $Q_5(V_G)$ calculated for $\lambda_2$ = 10, 12, and 13 μm. The calculations were performed for photodiodes with parameter values indicated in Table 2 as case 1.

The detectivity of the photodetector channel (Fig. 16a) at $\lambda_2$ = 10 μm is close to BLIP detectivity $D^*$; this detectivity is limited by the photodiode current noise $Q_3$ and at voltages $V_G$ >1.2 V it is almost independent of $V_G$. For $\lambda_2$ = 12 μm (Fig. 16b) the maximum detectivity of the photodetector channel is $D^* \approx 8.34 \cdot 10^{10}$ cm $\cdot$Hz$^{1/2} \cdot$Wt$^{-1}$ (BLIP detectivity is $D^*= 1.77 \cdot 10^{11}$ cm $\cdot$Hz$^{1/2} \cdot$Wt$^{-1}$). For $\lambda_2$ = 13 μm (Fig. 16c) the maximum detectivity of the photodetector channel is $D^* \approx 4.7 \cdot 10^{10}$ cm $\cdot$Hz$^{1/2} \cdot$Wt$^{-1}$ (BLIP detectivity is $D^*= 1.74 \cdot 10^{11}$ cm $\cdot$Hz$^{1/2} \cdot$Wt$^{-1}$). The degradation of detectivity at voltages $V_G > 1.225$ V is primarily defined by the increase in the photodiode 1/f-noise level ($Q_4$ in Fig. 16b, c).

An increase of the direct injection storage capacitance and the related increase of the accumulation time both lead to a greater contribution made by the 1/f-noise. Figure 17 shows the curves $D^*(V_G)$ and the dependences $Q_1(V_G)$, $Q_2(V_G)$, $Q_3(V_G)$, and $Q_4(V_G)$ similar to those shown in Fig. 16b yet calculated for direct injection storage capacity $Q_{in}$ = 2$\cdot 10^8$ electrons.

A comparison between the dependences in Figs. 16 and 17 shows that the maximal detectivity rather weakly depends on the storage capacity of the readout circuit. Yet, because of the increased contribution due to the 1/f noise of photodiodes, with increasing the input-gate voltage the dependence of $D^*$ on $V_G$ becomes more clearly manifested. That is why with increasing the storage capacity of direct injection readouts and, hence, with increasing the accumulation time, the requirements imposed on the standard deviation of threshold voltages become more stringent.

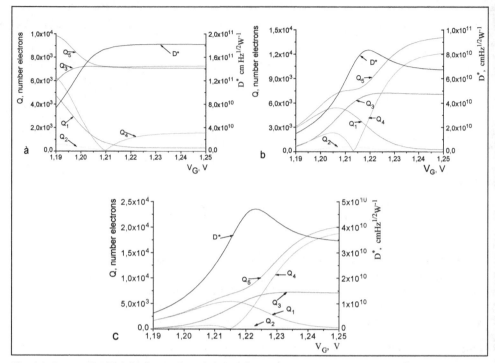

Fig. 16. Curves D*(V$_G$) (right axis) and dependences Q$_1$(V$_G$), Q$_2$(V$_G$), Q$_3$(V$_G$), Q$_4$(V$_G$), and Q$_5$(V$_G$) (left axis). Temperature 77 K, direct injection charge capacity Q$_{in}$=5·10$^7$ electrons; a – λ$_2$=10 μm, b – λ$_2$ = 12 μm, c – λ$_2$ = 13 μm.

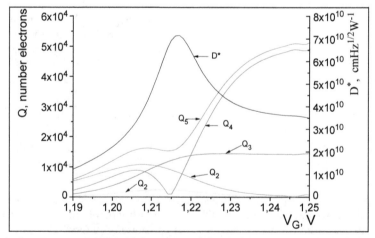

Fig. 17. Calculated curves D*(V$_G$) (right axis) and dependences Q$_1$(V$_G$), Q$_2$(V$_G$), Q$_3$(V$_G$), Q$_4$(V$_G$), and Q$_5$(V$_G$) (left axis). The photodiode temperature is 77 K; the storage capacity of direct injection readout is Q$_{in}$=2·10$^8$ electrons; λ$_2$ = 12 μm. The values of electrophysical parameters of photodiodes are given in Table 2.2, case 1.

## 4. Temperature resolution of thermography systems based on multi-element IR FPAs with direct injection readouts

For thermography systems based on multi-element photodetectors, an important figure of merit is the noise-equivalent temperature difference (NETD) (Rogalski, 2000). The NEDT value can be calculated by the formula (Taubkin, 1993).

$$\text{NETD} = \frac{\sqrt{1 / 2T_{in}}}{\dfrac{F^2}{4f^2}(A_{PD})^{1/2} G_{Op} \int\limits_{\lambda 1}^{\lambda 2} \dfrac{d(dR / d\lambda)}{dT} S(\lambda) D_{\lambda}^{*}\left(\lambda_1, \lambda_2, T\right) d\lambda} \tag{13}$$

Here, $dR/d\lambda$ is the blackbody spectral luminous exitance, $F/f$ is the relative aperture of the optical system, f is the focal distance, $G_{Op}$ is the optical transmission of the system, and $S(\lambda)$ is the atmospheric transmission.

An analysis of NETD starts with calculating the current-voltage characteristics of photodiodes and the background radiation level. Then, currents integrated in the readout circuit and detectivity are calculated.

Figure 18 shows the calculated curves of NETD versus the cutoff wavelength $\lambda_2$.

For an "ideal" thermography system (curve 1), in which the integration time is equal to the frame time, $T_{in}$ =20 ms in the present calculations, and the detectivity of photodetector channels is close to BLIP detectivity, we obtain the well-known dependences NETD($\lambda$) that show that the temperature resolution improves with increasing the wavelength $\lambda_2$. The dependencies in Fig. 18 reveal the causes of the losses in temperature resolution owing to insufficiently high dynamic resistance of photodiodes or limited storage capacitance of silicon direct injection readout circuits of thermography systems in comparison with the theoretical limit. It can be inferred from the dependences $D_{\lambda}^{*}(\lambda_2)$ (Fig. 15) and NETD($\lambda_2$) (Fig. 18) that the temperature resolution of thermography systems based on the photodiode – direct injection readout system attains a maximum value at some wavelength $\lambda_2$ and, then, decreases (see curves 2 and 3 in Fig. 18). The position of NEDT maximum is defined by electrophysical parameters of HgCdTe photodiodes, by the optical transmission of the system, and by the photodiode temperature. Yet, as it follows from the calculated curves of NETD in Fig. 18 (curves 4-7) the main factor limiting the temperature resolution of the thermography systems is the storage capacitance of silicon readout circuits.

With increasing the cutoff wavelength $\lambda_2$, the intensity of background radiation coming to photodetector also increases and, as a result, the integration time $T_H$ decreases, with simultaneous shift of the maximum temperature resolution of the thermography system towards short wavelengths in comparison with curves 2 and 3 in Fig. 18. The charge capacity of readout circuits $Q_{in}$ = $5 \cdot 10^7$ electrons presents a maximal value reported in literature for multi-element matrix IR FPAs (at 30-µm photocell pitch) (Rogalski, 2000). At such values of integration capacitance of readout circuits the temperature resolution of thermography systems (curve 5) already with $\lambda_2$ = 8 µm is more than one order of magnitude and with $\lambda_2$ = 14 µm, two orders of magnitude lower than that of the "ideal" thermography system (curve 1). The storage capacitance can be increased through implementing a new architecture of readout circuits (curves 6 and 7 in Fig. 18) (Lee, 2010). A cardinal solution here is implementation of ADC in each pixel of the readout circuit (Zhou, 1996; Martijn, 2000; Fowler, 2000). In (Bisotto, 2010), each pixel was provided with a 15-bit ADC, and the effective storage capacitance was in excess of $3 \cdot 10^9$ electrons. This has enabled an increase in integration time up to the frame time and, in this way, allowed reaching a NETD of LWIR thermography systems based on multi-element IR FPAs amounting to 2 mK, the latter value being close to the maximum theoretically possible NETD.

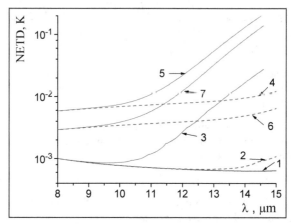

Fig. 18. Calculated curves NETD($\lambda_2$). In the calculations, the values $F_p/f = 0.5$ and $\lambda_1 = 5$ μm were adopted. Curve 1 was calculated for $D^*_\lambda = D^*_{\lambda BLIP}$ and $T_H = 20$ ms; curves 2 and 3 were calculated with allowance for the dependence of $D^*_\lambda$ on the long-wave spectral-response cutoff, integration time $T_{in} = 20$ ms; curves 4-7 were calculated for conditions with limited storage capacitance of direct injection readouts (curves 4 and 5 - $Q_{in} = 5 \cdot 10^7$ electrons, curves 6 and 7 - $Q_{in} = 2 \cdot 10^8$ electrons); curves 2, 4, and 6 – photodiode temperature 60 K, curves 3, 5, and 7 – photodiode temperature 77 K. The electrophysical parameters of photodiodes are given in Table 2, case 1.

The dependences $D^*(V_G)$ and NETD($\lambda_2$) calculated for different values of the storage capacity of readout circuits on the input-gate voltage $V_G$ and shown in Figs. 15-18 define the ultimate performance characteristics of thermography systems in the spectral range 8-14 μm. In fact, these dependences are will also be observed for a single-element FPA based on the system 'Hg$_{1-x}$Cd$_x$Te photodiode – direct injection readout circuit' since they were plotted using just the maximum values of $D^*$. For multi-element IR FPAs, the non-uniformity of input-FET threshold voltages leads to an increased fixed-pattern noise level and, for some part of photodetector channels, to a considerable reduction of $D^*$ and NETD in comparison with the dependences shown in Figs. 15 and 18. An additional factor causing an increase in the fixed-pattern noise, a decrease of detectivity, and worsened temperature resolution of multi-element IR FPAs is non-uniformity of the stoichiometric composition of the Hg$_{1-x}$Cd$_x$Te substrate.

For multi-element IR FPAs with $\lambda_2 \le 10$ μm, detectivity rather weakly depends on the gate voltage $V_G$ and, hence, on the dispersion of threshold voltages (see Fig.16a). The fixed-pattern noise level and the spread of NETD values of multi-channel IR FPAs are primarily defined by the scatter of the long-wave photosensitivity cutoff owing to non-uniform stoichiometric composition of the Hg$_{1-x}$Cd$_x$Te substrate.

Figure 19 shows calculated histograms of $D^*$, $I_{in}$ and NETD values of thermography systems for $Hg_{1-x}Cd_xTe$ photodiodes with photoelectric parameters given in Table 2, case 1. In the calculations, it was assumed that the dispersion of threshold voltages and the non-uniformity of the stoichiometric composition of substrate material obey normal distribution laws with parameter values indicated in the caption to the figure (the total number of realizations in the calculations was 200).

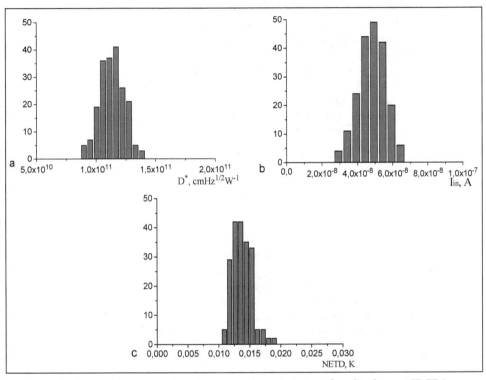

Fig. 19. Calculated histograms of performance characteristics of multi-element IR FPAs; a - detectivity $D^*$, b - currents $I_{in}$ integrated in the readout circuits, c - NETD. Standard deviation of input-FET threshold voltages - $\sigma(V_{th})$=5 mV. Standard deviation of the stoichiometric composition of $Hg_{1-x}Cd_xTe$ substrate - $\sigma(x)$ = 0.1%. Average stoichiometric composition - x=0.2167, T=77 K, storage capacitance of readout circuit $Q_{in}$=5·$10^7$ electrons.

The photocurrent level was calculated with allowance for the long-wave cutoff of photosensitivity individually for each photodiode. The integration time was defined by the magnitude of the storage capacitance of the readout circuit and by the maximum level of photodiode current over 200 realizations. At T=77 K, the mean stoichiometric composition x=0.2167 ensures a long-wave photosensitivity cutoff $\lambda_2$=11 μm at standard dispersion of stoichiometric composition $\sigma(x)$ = 0.1%, typical of the present-day state-of-the-art technology level (Rogalski, 2000).

For a single channel with c $\lambda_2$=11 μm, at the same values of electrophysical and design parameters of photodiodes and readout circuits the maximum sensitivity $D^*$ is 1.4·$10^{11}$

cm·Hz$^{1/2}$·W$^{-1}$, D$_{BLIP}$* =1.8·10$^{11}$ cm·Hz$^{1/2}$·W$^{-1}$, and NETD=10.8 mK (see curve 5 in Fig. 18). The dispersion of the stoichiometric composition of substrate σ(x) = 0.1% results in a spread of long-wave cutoff wavelengths in the interval from 11 to 11.8 μm, this being the main factor causing NETD degradation.

With increasing the long-wave cutoff wavelength λ$_2$, requirements to the uniformity of threshold voltages under the input gates of direct injection readout circuits and requirements to the uniformity of the stoichiometric composition of substrate both become more stringent. Figure 20 shows calculated histograms of currents I$_{in}$, detectivities D* and NETD values of thermography systems for Hg$_{1-x}$Cd$_x$Te photodiodes with photoelectric parameters indicated in Table 2 as case 1. At mean stoichiometric composition x=0.21055, σ(x) = 0.1% and T=77 K the non-uniformity of the stoichiometric composition of substrate material results in a spread of long-wave photosensitivity cutoffs in the interval from 12 to 13.2 μm. The input-gate voltage value V$_G$=1.23 V appears to be optimal for the radiation environment conditions and values of electrophysical and design parameters of photodiodes and readout circuits adopted in the calculations.

Note that for a single channel at λ$_2$=12 μm the maximum detectivity D* is 8.4·10$^{10}$ cm·Hz$^{1/2}$·W$^{-1}$ and NETD = 0.022 K (Fig. 18, curve 5). Taking the dispersion of threshold voltages with σ(V$_{th}$) = 5 mV and substrate stoichiometric composition with σ(x) = 0.1% into account results to two-three-fold degradation of NETD in a considerable fraction of photodetector channels.

Figure 21 shows calculated histograms of currents I$_{in}$, detectivities D* and NETD values of thermography systems for Hg$_{1-x}$Cd$_x$Te photodiodes with photoelectric-parameter values adopted in Fig. 20 yet under more stringent conditions in terms of the dispersion of FET threshold voltages and stoichiometric composition of Hg$_{1-x}$Cd$_x$Te substrate, σ(V$_{th}$) = 2 mV and σ(x) = 0.03%. A comparison between the histograms in Figs. 20 and 21 shows that more stringent requirements imposed on the uniformity of threshold voltages and stoichiometric composition of substrate allow a substantial reduction of the fixed-pattern noise and a considerable improvement of NEDT values of thermography systems.

With the adopted values of electrophysical parameters of Si readout circuits and Hg$_{1-x}$Cd$_x$Te photodiodes (Tables 1 and 2), thermography systems based on multi-element IR FPAs intended for operation at liquid-nitrogen temperature in the spectral range up to 13-14 μm with maximum possible NETD values (Fig. 18, curve 5) can be implemented at an acceptable level of fixed-pattern noise by:

-   increasing the dynamic resistance of Hg$_{1-x}$Cd$_x$Te photodiodes;
-   adhering to more stringent requirements in terms of uniformity of input-FET threshold voltages and stoichiometric-composition uniformity of the substrate.

However, requirements to σ(V$_{th}$) and σ(x) more stringent than the requirements that were adopted in calculating data in Fig. 21 presently cannot be met by silicon CMOS technology and synthesis processes of epitaxial Hg$_{1-x}$Cd$_x$Te layers (Phillips, 2002).

Achieving photosensitivity in the spectral region 12-14 μm necessitates cooling the hybrid IR FPA assembly to a temperature below liquid-nitrogen temperature, see curves 4 and 6 in Fig. 19.

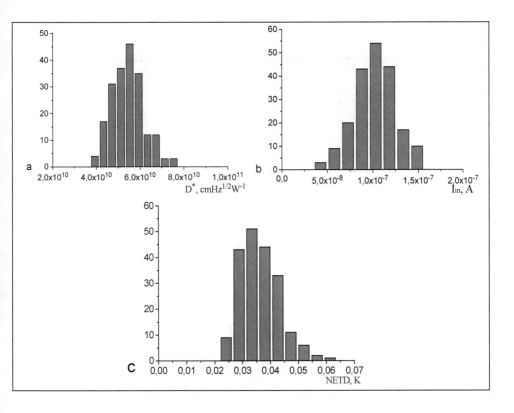

Fig. 20. Calculated histograms of performance characteristics of multi-element IR FPAs; a - detectivity $D^*$, b – currents $I_{in}$ integrated in the readout circuits, c - NETD. Standard deviation of input-FET threshold voltages - $\sigma(V_{th})$ = 5 mV. Standard deviation of the stoichiometric composition of $Hg_{1-x}Cd_xTe$ substrate $\sigma(x)$ = 0.1%. Average stoichiometric composition x=0.2105, T=77 K. Storage capacitance of readout circuit - $Q_{in}$=5 $\cdot10^7$ electrons.

Figure 22 shows calculated histograms of currents detectivities $D^*$, $I_{in}$ and NETD values of thermography systems for $Hg_{1-x}Cd_xTe$ photodiodes with photoelectric parameters indicated in Table 2 as case 1. The parameter values used in the calculations were the same as those in Figs. 19 and 20, and the photodiode temperature was assumed to be 60 K.

At mean stoichiometric composition x=0.2105, 0.1% dispersion of stoichiometric composition and temperature T=60 K, the long cutoff wavelength will fall into the wavelength interval from 12.6 to 14 μm.

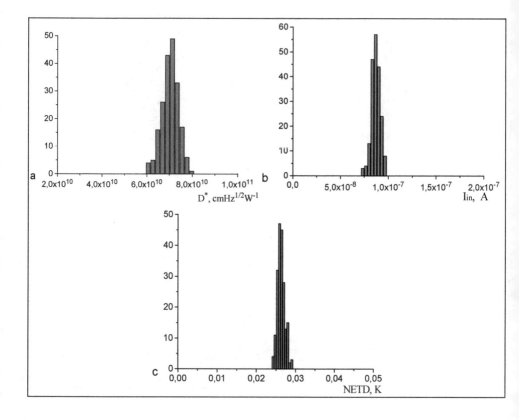

Fig. 21. Calculated histograms of performance characteristics of multi-element IR FPAs;
a - detectivity D*, b – currents $I_{in}$ integrated in the readout circuits, c - NETD. Standard
deviation of input-FET threshold voltages - $\sigma(V_{th})$ = 2 mV. Standard deviation of the
stoichiometric composition of $Hg_{1-x}Cd_xTe$ substrate $\sigma(x)$ = 0.03%. Average stoichiometric
composition x=0.2126, T=77 K.

As it follows from the calculated dependencies shown in Fig. 22, the cooling of the hybrid
assembly down to temperature 60 K will allow implementation of thermography systems
with NEDT values close to maximum possible figures, limited only by the value of the
storage capacitance of silicon readout circuits at an acceptable level of fixed-pattern noise.
The possibility of variation of the cooling temperature of hybrid assembly in the
calculations allows formulation of requirements to required accuracy in maintaining the
temperature of cooled hybrid assembly, i.e. requirements to be imposed on the cryostat
and cooling system.

Mathematical Modeling of Multi-Element Infrared Focal Plane Arrays Based on the System 'Photodiode – Direct-Injection Readout Circuit'

51

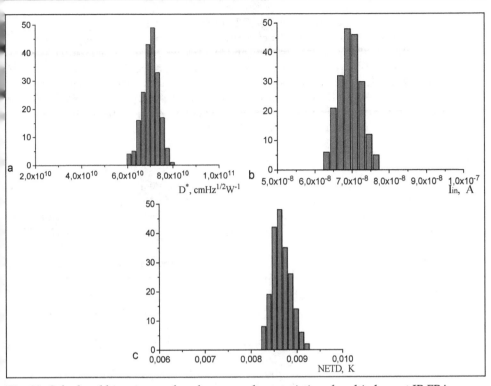

Fig. 22. Calculated histograms of performance characteristics of multi-element IR FPAs; a - detectivity $D^*$, b – currents $I_{in}$ integrated in the readout circuits, c - NETD. Standard deviation of input-FET threshold voltages - $\sigma(V_{th})$ = 5 mV. Standard deviation of the stoichiometric composition of $Hg_{1-x}Cd_xTe$ substrate $\sigma(x)$ = 0.1%. $V_G$ = 1.26 V. Average stoichiometric composition x=0.2105, T=60 K.

## 5. Conclusion

A mathematical model was developed to analyze the performance characteristics of IR FPAs based on the system 'IR photodiode – direct injection readout circuit'. The proposed mathematical model is based on the solution of the self-consistent problem for the current through photodiode and for the current integrated in the readout circuit, and also on the calculation of the noise charge Q(t) in terms of noise-current spectral density $S_i(\omega)$. Such an approach allows one to determine the main performance characteristics of thermography systems based on multi-element IR FPAs versus the gate voltage of photodetector channels, to calculate histograms $D^*(V_G)$, $I_{in}(V_G)$, and NETD as functions of non-uniformity of photoelectric parameters of photodiodes and silicon readout circuits, and to compare predicted performance characteristics of IR FPAs with experimental results.

The analysis of performance characteristic of multi-element FPAs can be performed with setting photodiode current-voltage characteristics either in analytical form ("classical" photodiode model, analytical model of $Hg_{1-x}Cd_xTe$ photodiodes) or as an approximation of experimental current-voltage characteristics of photodiodes. The proposed approach will be helpful in identifying the main factors that limit the performance characteristics of multi-element IR FPAs, and in formulating requirements to photoelectric characteristics of IR photodiodes, design and electrophysical parameters of readout circuits, and requirements to silicon technology used to fabricate direct injection readout circuits enabling the achievement of required performance characteristics of multi-element IR FPAs and thermography system based on such FPAs.

## 6. References

Anderson W. (1981). Tunnel contribution to $Hg_{1-x}Cd_xTe$ and $Pb_{1-x}Sn_xTe$ p-n junction diode characteristics. *Infrared Physics*, Vol. 20, (1981), pp 353-361

Anderson W. & Hoffman H. (1982). Field ionization of deep levels in semiconductors with applications to $Hg_{1-x}Cd_xTe$ p-n junctions. *J. Appl. Physics*, Vol. 53, N.12, (1982), pp.9130-9145

Bisotto S. et al. (2010). A 25µm pitch LWIR staring focal plane array with pixel-level 15-bit ADC ROIC achieving 2mK NETD. *Proc. of SPIE*, (2010) vol.7884, 7884oJ-1

Bluzer N. & Stehlik R. (1978). Buffered direct injection of photocurrents into charge-coupled devices. *IEEE Transaction on Electron Devices*, Vol.ED-25, No.2, (1978), pp.160-167

Buckingham M. (1983). Noise in Electronic Devices and Systems. *Ellis Horwood Ltd*, (1983)

D'Souza F. et al. (2002). 1/f noise in HgCdTe detectors. *Proc. SPIE "Infrared detectors and Focal plane arrays VII"*, No.4721, (2002), pp. 227-233

Felix P., et al. (1980). CCD readout of infrared hybrid focal plane arrays. *IEEE Transaction on Electron Devices*, Vol. ED-27, No.1, (1980), pp.175-188

Fowler B.l, Gamal A. & Yang D. (2000). Technical for pixel level analog to digital conversion. *Proc. SPIE*, Vol.3360, (2000), pp.124-138

Gopal V. (1996). Spatial noise limited NETD performance of a HgCdTe hybrid focal plane array. *Infrared physics and technology*, Vol.37, (1996), pp.313-320

Gumenjuk-Sichevska J. & Sizov F. (1999). Currents in narrow-gap photodiodes. *Semicond. Sci. Techn.* Vol. 14, (1999), pp.1124-1133

Gumenjuk-Sichevska J.; Karnaushenko D.; Lee I. & Polovinkin V. (2011). Infrared photodetectors based on the system $Hg_{1-x}Cd_xTe$ photodiode – direct-injection readout circuit. *Opto-electronics review* Vol.19, No.2, (2011), pp.63-70

Iwasa S. (1977). Direct coupling of five-micrometer (HgCd)Te photovoltaic detector and CCD multiplexer. *Optical engineering.* Vol. 16, N.3, (1977), pp. 233-236

Karnaushenko D.; Lee I. et al. (2010). Infrared focal plane arrays based on systems photodiode-direct injection readout circuits. *Opticheskii Zhurnal*, Vol. 77, No.9, (2010), pp.30-36

Krishnamurthy S. et al. (2006). Tunneling in long-wavelength infrared HgCdTe phodiodes. *Journal of Electronic Materials*, Vol. 35, No. 6, (2006), pp.1399-1402

Kunakbaeva G. & Lee I. (1996). Selection of the spectral range for infrared vision systems based on $Cd_xHg_{1-x}Te$ multielement photodiodes. *Optoelectronics, Instrumentation and Data Processing*, Vol.5, (1996), pp.19-26

Kunakbaeva G., Lee I. & Cherepov E. (1993). The system photodiode – direct-injection input CCD for multi-element FPAs. *Radiotekhnika i Electronika*, Vol.5, (1993), pp.922-930

Lee I. (2010). A new readout integrated circuit for long-wavelength IR FPA. *Infrared Physics and Technology*, Vol. 53, Issue 2, (2010), pp. 140-145

Longo J.T., et al.. (1978). Infrared focal plane in intrinsic semiconductors. *IEEE J. Solid State Circuits*, Vol. SC-13, N.1, (1978), pp.139-157

Martijn H.& Andersson J. (2000). On-chip analog to digital conversion for cooled infrared detector arrays. *Proc. SPIE "Infrared detectors and Focal plane arrays VI"*, Vol.4028, (2000), pp. 183-191

Mikoshima H. (1982). 1/f noise in n-channel silicon –gate MOS transistors. *IEEE Transaction on Electron Devices*, ED-29 – (1982), pp. 965-970

Overstraeten R., Declerck G. & Muls P. (1975). Theory of the MOS transistor in weak inversion – new method to determine the number of surface states. *IEEE Transaction on Electron Devices*, ED-22, (1975), pp. 282-288

Phillips J., Edwall D. & Lee D. (2002). Control of Very-Long-Wavelength Infrared HgCdTe Detector-Cutoff Wavelength. *Journal of Electronic Materials*, Vol. 31, No. 7, (2002), pp.664-668

Reimbold G. (1984). Modified 1/f trapping noise theory and experiments in MOS transistors biased from weak to strong inversion – influence of interface states. *IEEE Transaction on Electron Devices*, ED-31, (1984), pp. 1190-1198

Reimbold G. (1985). Noise associated with charge injection into a CCD by current integration through a MOS transistor. *IEEE Transaction on Electron Devices*, ED-32, (1985), pp. 871-873

Rogalski A. (2000). Infrared detectors. *Gordon and breach science publishers*, Canada, 2000

Sizov F. et al. (2006). Gamma radiation exposure of MCT diode arrays. *Semicond. Sci. Technol.*, Vol. 21, (2006), pp. 356-363

Steckl A.& Koehler T. (1973). Theoretical analysis of directly coupled 8-12 μm hybrid IR CCD serial scanning. *Proc. Int. Conf. Application of CCD's*, (1973), pp.247-258

Steckl A. (1976). Infrared charge coupled devices. *Infrared Physics*. Vol. 16, (1976), pp. 65.

Takigawa H., Dohi M. & Ueda R. (1980). Hybrid IR CCD Imaging Arrays. *IEEE Transaction on Electron Devices*, Vol. ED-27, No.1, (1980), pp.146-150

Taubkin I. & Trishenkov M. (1993). Minimum temperature difference resolvable with the IR imaging method. *Opticheskii Zhurnal*, Vol. 5, (1993), pp. 20-23

Tobin S., Iwasa S. & Tredwell T. (1980). 1/f noise in (Hg, Cd)Te photodiodes. *IEEE Transaction on Electron Devices*. ED-27 (1980), pp. 43-48

Vasilyev V. et al. (2010). 320.256 HgCdTe IR FPA with a built-in shortwave cut-off filter. *Opto–Electron. Rev.*, Vol.18, No. 3, (2010), pp.236-240

Yoshino J. et al. (1999). Studies of relationship between deep levels and RA product in mesa type HgCdTe devices. *Opto-Electronics Review*, Vol. 7, (1999), pp.361-367

Zhou Z., Pain B., et al. (1996). On-focal-plane ADC: Recent progress at JPL. *Proc. SPIE,* *"Infrared readout electronics III"*, Vol. 2745, (1996), pp.111-122

# 3

# Multiscale Computer Aided Design of Microwave-Band P-I-N Photodetectors

Mikhail E. Belkin

*Moscow State Technical University of Radio-Engineering, Electronics and Automation, Faculty of Electronics, Joint Research Laboratory "Microwave and Optoelectronic Devices", Moscow Russian Federation*

## 1. Introduction

Long wavelength InP-based p-i-n photodetectors (PD) are ubiquitous in modern optoelectronic circuits due to their inherent combination of ultra-high speed, high sensitivity in the most popular for modern telecom systems spectral range of 1.3-1.6 µm, and low bias voltages features that are impossible in principle for Si, GaAs or Ge counterparts. Typical material systems for the telecom spectral range are GaInAsP and GaInAs on InP substrate (Capasso et al., 1985). Now in the number of classical and present-day works (see, e.g. Bowers & Burrus, 1987; Beling & Campbell, 2009) is well-proved that compound semiconductor p-i-n photodetectors have the valid merits such as: high responsivity (up to 1 A/W), lowest dark current (below 10 fA), ultra-high bandwidth (up to 100 GHz and above), possibility of monolithic optoelectronic receiver module creation on common InP substrate.

Through a wide time period this type of PD found the most intensive application inside a receiver of the long-haul digital fiber-optic systems. But recently the p-i-n PDs for the wavelength of 1.3 and 1.55 µm are advancing in various RF and microwave photonic apparatus, particularly, in microwave-band optoelectronic oscillators (X.S.Yao, 2002), frequency converters (J. Yao, 2009), as well in so called photonic antennas for the base stations of radio over fiber (RoF) systems (Sauer et al., 2007). While designing a receiving module for the above devices and systems the stage of circuit schematic development referred to active and passive, electronic and photonic components' modeling would be most complicated and labour-intensive. Based on microwave IC design practice in this case the modeling is realized as a description of electrical and optical features of the integrated functional elements by means of mathematical equations, equivalent circuits or tables.

At present time computer-aided design (CAD) exploitation for innovative high-tech production R&D acceleration is a common way (Minixhofer, 2006). This is especially important for microwave semiconductor component base with measurement equipment and experimental work's cost being considerably more expensive compare to that in lower frequency bands. Using modern CAD software the next two approaches may be relevant for this case: by means of so-called physical models and by means of equivalent circuit models. A physical modeling is generally the most accurate but at the same time the most complicated. It is executed through a computer simulation of dedicated physical processes running in the semiconductor chip's epitaxial layers. On the other hand, equivalent circuit model is simpler and includes current and voltage sources as well as passive elements that subject to frequency band are built on a linear circuitry with lumped (for RF band) or distributed (for microwave band) parameters. Correct simulation of the real device features is ensured by the account of nonlinearity and lag effects in the sources and passive elements. Their parameters and characteristics are specified by the DC or low-frequency measurements also by the small- or large-signal mode measurements in operating band and adjusted due to parametric optimization process. The main advantage of the approach is in clarity of device operation scripts and in versatility for the same class components when the parameter spread could be simply taken in account by proper changing the numerical value of the separate element. Therefore, a simulation issue optimal decision based on criteria of accuracy and computation time would rely on multiscale combination of the physical and equivalent circuit models. For computations the means of one or a number of CAD tools could be applied. Such approach might be called as multiscale or end-to-end design.

Generally a transparent way for microwave photonic devices' computer design lies in simulating with a help of so called optoelectronic CAD (OE-CAD) tool (Lowery, 1997). Nevertheless, our review of modern commercial OE-CAD software has shown up their lack of development to simulate a microwave circuitry. An alternative approach is to model by two classes of CAD tools: so-called technology CAD or TCAD software (Blakey, 2008) and microwave-band electronic CAD or MW-ECAD software (Kielmeyer, 2008). By means of the first one a designer could realize the analysis procedures for a number of semiconductor chip physical, electrical and optical features and its DC characteristics; also simulate semiconductor laser, photodiode, transistor, or diode structure's small-signal frequency response. But for designing a number of relevant microwave photonics devices, e. g. optically controlled microwave amplifiers and switches, optoelectronic oscillators, transmitter and receiver optoelectronic modules, and so on one needs to know quite a number of the active semiconductor device parameters specifying its functioning in the operating bandwidth and in large-signal mode: output intercept point, harmonic and intermodulation distortions, large-signal output reflection coefficient. It is impossible to simulate the above features by the means of a TCAD tool. One more significant drawback of the tool is that the account of device package and assembling parasitic elements impacts is also impractical. This matter results in the solution error up to 30% even at the lower part of the microwave band that is growing along with a modulation frequency is being increased (Belkin & Dzichkovski, 2009).

The both issues are overcome by the application of ECAD tool optimizing for microwave band operation. The MW-ECAD tool special preference is in powerful electromagnetic (3D) analysis and optimization resources of microwave electronic circuitry and layouts design, for instance low-noise and power amplifiers, highly complex receiver optoelectronic modules. So introducing a convenient interface between these CAD tools one could realize multiscale end-to-end design of the device or module with microwave bandwidth. Usually for this goal well-known HSPICE interface is feasible. In this case, input data are multilayer semiconductor heterostructure with the certain cross section, and output data are a whole number of the outer parameters that totally and correctly characterize its features in microwave band. Also it might be a layout drawing of the module in development.

Thus, there exists a need for a reliable circuit-level photodetector model suitable for a designer of receiving modules of modern optical fiber-based systems and microwave photonics devices. In this chapter our recent results of the combined TCAD&ECAD and solitary MW-ECAD simulations for designing a long wavelength microwave-band p-i-n photodetector as a basic component of a multichannel analog system are presented.

## 2. General review

Generically, a photodiode as any electronic diode is a nonlinear device with a saturation effect and when it works in large-signal mode the fundamental signal is usually accompanied by a remarkable level of nonlinear distortions (Yu & Wu, 2002). The main distortion sources of the PD's conversion characteristic include absorption saturation, transit-time effects, and electric field screening arising from the space charge in the depletion layer (Williams et al., 1996). The latter effect also arises due to series resistance outside a junction region of a PD.

Nevertheless, in digital optical fiber systems handling with typical receiving power levels lower than 1 mW a photodiode is able to represent as linear device. On the other hand, in order to secure the needed spurious free dynamic range (SFDR) in the above microwave photonics devices and RoF systems the power levels may be up to 100 mW. For example a simple analysis of an opto-electronic oscillator model using quasi-linear interpretation results in the next expression for self-oscillating regime (X.S.Yao, 2002):

$$I_{ph}R \geq \frac{V_\pi}{\pi},$$ (1)

where $I_{ph}$ is the PD photocurrent; $R$ is the PD load impedance; $V_\pi$ is so called half-wave voltage of Mach-Zehnder modulator. Relationship between the photocurrent $I_{ph}$ and the PD optical power $P_0$ is addressed to the well-known formula:

$$I_{ph} = P_0 S_0,$$ (2)

where $S_0$ is the PD current responsivity. From (1) and (2) one can clearly conclude that self-oscillating condition could be simplified through increasing optical power at PD input. Namely, for typical values of $R$=50 $\Omega$, $V_\pi$=5 V, $S_0$=0.5 A/W it would be more than 64 mW. Naturally, it is unpractical to represent photodiode in such a regime as a linear element.

Large-signal operation of PD in the above-mentioned devices and system called out an additional issue that a pre-amplifier following the PD must concurrently have low noise figure and enough high linearity. A good chance to circumvent this issue is in utilizing a special type high-linear PD and a scheme without pre-amplifier (Joshi & Datta, 2009). Note that operating the photodiodes at larger photocurrent has added benefit of reducing overall noise figure of the device or link. But an employment of special type PDs that are structurally more complicate than standard ones might degrade the economical features of optical receivers. Thus, a direct linearity comparison of the widespread top-illuminated PD and middle-power microwave transistor amplifier is a subject of practical interest.

As known from RF and microwave amplifier design experience (Kenington, 2000), device linearity in large-signal mode is regularly characterized by so-called harmonic and inter-modulation distortions (IMD). Among them a two-tone third-order IMD (IMD3) is the most severe issue in design practice. The reason is in their closer positioning to carrier signals to allow rejecting by filters. Also for super wideband microwave photonic devices and systems when the bandwidth is more than an octave the two-tone second-order IMD (IMD2) is in great importance too. Generally, two-tone IMD is a product of passing two unmodulated carriers at the frequencies of $f_1$ and $f_2$ through a nonlinear device. When viewed in the frequency domain, output spectrum of the circuit includes besides $f_1$ and $f_2$ a lot of overtones. In the general case of distortions created by any order nonlinear circuit new frequencies will be generated in the form:

$$f_i = mf_1 \pm nf_2 \tag{3}$$

where $m$ and $n$ are positive integers (including zero), $m+n$ is equal to the order of the distortion. Then in the case of second-order nonlinearity each of the two tones will have a second harmonic and additional sum and difference beatings will occur at frequencies of $f_1 \pm f_2$. Also in the case of third-order nonlinearity each of the two tones will have a third harmonic and additional sum and difference beatings will occur at frequencies of $2f_1 \pm f_2$ and $2f_2 \pm f_1$. These overtones are known as second-order and third order intermodulation products correspondingly. One can calculate the above distortions through carrier transport equations (Williams et al., 1996) but the most simple and accurate way lies in measuring them in microwave band.

An obvious methodological drawback of IMD is referred to the dependence from powers of input signals that make impossible to compare devices at different input powers. To cope with it another well-known figure of merit for microwave amplifier linearity so called output intercept point (OIP) (Kenington, 2000) is widespread for microwave-band optoelectronic devices acting in large-signal mode. Following RF and microwave amplifiers, a rough estimate of the photodiode's 2-order OIP (OIP2) and the 3-order OIP (OIP3) can be produced by the following simple equations.

$$OIP2 = P_{out} - IMD2, \tag{4}$$

$$OIP3 = P_{out} - IMD3 / 2, \tag{5}$$

where $P_{out}$ is output AC power of the fundamental tone in the photodiode load. Each parameter is usually expressed as a power in dBm. However, a common practice to define

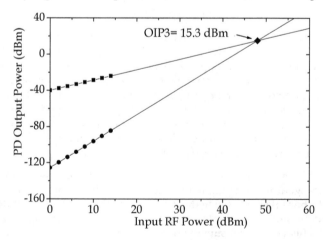

Fig. 1. A typical example of OIP3 determination for a photodiode

| No | Description of the PD under test | Small-signal bandwidth | Output Intercept Point | Reference |
|---|---|---|---|---|
| 1 | Large optical cavity p-i-n waveguide InGaAs photodiode | 20 GHz | OIP3 (-9 V, 1 GHz)=15.3 dBm<br>OIP3 (-9 V, 18 GHz)=2.4 dBm | Jiang et al., 2000 |
| 2 | Uni-traveling carrier waveguide integrated InGaAs photodiode | 20 GHz | OIP3 (-5 V, 1 GHz)=24 dBm<br>OIP3 (-5 V, 18 GHz)=8 dBm | Liao et al., 2003 |
| 3 | Backside illuminated uni-traveling carrier InGaAs photodiode | 10 GHz | OIP3 (-3 V, 10 GHz)=32 dBm<br>OIP3 (-5 V, 10 GHz)=40 dBm | Gustavsson et al., 2005 |
| 4 | GRIN lens-coupled top-illuminated InGaAs photodiode | 6.5 GHz | OIP3 (-5 V, 0.83 GHz)=43 dBm<br>OIP3 (-9 V, 0.83 GHz)=49 dBm | Joshi et al., 2008 |
| 5 | P-i-n waveguide InGaAs photodiode | 20 GHz | OIP3 (-4 V, 1.1 GHz)=32.6 dBm | Draa et al., 2008 |
| 6 | Charge compensated modified uni-traveling carrier InGaAs photodiode | 23 GHz | OIP3 (-7 V, 0.31 GHz)=52 dBm | Beling et al., 2008 |
| 7 | Partially-depleted absorber back-illuminated mesa structured photodiode | 10 GHz | OIP3 (-1 V, 1 GHz)=30 dBm<br>OIP3 (-3 V, 1 GHz)=38 dBm<br>OIP3 (-5 V, 1 GHz)=43 dBm | Tulchinsky et al., 2008 |
| 8 | Top-illuminated InGaAs p-i-n photodiode | 8 GHz | OIP3 (-2 V, 1 GHz)=17 dBm<br>OIP3 (-8 V, 1 GHz)=30 dBm<br>OIP2 (-2 V, 1 GHz)=33 dBm<br>OIP2 (-8 V, 1 GHz)=46 dBm | Godinez et al., 2008 |
| 9 | Top-illuminated InGaAs planar p-i-n photodiode | 3.2 GHz | OIP3 (-16 V, 0.83 GHz)=49 dBm | Joshi & Datta, 2009 |
| 10 | Ge n-i-p on silicon-on-insulator substrate photodiode | 4.5 GHz | OIP3 (-5 V, 1 GHz)=20 dBm<br>OIP2 (-5 V, 1 GHz)=40 dBm | Ramaswamy et al., 2009 |
| 11 | Backside illuminated InGaAs p-i-n photodiode array | 8 GHz | OIP3 (-2 V, 5 GHz)=32.5 dBm | Itakura et al., 2010 |
| 12 | Modified uni-traveling carrier InGaAs photodiode with cliff layer | 24 GHz | OIP3 (-5 V, 0.32 GHz)=45 dBm<br>OIP3 (-9 V, 0.32 GHz)=50 dBm | Li et al., 2010 |

Table 1. Results of photodetector's output intercept point measurements

OIP relies on a graphical representation of measured amplitude curves for fundamental and second- or third-order IMD tones. Following it Figure 1 illustrates a typical example of OIP3 determination for a photodiode (Jiang et al., 2000) extrapolated from the measured data of fundamental (squares) and IMD3 (circles) plots. In the Figure's logarithmic scale on the both axes the graphs represent straight lines where the fundamental tone has a slope of 1:1 and IMD3 tone has a slope of 1:3.

Up-to-date there is a series of papers referred to OIP measurements. Table 1 lists the results of some of them.

At a glance the values of photodiode OIP were eventually increased and for today the highest OIP of more than 50 dBm possesses charge-compensated modified uni-traveling carrier InGaAs photodiode (Beling et al., 2008). Standard vertical-illuminated p-i-n PDs have some lower OIP values up to 46-49 dBm. The general tendencies for all types of PDs consist in decreasing OIP as the test frequency and the negative voltage bias are increased.

Another known property of nonlinear circuits is that vector addition of the output fundamental and harmonic components also determines a phase variation of the resultant output signal when the input level varies. This effect is a requisite of dynamic systems and as the figure of merit for microwave amplifier linearity is called amplitude-to-phase (AM-to-PM) conversion for single-tone mode that is usually expressed as a certain phase deviation, in degrees or radians at predetermined input power (Pedro & Carvalho, 2003). In the case of photodetectors the similar parameter in terms of power-to-phase conversion (PPC) is being investigate by a joint research team (Tailor et al., 2011) for emerging microwave photonics-based timing applications in the fields where it is important to precisely know the arrival time of light pulse such as RoF, phased-array radars, optical fiber coherent communications and so on. Our results of MW-ECAD's photodiode nonlinear modeling will be highlighted in section 4.

## 3. Combined TCAD&ECAD simulation

To illustrate effectiveness of multiscale approach, the end-to-end design of the photodiode modules with bandwidth near thirty gigahertz was fulfilled. At the first step following the modern CAD tools capabilities (see the Introduction), the simulation of the super high-speed pin-photodiode was carried out with the aid of Synopsys Sentaurus TCAD software tool[1]. Sentaurus package allows for correct modeling of the wide spectrum of semiconductor devices due to transport equations and various physical models combining. There is fundamental capacity of photodetector simulation too but up to date the examples of it are not frequently (see e. g. Zakhleniuk, 2007).

### 3.1 P-i-n photodiode heterostructure TCAD simulation

For the simulation a widespread mesa-construction with semi-insulator-buried and vertically-illuminated p-i-n photodiode heterostructure similar to partially-depleted absorber type (Tulchinsky et al., 2008) was selected (Figure 2). In the Figure: region 1 is p+-type contact layer, region 2 – p-type buffer layer, region 3 – i-type absorption layer, region 4

---

[1] http://synopsys.com/home

- n-type buffer layer, region 5 – n$^+$-type contact layer. The mesa's material composition, layers initial doping levels and thickness selected for simulation are: region 1 - Ga$_{0.47}$In$_{0.53}$As, $5 \cdot 10^{18}$ cm$^{-3}$, 0.5 µm; region 2 –InP, $3 \cdot 10^{17}$ cm$^{-3}$, 3 µm; region 3 - Ga$_{0.47}$In$_{0.53}$As, $3 \cdot 10^{16}$ cm$^{-3}$, 1 µm; region 4 – InP, $3 \cdot 10^{18}$ cm$^{-3}$, 0.3 µm; region 5 - InP, $8 \cdot 10^{18}$ cm$^{-3}$, 2 µm. The substrate's doping level and thickness are $2 \cdot 10^{18}$ cm$^{-3}$ and 300 µm correspondingly. The chip dimensions are 200x200 µm$^2$, the mesa's top diameter is 50 µm.

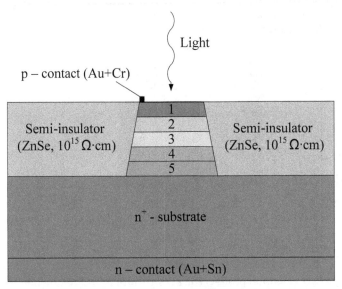

Fig. 2. The cross section of the photodetector in research

Such double-heterojunction p-i-n photodiode structure as will be shown below is lightly slower than more popular single-heterojunction one (Bowers & Burrus, 1987) but has a number of physics and technological advantages with minor deterioration of operating speed (Belkin et al., 2008):

- the possibility of the buffer and absorption layers precise doping;
- decrease of growth defects (growth pits, packing defects, absorption layer surface erosion);
- volatile components (such as As and P, and Zn doping agent) evaporation exclusion;
- the possibility of using InP "wide-gap window" effect that permits to collect photons on absorption layer without their surface recombination in the air-absorption layer interface;
- the possibility of desired form photodiode mesa-structure creation, using indium phosphide and solid solution etching selectivity, and small-area photodiodes forming selectivity;
- the possibility of compensating in part space-charge effect due to different thickness of the layers 2 and 4.

All the advantages above allow minimizing photodiode dark current ensuring maximum current responsivity.

The SYNOPSYS Sentaurus TCAD platform was used to model the Fig. 2's photodiode structure. The simulating procedure based on the following methodology. As the first step optical field distribution profile was inspected (Figure 3). In the result we determined that optical field is absorbed completely in the depletion layer 3 of Fig. 2 (dark horizontal stripe in Fig. 3). Then, by optical power-dependence reverse current-voltage characteristics computation we found that optimal photodiode operating voltage was -10 V (Belkin et al., 2008). The final step was AC small-signal current responsivity-frequency characteristics analysis and optimization in dependence of the absorption layer doping density, $N_D$, thickness, s and the photosensitive window width, w (in 2D-model).

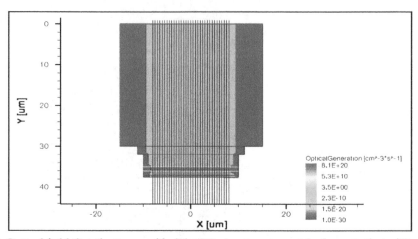

Fig. 3. Optical field distribution profile (The PD structure is upside-down to that of Fig. 2)

As known (Bowers & Burrus, 1987), the general limitations on the speed of p-i-n photodetectors are: 1) carrier drift time across depletion layer, 2) diode inherent capacitance charge/discharge time, 3) charge trapping at heterojunctions, 4) diffusing time out of undepleted regions, 5) package parasitic effects. Among them, in photodetector heterostructure with target bandwidth of 30 GHz the major effect is due to the first two factors because the hole trapping time even for abrupt p-InP/i-GaInAs heterojunction was estimated to be about 5 ps (Wey et al., 1995).

In the course of simulations by means of the program developed, some data for technology research and photodiode structure design, that are optimized for all the three above parameters, are received. First, optimal absorption layer doping density with s=1 μm and w=25 μm was investigated (Figure 4). Absorption region doping level analysis showed: 1) photodiode bandwidth was increased with the level diminishing, 2) there was no sense to provide lower than $N_D= 10^{16}$ cm$^{-3}$ as the bandwidth became almost invariable (The latter is important from the technology point), 3) to meet the above project goal of 30 GHz the simply realized by well-known liquid-phase epitaxy $N_D=1 \cdot 10^{16}$ cm$^{-3}$ is quite enough. That's why the above value was used in the future simulations.

Furthermore, optimal absorption layer thickness with $N_D=1 \cdot 10^{16}$ cm$^{-3}$ and w=25 μm was investigated (Figure 5). Absorption layer thickness analysis showed: 1) as for super high-speed p-i-n photodiode it is necessary not to diminish, according to known results (see e.g.

Agethen, et al., 2002), but to increase the thickness for bandwidth enhancement because of the need to consider not only time of carrier drift through depletion region, but diode's self-capacitance charge/discharge time too; 2) optimal thickness of absorption region for photodiode modeled is near s=1 μm as for the larger ones the bandwidth became almost invariable. That's why the above value was determined as optimal for the project goal achievement.

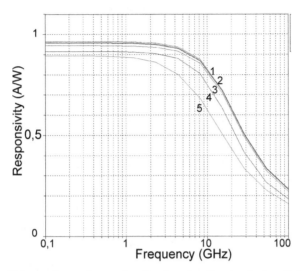

Fig. 4. The bandwidth vs. absorption region doping density simulation (1: $N_D$=8 ·$10^{15}$ cm$^{-3}$; 2: $N_D$=1 ·$10^{16}$ cm$^{-3}$; 3: $N_D$=2 ·$10^{16}$ cm$^{-3}$; 4: $N_D$=3 ·$10^{16}$ cm$^{-3}$; 5: $N_D$=4 ·$10^{16}$ cm$^{-3}$)

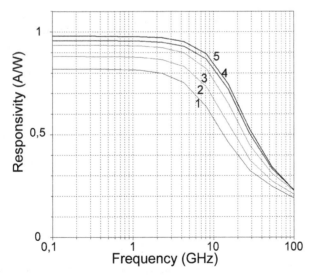

Fig. 5. The bandwidth vs. absorption region thickness simulation (1: s=0.3 μm; 2: s=0.55 μm; 3: s=0.8 μm; 4: s=1 μm; 5: s=1.3 μm)

Finally, the optimal photosensitive window width with $N_D=1 \cdot 10^{-16}$ cm$^{-3}$ and s=1 µm was investigated (Figure 6). Photosensitive window width analysis showed: 1) the photodiode bandwidth was increased with the width diminishing; 2) the project goal value of 30 GHz is arrived when w≤22 µm that relatively similar with the other results (Bowers & Burrus, 1987; Agethen, et al., 2002).

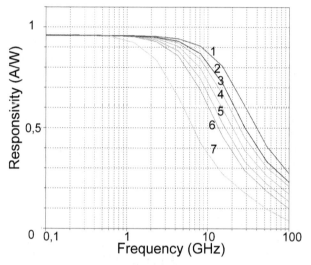

Fig. 6. The bandwidth vs. photosensitive window width simulation (1: w=18 µm; 2: w=20 µm; 3: w=22 µm; 4: w=24 µm; 5: w=26 µm; 6: w=28 µm; 7: w=30 µm)

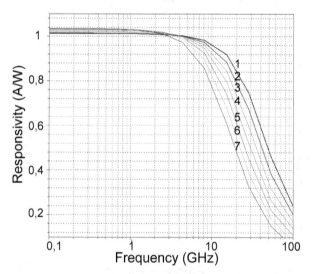

Fig. 7. The bandwidth vs. photosensitive window width simulation for a usual top-illuminated p-i-n photodiode structure with single heterojunction (1: w=18 µm; 2: w=20 µm; 3: w=22 µm; 4: w=24 µm; 5: w=26 µm; 6: w=28 µm; 7: w=30 µm)

Our last TCAD modeling was referred to the photodetector bandwidth estimation due to additional epitaxial layer introduction (Fig. 2, region 2). For this purpose the same photodiode structure response with closed region 2 was simulated (Figure 7). From the Figure one may conclude that the photosensitive window width needed for the project goal of near 30 GHz was increased slightly (up to 24 μm). So its effect has not to simplify substantively the mesa etching process.

### 3.2 P-i-n photodiode MW-ECAD simulation

As mention above for end-to-end design of a microwave-bandwidth photodiode an appropriate ECAD tool is required. The most convenient way for this microwave network research is the approximation by a physical equivalent circuit (Figure 8). In the Figure, $I_{ph}$ is the complex photocurrent source (frequency $F$ dependent), $R_d$, $C_d$, $R_s$ are the p-i-n photodiode heterostructure differential resistance, intrinsic capacitance and serial resistance respectively. Furthermore, $L_w$ and $C_c$ are the connecting wire inductance and case capacitance approximation.

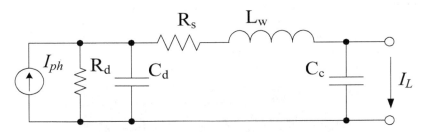

Fig. 8. The p-i-n photodiode electrical equivalent circuit

Based on the usual RF circuit simulation technique the photodiode transducer gain $k_{pd}$ can be determined from this scheme as:

$$k_{pd} = I_L / I_{ph} \qquad (6)$$

where $I_L$ is complex load current. Then from the Figure 8:

$$\dot{k}_{ph} = \frac{(1 + j\omega C_c z_L)^{-1}(1 + j\omega\tau)}{1 - \omega^2 L_w C_d + j\omega C_d R_s + z_L(1 + j\omega C_d z_L)^{-1}} \qquad (7)$$

where $\omega = 2\pi F$, $z_L$ is load impedance, $\tau$ is transit time.

The values of $I_{ph}(F)$, $\tau$, $R_d$=3.5 MΩ, $C_d$=80 fF, $R_s$=21 Ω were found analytically from the Sentaurus program. Also the values of $L_w$ and $C_c$=0.11 pF were determined experimentally by the microwave vector network analyzer. Figure 9 illustrates some results of the AWRDE [2] ECAD environment simulation. It follows from the graphs that the designer can realized twofold expansion of the photodiode 3 dB bandwidth (20 to 40 GHz) owing to the appropriate fitting of the connecting wire inductance $L_w$.

---

[2] http://web.awrcorp.com

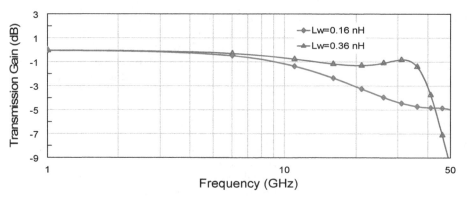

Fig. 9. The photodiode packaging effect simulation

### 3.3 Experimental verification

Based on the above simulation results some p-i-n photodiode samples were produced and mounted into the microwave microstrip test fixture. The typical time response of the photodetector sample measured by the two-section mode-locked laser and high-speed sampling oscilloscope is shown in the Figure 10. The pulse duration from the Figure is near 10 ps, so the accuracy of the above simulation results is verified.

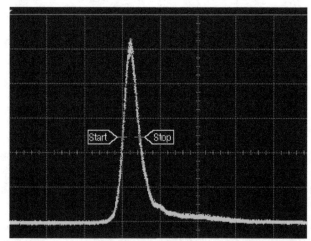

Fig. 10. Experimental photodiode impulse response (horizontal scale 25 ps/div.)

Thus, the methodology of the end-to-end computer-aided design for a p-i-n photodiode with microwave bandwidth is addressed. First, using the technology CAD tool a designer can define and optimize the dependence of the major photodiode structure parameters (absorption layer doping density and thickness, photosensitive window diameter, so on) on small-signal bandwidth and get the data for the following calculations. Then the electronic MW-CAD simulation based on the photodiode electrical equivalent circuits is performed that allow to take into account various effects in real receiver module with microwave bandwidth: LC package parasitics, matching networks, front-end amplifier and so on. To confirm the actuality and accuracy of the proposed methodology the specific p-i-n photodiode with bandwidth of near 30 GHz was simulated. The simulation results were experimentally verified with fine accuracy.

## 4. Solitary MW-ECAD simulation

The main concept of this approach can be formulated as follows: a designer of analog receiving modules usually does not know a lot about the PD heterostructure's specific parameters, usually has not an access to a specific cost-expensive TCAD tool but he has facilities suitable for measuring transmittance and reflectance (or S-parameters) of the photodiode chip under test in the optical system's modulation band. Then equipped with experimental results, and equivalent circuit interpretation as well as AWRDE or another MW-ECAD tool facilities he builds up a non-linear "optoelectronic" equivalent circuit and if necessary some linear or nonlinear electronic circuits behind it.

### 4.1 P-i-n photodiode nonlinear model

Figure 11 represents cascaded equivalent circuit of PD that does it quite effective for simulation by means of a highly-developed MW-ECAD tool. There are two sub-circuits inside it: linear (right) and nonlinear (left). The linear sub-circuit simulating the frequency distortions due to parasitic elements impact agrees with small-signal PD model of Fig. 8. The nonlinear sub-circuit includes a diode D1 and a capacitance of p-n junction that is changed in according with applied reverse voltage (element PNCAP). To supply the nonlinear elements a DC source V2 is introduced. Inductance L1 is assigned for DC/AC isolation; resistance R2 models PD's DC termination. Non-ideality, phase shift, and delay of the optical-to-electrical conversion process are modeled by voltage-controlled current source U1.

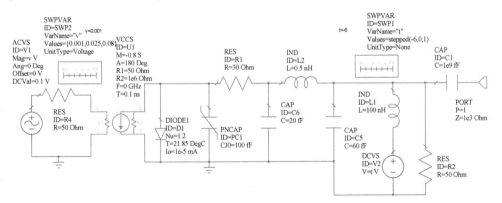

Fig. 11. MW-ECAD nonlinear p-i-n photodiode model based on electrical equivalent circuit

For the simulation an equivalent electrical AC source V1 is placed at the circuit input. A frequency of this source is tuned through the microwave-band modulation bandwidth, and its amplitude is in exact numerical agreement with a power of intensity-modulated optical signal. As it follows from Figure 11 for the given model experiment the next values are assumed: an equivalent power of optical carrier is 100 mW (DC voltage 0.1 V), equivalent modulated optical power is changed in the range of 10-100 mW (0.01-0.1 V) that meets the optical modulation rate between 10 and 100 percentage. A reasonableness of the optoelectronic circuit's simulation by ECAD tool is founded on the proposed method of equivalent voltage which uses identical linear dependences of a current vs. voltage in an electrical circuit and a photocurrent vs. optical power in an optoelectronic circuit of a photoreceiver.

Thus, the next linear or nonlinear effects are taken into account in the given model: (i) nonlinearity of depletion layer capacity; (ii) frequency distortions due to parasitic elements of the PD package and receiver circuitry; (iii) reflections from photosensitive window; (iv) dark current and its dependence of temperature and bias voltage; (v) DC photocurrent generation due to optical carrier impact; (vi) deviation of the optical signal modulation rate; (vii) electric field screening due to series resistance.

### 4.2 Simulation results and discussion

At the first step, to calibrate the model the device-under-test small-signal frequency characteristics are simulated. As the result, 3-dB bandwidth is limited in the range of 20-30 GHz depending on reverse bias voltage and agreed well with the Figure 9 at the bias of 10 V. Then, the above-described three nonlinear distortion parameters depending on equivalent input optical power and reverse bias (Ub) are simulated through one-button operations. Figure 12 shows an example of equivalent AM-to-PM distortion characteristics at the fundamental frequency 15 GHz.

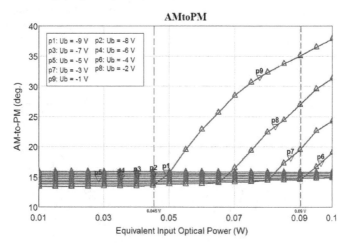

Fig. 12. Photodiode power to phase conversion characteristics

Also Figure 13 depicts an example of OIP2 characteristics for the input tones of 5 GHz and 6 GHz, and distortion tone of 11 GHz. At last Figure 14 shows an example of OIP3 characteristics for the input tones of 5 GHz and 5.1 GHz, and distortion tone of 4.9 GHz.

A combined outcome from Figures 12-14 follows that large-signal conversion features of a p-i-n photodetector are notably degraded as far as negative reverse DC bias is increased that is quite coincided with the experimental data of Table 1. In particular, Figure 12 depicts that the PPC linearity threshold is twofold as the PD bias is decreased from -1 to -4 V. Also from Figure 13 one can see that second-order output intercept points are almost constant vs. optical power and their numerical deviations vs. bias are near 26 dB (from 47 dBm at -8 V to 21 dBm at -1 V). In spite of this, shown in Figure 14 third-order output intercept points begin to degrade with optical power at the bias more than -4 V, and their numerical deviations vs. bias are near 17 dB (from 44 dBm at -8 V to 27 dBm at -1 V). Note that the results of power-to-phase conversion simulation are comparable in principle with recent experimental data (Tailor et al., 2011), also the results of OIP2 and OIP3 simulation are closed to recent experimental data of Table 1.

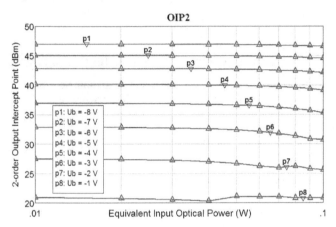

Fig. 13. Output second-order intercept point characteristics

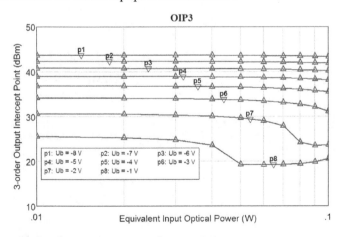

Fig. 14. Output third-order intercept point characteristics

Generally, based on our model experiments we are able to state that simple in design and producing cost-effective top-illuminated photodetector has been inspected holds somewhat (3-5 dB) lower linearity as compared to a special high-linear charge-compensated modified uni-traveling carrier photodiode (see Table 1). But the latter has at the same time significantly higher bandwidth running up to THz band (Nagatsuma et al., 2007). Nevertheless the researched type of photodetector thanks to inherent cost-effectiveness also has a potential for RoF systems and microwave photonics applications with moderate frequency and power requirements.

For example, the typical values of the allowable AM-to-PM distortion and OIP3 for middle-power reasonable-noise microwave amplifiers are not more than $10°$ as well near 30-35 dBm correspondently. Following it we are able to conclude: (i) even a standard p-i-n photodetector is more linear circuit element than pre-amplifier and can be used without it in the most photoreceiving schemes of microwave photonics devices and RoF system's photonic antennas; (ii) to secure a comparable p-i-n photodetector linearity the reverse bias voltage must be less than -4 V at optical power up to 100 mW.

## 5. Comparison of OE-CAD and MW-ECAD photodiode simulations

As we communicated in Introduction, the available optoelectronic CAD tools are unpractical for designing nonlinear device operating in microwave band. That's why it is in essential importance to state the numerical value of error while simulating for instance noise features of a photodetector in microwave band using MW-ECAD and OE-CAD software. Our direct comparison was fulfilled by AWRDE and advanced OE-CAD tool VPItransmission Maker™ version 8.6.

In the case of VPI software a library p-i-n photodiode model considering thermal and shot noise sources and dark current of 0.15 nA was selected. On the other hand, for photodiode MW-ECAD simulation the equivalent circuit model of Fig. 11 was used. For the both simulating experiments a broadband Gaussian-distributed white noise generator model with identical power levels was employed as a source. Varying its levels output noise power was measured in the band from 1 GHz to 50 GHz by the noise analyser model in the bandwidth of 1 Hz. Figure 15 depicts the simulation results.

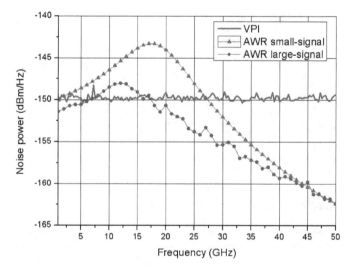

Fig. 15. OE-CAD tool vs. MW-ECAD tool simulation

As shown from Figure, simulated by VPI (solid curve) spectral density of average output noise power represents frequency and noise source power independent horizontal line that is unrealistic. On the other hand, simulated by AWRDE results follow the real frequency characteristic of the photodiode under test and differ subject to small-signal (triangles) or large-signal (circles) power regime of the noise source. The numerical value of error is up to 10 dB.

## 6. Conclusion

This chapter presented a multiscale approach to the combined TCAD&ECAD and solitary MW-ECAD simulations for designing a microwave-band p-i-n photodetector as a basic component of the receiving modules for radio-over-fiber systems and microwave photonics devices. Within the limits of the first approach, primarily, using a TCAD tool a designer can define and optimize the dependence of the major photodiode structure parameters on small-signal bandwidth. Then an electronic MW-CAD simulation based on the photodiode electrical equivalent circuits is performed that allow to take into account various effects in real receiver module with microwave bandwidth. To confirm the actuality and accuracy of the proposed methodology the specific p-i-n photodiode with bandwidth of near 30 GHz was simulated. The main concept of the second approach can be formulated as follows: a designer of analog receiving modules usually does not know a lot about the PD heterostructure's specific parameters, usually has not an access to a specific cost-expensive TCAD tool but he has facilities suitable for measuring transmittance and reflectance of the photodiode chip under test in the optical system's modulation band. Then equipped with experimental results, and equivalent circuit interpretation as well as AWRDE or another MW-ECAD tool facilities he builds up a non-linear "optoelectronic" equivalent circuit and if necessary some linear or nonlinear electronic circuits behind it. Following the simulation results we have concluded: (i) even a standard p-i-n photodetector is more linear circuit element than pre-amplifier and can be used without it in the most photoreceiving schemes of microwave photonics devices and RoF system's photonic antennas; (ii) to secure a comparable p-i-n photodetector linearity the reverse bias voltage must be less than -4 V at optical power up to 100 mW.

The modeling results are validated by the experimental data for TCAD&ECAD similation and by close agreement with the known measured data for MW-ECAD large-signal simulation.

## 7. Acknowledgment

This work was supported by the Russian Ministry of Education and Science program "Progress of scientific potential of the higher school (2009-2011)". Author wishes to thank Dr. E. Portnoi from Ioffe Physico-Technical Institute RAS for the mode-locked laser's photodiode measurement.

## 8. References

Agethen, M., et.al. (2002). InGaAs pin Detectors for Frequencies above 100 GHz, *IEEE Indium Phosphide and Relative Materials Conference, IPRM'2002*, A8-2, pp. 673-676

Beling, A., Pan, H., Chen, H. & Campbell, J.C. (2008). Measurement and modeling of high-linearity modified uni-traveling carrier photodiode. *IEEE Photonics Technology Letters*, Vol.20, No.14, (July 2008), pp. 1219-1221, ISSN 1041-1135

Beling, A. & Campbell, J.C. (2009). InP-Based High Speed Photodetectors, *IEEE Journal of Lightwave Technology*, Vol. 27, No 3, (February 2009), pp. 343-355

Belkin, M.E., Dzichkovski, N.A. & Indrishenok.V.I. (2008). Super high-speed p-i-n-photodiode heterostructures computer simulation. *Journal of Nano and Microsystem Techniques*, No 10 (99), (October 2008) pp. 23 – 28, (in Russian)

Belkin, M.E. & Dzichkovski, N.A. ( 2009). Research of microwave-bandwidth p-i-n photodetectors, *Proceedings of EUROCON* 2009, p.193-196, St. Peterburg, Russia, May 2009

Blakey, P.A. (2008). Technology CAD. In: *RF and Microwave Circuits, Measurements, and Modeling*, M. Golio, J. Golio, (Ed.), Boca Raton, 34-1 – 34-11, ISBN 13-978-0-8493-7218-6, London

Bowers , J. E. & Burrus C.A. Ultrawide-Band Long-Wavelength p-i-n Photodetectors, *IEEE Journal of Lightwave Technology*, vol. 5, No 10,( October 1987), pp. 1339-1350

Capasso, F., et. al. (1985). Photodetectors, In: *Lightwave Communications Technology*, Vol. 22, Part D, W.T. Tsang, ( Volume Ed.), Academic Press Inc.

Draa, M.N., et al. (2008). Frequency behaviors of third order intercept point for a waveguide photodiode using three laser two-tone setup. *Proceedings of 21st Annual Meeting of the IEEE Lasers and Electro-Optics Society. LEOS 2008*, pp. 284-285, Acapulco, Mexico, November 9-13, 2008

Godinez, M.E., et al. (2008). RF Characterization of Zero-Biased Photodiodes. *IEEE Journal of Lightwave Technology*, Vol. 26, No 24, (December 2008), pp. 3829-3834

Gustavsson, M., Hedekvist, P.O. & Andrekson, P.A. (2005). Uni-Travelling-Carrier Photodiode Performance with X-band Modulation at High Optical Power. *IEEE Microwave and Wireless Components Letters*, v.15, No.5, (May 2005), pp. 297-299, ISSN 1531-1309

Itakura, S., et al. (2010). High-Current Backside-Illuminated Photodiode Array Module for Optical Analog Links. *IEEE Journal of Lightwave Technology*, Vol. 28, No 6, (March 2010), pp. 965-971, ISSN 0733-8724

Jiang, H. et al. (2000). The Frequency Behavior of the Third-Order Intercept Point in a Waveguide Photodiode. *IEEE Photonic Technology Letters*, Vol.12, No.5, (May 2000), pp. 540-542

Joshi, A., Datta, S. & Becker, D. (2008). GRIN Lens Coupled Top-Illuminated Highly Linear InGaAs Photodiodes. *Photonics Technology Letters*, Vol.20, No.17, (September 2008), pp. 1500-1502, ISSN 1041-1135

Joshi, A. & Datta, S. (2009). Highly Linear, High Power Handling Photodiode for RF Photonic Links, *Proceedings of ECOC 2009*, p. 9.2.4, Vienna, Austria , 20-24 September 2009

Kenington, P.B. (2000). *High-Linearity RF Amplifier Design*. Artech House, ISBN 1-58053-143-1, Norwood, MA

Kielmeyer, R. (2008). Computer Aided Design (CAD) of Microwave Circuitry. In: *RF and Microwave Circuits, Measurements, and Modeling*, M. Golio, J. Golio,, (Ed.), Boca Raton, 31-1 – 31-8, ISBN 13-978-0-8493-7218-6, London

Li, Z., et al. (2010). High-Saturation-Current Modified Uni-Traveling-Carrier Photodiode With Cliff Layer. *IEEE Journal of Quantum Elecronics*, Vol.46, No.5, (May 2010),pp. 626-632, ISSN 0018-9197

Liao, T.S. et al. (2003). Investigation of the High Power Integrated Uni-Traveling Carrier and Waveguide Integrated Photodiode, *IEEE 2003 MTT-S International Microwave Symposium Digest*, pp. 155-158, vol.1, Philadelphia, PA, 8-13 June, 2003

Lowery, A.J. (1997). Computer-aided photonics design. *IEEE Spectrum*, Vol. 34, No 4 (April 1997), pp. 26-31, ISSN: 0018-9235

Minixhofer, R. TCAD as an integral part of the semiconductor manufacturing environment. (2006), *Proceedings of International Conference on Simulation of Semiconductor Processes and Devices, SISPAD*, pp. 9 – 16. Monterey, California, September 6-8, 2006

Nagatsuma, T., Ito, H. & Ishibashi, T. (2007). Photonic THz Sources Using Uni-Traveling-Carrier Photodiode Technologies. *Proceedings of LEOS 2007, International Topical Meeting of Laser and Electro-Optic Society*, pp. 792-793.

Pedro, J. C. & Carvalho, N. B. (2003). *Intermodulation Distortion in Microwave and Wireless Circuits*, Artech House, ISBN 1580533566, Boston, London

Ramaswamy, A. et al. (2009). Experimental analysis of two measurement techniques to characterize photodiode linearity. *Proceedings of MWP '09, International Topical Meeting on Microwave Photonics*, pp. 1-4, ISBN 978-1-4244-4788-6. Valencia, Spain, October 14-16, 2009

Sauer, M., Kobyakov, A. & George, J. (2007). Radio over Fiber for Picocellular Network Architectures, *IEEE Journal of Lightwave Technology*, vol. 25, no. 11, (November 2007), pp. 3301-3320

Taylor, J.A., et al. (2011). Characterization of Power-to-Phase Conversion in High-Speed P-I-N Photodiodes. *IEEE Photonics Journal*, Vol. 3, No.1, (February 2011), pp. 140 – 151, ISSN 1943-0655

Tulchinsky, D.A. et al. (2008). High Current Photodetectors as Efficient, Linear and High-Power RF Ouput Stages. *IEEE Journal of Lightwave Technology*, Vol. 26, No.4, (February 2008), pp. 408–416

Wey, Y.-G. et al. (1995). 110-GHz GaInAs/InP Double Heterostructure p-i-n Photodetectors, *IEEE Journal of Lightwave Technology*, vol. 13, No 7,( July 1995), pp. 1490-1499

Williams, K. J., Esman, R. D. & Dagenais, M. (1996). Nonlinearities in p-i-n Microwave Photodetectors., *IEEE Journal of Lightwave Technology*, vol. 14, No 1,( January 1996), pp. 84-96

Yao, J.P. (2009). Microwave Photonics, *IEEE Journal of Lightwave Technology*, vol. 27, No 3, (February 2009), pp. 314-335

Yao, X.S. ( 2002). Opto-electronic Oscillators, In: *RF Photonic Technology in Optical Fiber Links*, W. S. Chang(Ed.), 255-292, Cambridge University Press, Cambridge, UK

Yu, P. K. L. & Wu, M. C. (2002). Photodiodes for high performance analog links, In: *RF photonic technology in optical fiber links*, W. S. Chang, (Ed), 231-254, Cambridge university press, Cambridge, UK

Zakhleniuk, N.A. (2007). Theory and Modelling of High-Field Carrier Transport in High-Speed Photodetectors. *Proceedings of NUSOD'07 Numerical Simulation of Optoelectronic Devices*, pp. 77-78, ISBN 978-1-4244-1431-4, Newark, DE, September 24-28, 2007

# Part 2

# Photodetection Systems

# 4

# Photoconductors for Measuring Speckle Motion to Monitor Ultrasonic Vibrations

Jonathan Bessette and Elsa Garmire
*Dartmouth College*
*USA*

## 1. Introduction

This chapter explains a unique miniature photoconductive detector design that has been specifically engineered with a speckle-monitoring application in mind. Small arrays of these detectors have been fabricated with silicon-on-insulator (SOI) technology and in semi-insulating GaAs, and have been put to work measuring surface vibrations in laboratory experiments. The important characteristics of these photodetectors are their small physical dimensions, on the order of 10 $\mu$m, with an ability to measure ultrasonic bandwidths extending into the megahertz regime, and with high internal gain. The advantage of photo-conductors is their capability of generating multiple electrons of photocurrent for each photon captured.[1]

## 2. Performance of a general photodetector

At the most fundamental level, electronic photodetection is a method of counting photons by converting them into some form of electronic current. There are many ways to do this, including all the familiar detector technologies such as photodiodes, photomultiplier tubes, and photoconductors, to name a few. For the purposes of this discussion, all these detectors can be abstracted into a simple detector model consisting of three primary parameters: gain, noise, and bandwidth.

Gain can be neatly expressed as the number of electrons per unit time of signal current either generated or modulated by the detector divided by the number of photons per unit time incident upon it. A gain of one, for example, indicates that one electron flows for every photon collected.

Noise is a measure of the electronic current generated or modulated due to random fluctuations in either the arrival or conversion rate of photons, or due to the random generation of the flow of electrons within the detection circuit not caused by photons at all.

Bandwidth is a measure of the maximum frequency response of the detector to a photonic signal. The higher the bandwidth, the faster the detector can respond to changes in the level

---

[1] Much of this chapter comes from the PhD thesis of Jon Bessette: "Silicon-on-Insulator Photoconducting Mesas For High-Speed Laser Speckle Monitoring Applications," Dartmouth College, June, 2010.

of illumination. Bandwidth limits may be imposed by the internal physics of the device or additionally by a combination of factors in the circuit implementation.

Gain, noise, and bandwidth are always intimately connected in any detection system and improved performance in one parameter often comes at the price of decreased performance in another. It is important therefore to understand the connections between them for any given detector technology and the detector application.

A simple model combines these three elements into a description of the signal from a generic photodetector in the useful form of a signal-to-noise ratio (SNR). This ratio indicates a lower bound of sensitivity for which a detector is useful in a given operating regime. The convention adopted here is to express the SNR as the ratio of signal power to noise power:

$$SNR_{gen} = \frac{(e\Gamma A\Delta I)^2}{\sigma^2_{noise}(\Gamma, B)} \tag{1}$$

The numerator is given in units of current squared, where $e$ is the electron charge, $\Gamma$ is the detector gain, $A$ is the detector area, and $\Delta I$ is the change in optical intensity incident on the detector during the time $2/B$. The noise term $\sigma^2$ is expressed in units of current squared and is generally a function of the gain $\Gamma$, bandwidth $B$, and the specifics of the detector. There are multiple sources of noise in the signal of any detector that measures changes in incident optical power. The predominant forms of noise in the simple model presented here are photon shot noise, current shot noise, and circuit thermal noise. These noise sources are all well understood and characterized by circuit designers and so will only be summarized here:

- Photon shot noise is caused by the random arrival of discrete photons in a given photon stream. The power in the photon shot noise is proportional to the average incident optical power, and the variance can be given in units of current as $\sigma^2_\varphi = 2e^2\Gamma^2 FAI_oB/h\nu$, where $F$ is an excess noise factor created by probabilistic gain mechanisms and $I$ is the intensity incident on the detector (Liu, 2005).
- Current shot noise is caused by the random arrival of discrete electrons in a given electric current. Its power is proportional to the average background current $i$ and its variance is given as $\sigma_i^2 = 2ei\Gamma^2 FB$.
- Circuit thermal noise is created by random thermal motion of the electrons in a conductor. It is independent of the current flowing through the load resistor $R_L$ and is present even when no electric potential is applied. The thermal noise variance is approximately $\sigma_K^2 = 4k_BTB/R_L$.

In general, the different noise sources are assumed to be independent of each other. This means that the variance of the combined noise amplitude is the sum of the variances of the individual noise sources, i.e. $\sigma_{noise}^2 = \sigma_\varphi^2 + \sigma_i^2 + \sigma_K^2$. To quantify the expected performance of a given detector in a particular operating regime, we must explore the magnitude of the various gain and noise terms for different classes of detectors. Because rather small area detectors are required for our application, we will restrict ourselves to the discussion of semiconductor-based detectors. These can be manufactured on the appropriate scales and easily configured for array operation, which can greatly enhance their effectiveness.

## 2.1 Gain and noise

The fundamental operating principle of all solid-state photosensitive devices is the same. A photon with sufficient energy interacts with a semiconductor crystal, temporarily changing the distribution of electron energies within the crystal. One electron gains enough energy to attain an energetic conductive state, where it is free to move about within the crystal. The promoted electron leaves behind a vacancy, called a hole, which can also move about within the crystal. Together, these are referred to as an electron-hole pair, and in an ideal photodetector, one such pair is created for every absorbed photon. Eventually, if left in the crystal long enough, the electron-hole pair will recombine, giving up the extra energy in the form of heat. This happens on a characteristic time scale called the recombination lifetime, $\tau_r$.

While excited, the electron and hole will drift in the presence of an electric field, creating an electric current. This current can be detected by connecting the active area of the semiconductor into an electronic circuit. How the electric field is applied and the nature of the electrical contacts defines the class of detector.

### 2.1.1 Photoconductors

Photoconductors are uniformly doped semiconductors whose conductance changes in proportion to the number of photons absorbed per unit time. When an electrical potential is placed across the photoconductor, the amount of current flowing through the photoconductor indicates the amount of light incident on it. The photoconductor is perhaps the simplest kind of semiconductor photodetector. It is perhaps surprising, then, that such a simple device may exhibit internal gain.

The photoconductor gain is the ratio of electron collection at the contacts to electron-hole pair generation by photons within the channel. The gain is the ratio of recombination time to the average transit time of charge carriers across the photoconductive channel. This ratio is, in turn, determined (at low applied voltages) by the photoconductor's effective combined carrier mobility $\mu = (\mu_e + \mu_h)$, its carrier recombination time $\tau_r$, its channel length $L$ and the applied voltage $V$ according to

$$\Gamma = V\mu\tau/L^2 \tag{2}$$

(for a square detector, $\Gamma = V\mu\tau/A$ where $A$ is area). This gain process is fundamentally probabilistic for each individual photon absorbed, and so necessarily the mechanism produces an excess noise factor. The excess noise factor for a uniformly illuminated photoconductor with such a gain, governed by Poisson statistics of a uniform random variable, is 2 (Liu, 2005).

It is important to note that the gain does not scale linearly with voltage indefinitely. At least two effects may limit the maximum gain seen inside a photoconductive element. The first of these is drift velocity saturation, which occurs when the electric field is great enough to yield carrier drift velocities on the order of the carrier thermal velocity. Increasing the electric field beyond this saturation value - about 30 kV/cm for silicon - does not cause the carriers to drift any faster (Neamen, 2004). Consequently, the transit time is not reduced with greater field strength and gain saturates.

A second mechanism that limits the internal gain is space-charge current limitation (Liu, 2005; Bube, 1960; Rose 1958), when charges build up on either electrode due to the effective capacitance of the device. When this charge exceeds the total number of free carriers within the photoconductor, the excess charge is injected into the device and carries a current that is limited by the dielectric relaxation of the capacitor (Rose, 1954). These so-called space-charge currents screen the photogenerated carriers from any additional electric field, so that increasing the applied potential difference beyond a certain threshold does not increase the photocurrent, and the gain saturates.

The threshold voltage that limits the gain is called the space-charge voltage. It is equal to the applied potential that causes the charge on the effective capacitor to exceed the number of free carriers within the photoconductive channel. For a photoconductor with ohmic contacts and charge carriers of both sign, this is explicitly

$$V_{SC} = \frac{\oint (\sigma_0 + \sigma_{ph}) \mathbf{dv}}{e \mu C} \tag{3}$$

where $\sigma_0$ is the dark conductivity, $\sigma_{ph}$ is the photogenerated conductivity, and the integral is over the volume of the photoconductor. The value given for capacitance depends on the exact geometry and material properties of the photoconductor, but for the simplest model of parallel electrodes and a block-shaped semiconductor, the capacitance is simply $C = WD\varepsilon/L$ and the space-charge voltage is (Liu, 2005)

$$V_{SC} = \frac{(\sigma_0 + \sigma_{ph})L^2}{\epsilon \mu}. \tag{4}$$

The gain limit for a space-charge limited device becomes

$$\Gamma_{max} = \frac{\tau(\sigma_0 + \sigma_{ph})}{\epsilon}. \tag{5}$$

Note that the gain limit is a function of photoconductivity (i.e., depends on the amount of light being absorbed), but that this does not greatly affect the performance unless the photoconductivity is comparable to the dark current.

For the dimensions and material of devices investigated in this chapter, it is not the saturation velocity limit but rather the space-charge current that limits the photoconductive gain.

As far as noise is concerned, the dark current shot noise of a photoconductor can be significant in certain operating regimes. For an ideal photoconductor, this noise figure scales linearly with the applied voltage and the cross-sectional area-to-length ratio of the active region. This can be deduced by plugging in the ideal photoconductor dark current into the definition of shot noise

$$\sigma_{i,d}^2 = 2eV \frac{WD}{\rho_0 L} B, \tag{6}$$

where $\sigma_0$ is the dark resistivity of the photoconductor. We will see in a following section that this noise figure is dominated by the circuit thermal noise at high bandwidths.

## 2.1.2 Photodiodes

Photodiodes are rectifying junctions created by adjoining two regions of a semiconductor crystal that are doped with different impurities. The impurities change the equilibrium concentrations of conductive electrons and holes. When a semiconductor is $p$-doped, it has an excess of mobile positive charges (holes). Likewise when it is $n$-doped, the semiconductor has an excess of negative charges (electrons). Note that the total charge of the crystal remains neutral - this implies there are fixed negative charges in the $p$-doped regions and fixed positive charges in the $n$-doped regions.

When two regions of opposite doping are brought into contact, diffusion results in excess mobile charges from each region spreading into the neighboring region, leaving behind some fixed charge. The fixed charge is of opposite polarity on either side of the junction and together the two zones are called the space-charge region.

The space-charge region supports an electric field which points from the $n$-doped side of the junction to the $p$-doped side of the junction. When a photon is absorbed in this region, the resulting electron and hole are quickly swept out through opposite contacts as a result of the electric field. Reverse-bias is generally applied, which means a relatively positive potential is placed on the $n$-doped side of the junction. This widens the junction and increases the maximum field in the space-charge region.

Once mobilized by the absorption of the photon, the electron drifts toward the (positively charged) $n$-doped side of the space-charge region and the hole toward the (negatively charged) $p$-doped side of contact, and the circuit registers a blip of electric current. Once the carriers reach the end of the space charge region, they eventually recombine.

In contrast to the ohmic photoconductor contacts, the $pn$ junction barrier under reverse bias acts as a completely blocking contact for the drift of both charge carriers, so that no additional carriers are injected into the space-charge region. Since the photo-induced charge only transits the active region once, the photodiode produces only one unit of electronic current for every photon absorbed. The basic photodiode thus has a maximum of unity internal gain.

The dark current is generally very low for most photodiodes, on the order of pico-amperes for small diodes, and it only decreases as the detector area is reduced. Consequently, the dark current noise is dominated at higher bandwidths by the circuit thermal noise.

## 2.1.3 Metal-semiconductor-metal photodiodes

Metal-semiconductor-metal photodiodes are created by making contact with a semiconductor using a metal that forms a Schottky potential barrier. Each metal-semiconductor junction exhibits the rectifying properties of a diode. Charge carriers diffuse across the junction to establish a space-charge layer at each contact, similar to $pn$ junction diodes. However, unlike $pn$ diodes, the current across the Schottky junction barrier is carried by the drift of semiconductor majority-charge carriers and not minority carrier diffusion.

When a bias is placed across the contacts, one of the junctions is forward-biased and one is reverse-biased. It is the reverse-biased contact that limits the current in the rectifying regime.

Another distinction that separates MSM photodiodes from *pn* junction diodes is that they can exhibit photoconductive gain greater than unity, despite the non-ohmic contacts and rectifying Schottky barriers. The source of this gain is the same mechanism as for photoconductors, namely multiple charge carriers make a transit across the semiconductor region for every photon absorbed, despite the presence of the Schottky barriers at the contacts. This is possible because carriers can be injected via thermionic emission over the Schottky barrier (Neamen, 2003; Soares, 1992; Burm, 1996) to maintain charge neutrality. That is to say, the Schottky barrier is not a completely blocking contact - it merely presents an additional potential barrier that increases the device resistance. This is a key difference between *pn* junction photodiodes and Schottky barrier photodiodes.

The dark current in MSM photodiodes can be many orders of magnitude greater than the dark current in reverse-biased photodiodes. Consequently, the dark current noise in MSM photodiodes is of greater concern. The dark current can be estimated by the following formula for the reverse-bias saturation current density in units of current per cross-sectional area (Neamen, 2003):

$$J_{sT} = A^* T^2 \exp(\frac{-e\Phi_B}{k_B T}) \tag{7}$$

where $A^*$ is called the Richardson constant with units of AK$^{-2}$cm$^{-2}$ (A is amperes) and is specific to the semiconductor, T is temperature in Kelvin, and $\Phi_B$ is the Schottky barrier height in units of eV.

As an example, consider a silicon Schottky diode with chromium contacts, similar to the SOI prototype construction discussed later in this chapter. The Richardson constant for silicon is approximately $A^*$ = 114 AK$^{-2}$cm$^{-2}$ and the barrier height is approximately 0.49 V. Thus, at room temperature, the reverse saturation (dark) current is approximately $J_{sT}$ ≈ (114 AK$^{-2}$cm$^{-2}$) × (300 K)$^2$ × exp(-0.49 eV/0.026 eV) = 67 mA/cm$^2$. For a cross sectional area of 10 $\mu$m × 10 $\mu$m, this gives about 67 nA of dark current, which has a significant impact on noise at low bandwidths.

It should be noted that the Schottky barrier height can be a significant function of both the applied voltage and the level of illumination. This occurs through several mechanisms, including image force lowering (the Schottky effect), (Neamen, 2003; Soares, 1992) charge tunneling from the metal across the barrier (Soares, 1992), and photogenerated charge accumulation (Soares, 1992; Carrano, 1998). By these processes, charges are pulled to either contact where they establish compensating electric fields. The effect is to lower the potential barrier with additional applied voltage and consequently modify the current-voltage characteristics of the diode - as the reverse bias voltage increases, so does the reverse saturation current.

### 2.2 Small area detectors in the thermal noise limit

By way of example, let us compare the predicted performance of photoconductors and photodiodes in the face of noise for a megahertz bandwidth and low detector area situation. The parasitic capacitance for both photodiodes and photoconductors for small detectors is usually somewhere in the 10 pF range or greater. If we want to operate with a 1 MHz bandwidth, we require an $RC$ time constant such that $R_LC = 1/(2\pi \times 1\text{MHz}) = 160$ ns. This means we need to use a load resistance no greater than $RL = 16\ \Omega$, and the thermal noise generated by such a load at room temperature $T$ will be

$$\sigma_K = \sqrt{4k_BTB/R_L} \approx 1\ \text{nA} \tag{8}$$

This noise figure is quite significant and dominates the other primary sources of noise, photon shot noise and dark current noise. Take the photon shot noise for example. A detector with an active area of 100 $\mu$m$^2$ under 1 mW/cm$^2$ illumination sees a total of 1 nW optical power $P_0$. At an optical wavelength of 1 $\mu$m, this corresponds to a photon shot noise of

$$\sigma_\Phi = (\sqrt{2P_0F/h\nu})e\Gamma B \approx \Gamma \times 10\ \text{pA}.$$

This is less than thermal noise for gains less than 100.

The dark current noise is often negligible as well. For small area photodiodes, the dark current itself is negligible compared to the thermal noise. (Consider that the dark current scales with detector area and 1 mm$^2$ detectors have dark currents in the pA range.) Photoconductors have considerably more dark current, but in most cases the noise associated with it will not be more significant than the thermal noise. For instance, according to Eq. 6, the dark current noise of an ideal photoconductor with a dark resistivity of 10 k$\Omega$-cm that is 10 $\mu$m thick, 10 $\mu$m wide and 10 $\mu$m long, and operated with 1 V applied at 1 MHz bandwidth is

$$\sigma_{i,d} = \sqrt{2eVWDB/\rho_0L} \approx 200\ \text{pA}$$

This is still less than the thermal noise, even for the highest allowable resistor load.

If we examine the SNR for photoconductors and photodiodes using the above approximations and Eq. 1, we obtain

$$SNR = \frac{(e\Gamma A\Delta I)^2}{4k_BT/R_L} \tag{9}$$

where the gain $\Gamma_{pd} = 1$ for photodiodes and the gain $\Gamma_{pc}$ is a function of geometry and applied voltage for photoconductors. In this approximation, the SNR scales with the square of the internal gain. This implies that the performance of a photoconductor will exceed the performance of a photodiode whenever the gain is greater than one.

In Fig. 1, a direct comparison between the SNR for photoconductors and photodiodes is plotted as a function of detector area, $WD$, assuming a square detector, load resistance of $RL$ = 1 k$\Omega$, a bandwidth of $B$ = 1 MHz, a photoconductor recombination time $\tau_R$ = 10 ns, a photoconductor mobility of 1850 cm$^2$=V•s, and an applied photo-conductor voltage of 1 V. The signal is assumed to be a 50% modulation of uniform 1 mW/cm$^2$ illumination. The conductivity is 10$^{-3}$ $\Omega^{-1}$cm$^{-1}$, giving a maximum photoconductive gain of $\Gamma_{max}$ = 9.6. For this example, the threshold for enhanced photoconductor performance is at 43 $\mu$m, and gain saturation occurs at widths below 13 $\mu$m.

Below a certain threshold of detector area, the photoconductor exhibits a superior SNR to the photodiode. This is a direct consequence of the internal gain mechanism of the photoconductor: the noise is the same for both detectors, but the gain is for a photodiode is limited to one, whereas the gain for the photoconductor can be significantly greater than one.

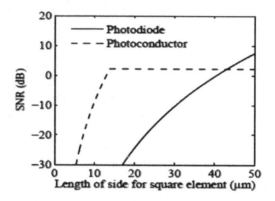

Fig. 1. Plot of SNR for a photoconductor and SNR of a photo-diode of the same square size as a function of the side-length.

## 2.3 Summary

In the limiting case of the thermal noise regime, where photon shot noise and current shot noise are dominated by the thermal noise of the load resistance, a detector with high internal gain is preferred. This thermal noise limit is approached at low light levels and at higher bandwidths, precisely the operating conditions required by speckle monitoring detectors for ultrasound vibration measurement. Both photoconductors and metal-semiconductor-metal detectors can exhibit internal gain at megahertz bandwidths with detector areas appropriate for direct speckle monitoring, and so merit further investigation.

## 3. Photoconductor development

This section describes the design and development process for the photoconductor prototypes built and tested at Dartmouth College for use in speckle monitoring applications. It also discusses several of the device characterization experiments performed on the resulting devices.

### 3.1 Gallium arsenide photoconductors

The first prototypes for speckle monitoring were constructed from GaAs (Heinz, 2004; Heinz, 2007; Heinz, 2003). They consisted of a raised mesa of GaAs 100 $\mu$m wide by several hundred $\mu$m long divided into four active areas defined and separated by superposed contacts. Each active area was approximately 100 $\mu$m $\times$ 40 $\mu$m.

Although quite sensitive to speckle motion, these devices suffered from a lack of bandwidth beyond some tens of kilohertz, and so were unsuitable for systems requiring ultrasonic bandwidths into the megahertz regime. It was determined that the limiting mechanism was most likely space-charge buildup, a result of high contrast in conductivity between the illuminated active regions and the shadowed contact regions.

Two potential solutions were investigated to reduce the effect of space-charge in the GaAs photoconductors and increase the individual detector bandwidth. The first of these

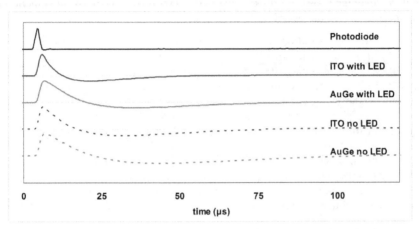

Fig. 2. Plot of GaAs photoconductor response to a 3 µs long laser diode pulse for various devices with and without LED background illumination. The response from a silicon photodiode is shown for comparison. The values on the y-axis are normalized (arbitrary units) (Bessette, 2006).

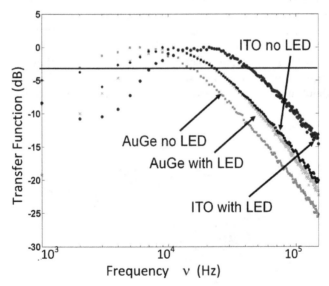

Fig. 3. Fast Fourier transform of AuGe and ITO device impulse response with and without red LED background illumination. The graph shows that the ITO devices are inherently faster than their AuGe counterparts with and without the 32 mW/cm² background illumination. The -3 dB rolloff for the un-illuminated AuGe device is around 10 kHz, while the ITO device rolls off at 20 kHz. With background illumination, the ITO device rolls off around 40 kHz (Bessette, 2006).

solutions was to use a transparent electrical contact, thereby lessening contact shadowing and reducing the impact of the resulting space-charge. To that end, a new GaAs prototype design was developed and produced (Bessette, 2006). The major improvement in this detector generation was the use of indium-tin oxide (ITO) for electrical contacts.

The second proposed solution for improving bandwidth was to provide a uniform background illumination with an LED during device operation. This would reduce space-charge buildup by shortening the space-charge relaxation time via increased average conductivity. Both of these measures were evaluated for effectiveness in a single impulse-response style experiment. Figure 2 shows the oscilloscope traces from different versions of the detector in response to an impulse-modulated laser diode. The settle time is seen to improve significantly with the presence of LED background illumination in both ITO and AuGe contacted devices. The settle time is also remarkably shorter for the ITO device, compared to the AuGe device. The photodiode response is limited by the diode laser driver which has a 200 kHz rolloff frequency. There was indeed an improvement in response time for the devices when using transparent contacts and background illumination, but as seen in Fig. 3, the gain in frequency response was still not enough to obtain ultrasonic bandwidth.

## 3.2 Silicon-on-insulators as a solution

To altogether eliminate the problems associated with persistent space-charge, a photoconductor array made from a relaxation-limited semiconductor was proposed. Silicon is a logical choice, since it is relaxation time-limited, responsive to visible and near-infrared wavelengths, and the most commonly used semiconductor material. The challenge with silicon is that in bulk form, the carrier diffusion length, or the distance on average that an unconstrained, excited electron or hole will diffuse before recombining, can be on the order of centimeters (Neamen, 2003). This rules out the kind of mesa architecture utilized in the GaAs array design, since the photo-generated carriers would escape from each element and diffuse throughout the crystal, destroying the spatial resolution required for speckle monitoring.

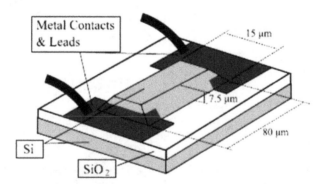

Fig. 4. Each detector element consists of an island of silicon bonded to an insulating layer of silicon dioxide grown on the silicon substrate. These structures are achieved by anisotropic etching through photolithographic masks. Electrical connection to each island is made by metal contacts, also defined by photolithography.

A new design was proposed incorporating silicon-on-insulator (SOI) technology, which consists of a thin layer of silicon crystal, on the order of 10 μm thick, fusion-bonded to a layer of insulating amorphous silicon dioxide grown on a supporting silicon wafer. Photoconductive islands are defined by etching through the thin top layer of silicon on every side, as is shown in Fig. 4. The mutual isolation of the islands in the array ensures that carrier diffusion will not compromise the spatial resolution of each detector element.

### 3.3 Fabrication details

The fabrication of SOI detector arrays, although they are comprised of only a few component layers, is not a trivial procedure to develop or execute. The difficulty arises as a consequence of both the small feature-size and the high aspect-ratio of individual photoconducting islands. The structures must be considered as fully three-dimensional, which makes processing in a top-down manner, the traditional approach to two-dimensional semiconductor wafer processing, more difficult.

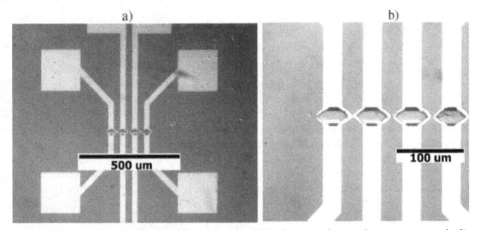

Fig. 5. a) Photograph of a finished four-element SOI photoconductor device array, including metal contact pads. In this design, each element can be contacted individually; b) Close-up of a finished four-element array. This sample was presumably underexposed during the contact layer photolithograpy step, resulting in a short circuit around the perimeter of each element after the metal etch step. This array was not actually used in experiments.

The following is an overview of the most current fabrication protocol for the SOI devices as designed for this project and executed in the available microfabrication facilities. An outline of microfabrication protocol:

1. RCA cleaning routine to remove surface impurities from wafer
2. Oxide layer grown on outer silicon layer
3. Photolithography step to define oxide etch
4. Oxide etch with HF, leaving oxide as a hard mask
5. Silicon etch with KOH
6. Metal contaminant removal and wafer cleanup
7. Optional removal of remaining oxide mask layer and optional oxide under-etch filling with hard-baked photoresist

8.  Thermal deposition of chromium and gold layers to provide adhesive conducting layer
9.  Photolithography step to define electrical contacts
10. Gold etch
11. Chrome etch
12. Mount wafer dice into electronic DIP package, connect leads to contacts

Optical micrographs of a finished sample that incorporates four individually addressable elements into a small array are shown in Fig. 5.

### 3.3.1 Anisotropic etching of silicon device layer

In order to facilitate the required electrical connections using top-down deposition of electrical contacts, a potassium hydroxide (KOH) anisotropic silicon etch step was chosen to define the mesa structures from the thin SOI device layer. KOH attacks the different crystal planes of silicon at different rates. Chiefly, there is a high degree of selectivity between the etch rate in the (100) direction of the crystal and the etch rate in the (111) direction. The (100) direction etches hundreds of times faster than the (111) direction. The result of etching with a mask edge aligned parallel with the <110> plane (that is, generally speaking, a mask edge aligned with the wafer flat of a (100) oriented wafer) is a sloped vertical etch exposing the <111> crystal plane, as is seen in Fig. 6(a).

This is in contrast to an isotropic chemical etch such as with nitric acid (HNO₃) based etchants, which would result an under-etched mesa, as in Fig 6(b). Under-etching is bad in this context for two reasons. Firstly, it would be very difficult to connect these devices electrically without removing the top oxide mask. Removing the oxide mask is undesirable since such a step would also partially etch the SOI buried oxide layer, further complicating electrical connection. Secondly, since the devices are almost as thick as they are wide, isotropic etching would eliminate the majority of the actual device. This problem only gets worse as the device dimensions are decreased. For these reasons the anisotropic etch was chosen.

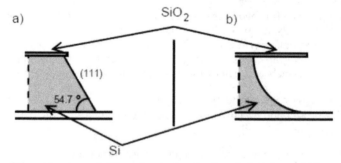

Fig. 6. Sketch of the side-view of silicon facets due to chemical etching a) Sloped vertical facet resulting from anistropic KOH etching of (100) oriented silicon. b) Under-etching resulting from an isotropic etch in which the device is etched laterally as much as it is etched vertically.

One difficulty encountered in utilizing such an anisotropic etch on these structures is the undesired etching of convex corners. There are many other crystal directions besides (110) which etch faster than the (111) direction. When convex corners are exposed to the etchant, all of these other directions get etched, in addition to the desired downward etching (Lee, 1969; Wu, 1989). This results in convex corners that are etched inward, rapidly degenerating the structure from that direction. Fig. 7 shows the result of this effect, and demonstrates how this undesirable effect is exacerbated as device dimensions are reduced.

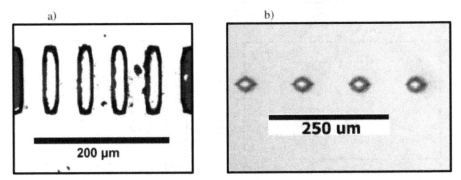

Fig. 7. The devices in prototype (a) are much longer than wide and the device layer thickness of silicon is only 7.5 $\mu$m, so the convex corner etching, while apparent, is not as significant. The devices in prototype (b) were meant to be more square, and the device layer thickness is 10 $\mu$m, which exaggerates the effect of the convex corner etching.

Convex corner etching is particularly problematic for these three dimensional structures, whose height is on the same order as their width. A common solution to this problem is to extend the etch mask over these corners (Wu, 1989; Biswas, 2006). The idea is to add extra sacrificial material that must be etched in the fast etching directions of the corner, thereby giving enough time for the vertical etch to finish before the protected corner material begins to etch. See Fig. 8 for a diagram of such a mask solution, after (Wu, 1989). Certainly, any new prototype including the anisotropic etch step during fabrication should include some form of convex corner protection in the mask design.

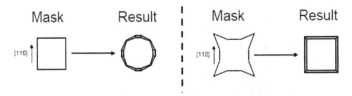

Fig. 8. Sketch of typical results of mask patterns for anisotropically etched structures with convex corners. The triangular overhangs on the mask help protect the desired corner integrity.

### 3.3.2 Electrical contacts

Early versions of the prototype used only a sputtered gold contact layer, but this was found to be insufficiently adhesive. In later prototypes, the contact layer was deposited via thermal deposition of a thin layer of chromium (to promote adhesion) of approximately 30 Å thickness followed by thermal deposition of a conductive gold layer of approximately 2000 Å. These contacts proved to be robust enough to avoid excessive flaking, but they are not ohmic, as illustrated by the current versus applied voltage curve of a single device shown in Fig 9. The metal-semiconductor junctions formed by the contacts do exhibit rectifying behavior that typifies MSM photodiodes rather than a pure photoconductor. The next section discusses the implications of this result.

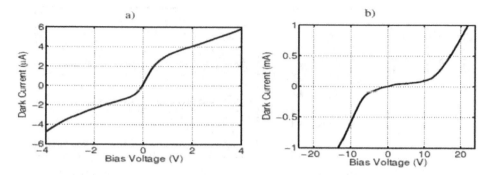

Fig. 9. Current versus bias voltage for an SOI detector with zero incident optical power a) near zero bias and b) extending into diode breakdown regions. The shape of the curve is typical of MSM photodiodes (Soares, 1992).

### 3.4 Characterization

After the prototypes were constructed, they were evaluated on different aspects of their performance, including the current-voltage relationship, gain, bandwidth, and noise.

### 3.4.1 Current versus voltage

Figure 9 shows the current versus voltage for one SOI device both near zero bias and over an extended range. The devices behave as metal-semiconductor-metal diodes already introduced in a previous section. Near zero bias (see Fig. 9(a)) there is a greater effective conductivity until one of the metal-semiconductor junctions is sufficiently reverse-biased to partially rectify the current, at which point the effective conductivity decreases. Note that the reverse bias current densities of these Schottky diodes are orders of magnitude greater than the equivalent $pn$ junction diodes. Additionally, the barrier-lowering effects of the applied electric field result in a continuously rising reverse current. At even greater biases, the reverse bias current increases at a much faster rate due to ionization multiplication effects and eventually junction breakdown (Neamen, 2003; Tu, 1992).

Even without improving the contacts to be more ohmic, SOI MSM devices can still have the internal gain mechanism that makes them ideal for speckle monitoring applications with higher bandwidths. This has been discussed earlier and is illustrated in the next section on device characterization.

To optimize these devices for an extended range of ohmic behavior, some experimentation with the contact materials and deposition procedure will be required. Ohmic contacts in silicon devices are generally achieved by introduction of a heavily doped silicon region near the contact, or by the use of silicide contacts (a silicon compound with a metal such as titanium, cobalt, tungsten, or platinum) rather than pure metals (Sze, 2007).

Additional doping would require at least one additional photolithographic and dopant implantation step, and is perhaps best to be avoided if possible. The formation of silicide contacts, however, requires only a slight modification of the contact deposition process and no additional photolithography. Silicide formation can be achieved in any one of four ways, including *direct reaction, chemical vapor deposition, co-evaporation,* or *co-sputtering* (Sze, 2007). Direct reaction is the easiest method to implement - it involves a metal deposition step followed by high temperature annealing. The metal component of the silicide is deposited (for instance by thermal evaporation or sputtering) onto the desired contact area. During the annealing step, some of the silicon device is consumed and becomes part of the contact silicide layer. A combination of careful literature review and experimentation should be sufficient to determine the appropriate metal choice and processing protocol for the formation of properly ohmic contacts with less rectifying effect on the current-voltage behavior of the photoconductors.

There is an apparent asymmetry in the biasing direction. This is possibly due to a difference in the metal-silicon junction size on either side of the element, which would make the reverse bias current in one direction greater than in the other. All elements in this array showed approximately the same shift with biasing applied in the same orientation.

### 3.4.2 Gain

The photo gain of an SOI device was measured as a function of bias voltage and compared to a silicon *pn* junction photodiode. The raw photocurrent measurements were converted to gain by calibration with a silicon photometer. In Fig. 10, we see that the silicon photodiode exhibits the expected flat gain throughout reverse bias. The gain value is approximately 0.85, indicating an internal quantum efficiency of about 85%.

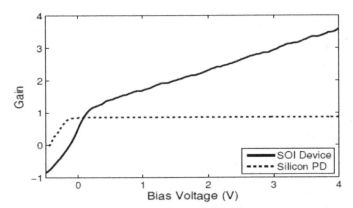

Fig. 10. Measured photo gain for an SOI photodetector (solid) compared to a reverse-biased silicon photodiode (dashed) as a function of bias voltage.

In contrast, the SOI device starts with a small gain near zero bias (due to the Schottky barrier induced fields). The gain increases sharply with applied voltage, quickly surpassing the *pn* junction diode gain, until the bias reaches 0.20 V. At this point, the gain continues to rise linearly, but at a reduced rate.

Two related phenomena were discovered in the behavior of the photodetectors that were not initially predicted. Firstly, the gain increases monotonically with bias voltage up to a maximum and then declines again. It was thought that the gain would saturate, but not decline. Secondly, the gain is not independent of incident illumination, but actually decreases with increased light levels.

Figure 11 shows the block diagram of an experiment performed to demonstrate this behavior. A sweeping DC bias voltage was applied to one SOI element under illumination from two diode lasers. The first laser was a near-IR InGaAs diode that provided a constant background level of illumination. The second laser was an AlGaAs diode operating at 661 nm that provided a low level AC signal (10 kHz). The resulting photocurrent was filtered with a lock-in amplifier and recorded as a function of the bias voltage to obtain the curves shown in Fig. 12.

The peaking phenomenon at some bias voltage and the reduction in gain at high illumination levels can be understood in the context of carrier traps within the photoconductor crystal (Rose, 1958; Macdonald, 2001). The presence of traps in silicon and SOI that affect carrier lifetimes is well noted (MacDonald, 2001; Celler, 2003).

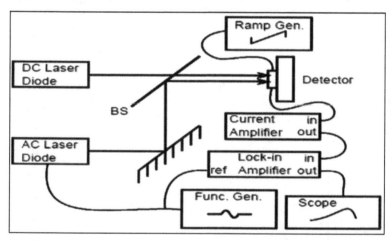

Fig. 11. Block diagram for experimental small-signal gain measurements as a function of bias voltage and background illumination.

Some trapping behavior can increase the effective lifetime of charge carriers; for instance, the presence of shallow traps in effective thermal equilibrium with the conduction band. These states capture some of the photogenerated carriers and prevent them from recombining and returning to the ground state. Since charges in these trapping states communicate with the conduction band, they can carry currents and contribute a photocurrent component with a long-lived response.

Another trap-related phenomenon that can contribute to the apparent long carrier-lifetimes is electron-hole separation due to the trapping of one of the charge carriers (MacDonald,

2001). For example, if holes are trapped in a state where electrons cannot recombine with them, a corresponding excess of free electrons will build up and contribute to the photoconductance (even if the holes do not contribute to the photoconductance).

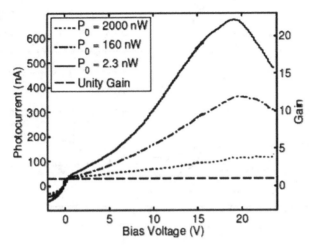

Fig. 12. SOI photodetector gain over a larger interval of bias voltage, for three levels of background illumination. The gain is a strong function of background illumination, presumably due to the sensitivity of the recombination time to trap-filling. Gain includes reflectivity losses.

The trapping states can become saturated by the two methods, as suggested by the curves in Fig. 12, both of which increase the density of excited charge carriers. The first is to raise the bias beyond the threshold to induce space-charge currents. The second is to expose the detectors to increasing levels of background illumination. In either case, the traps will begin to fill up and eventually become saturated.

At low carrier concentrations, these traps can significantly lengthen the effective carrier lifetime. Since the gain is proportional to lifetime (see Eq. 2), the gain is greatest when these traps are relatively empty and available to receive carriers. This phenomenon is also evident in the following impulse response and frequency measurements.

Since we know how much background optical power it takes to reduce the excess gain, we can make an estimate as to the density of these trapping states once we determine their effective lifetime. That will be determined in the next section.

### 3.4.3 Impulse response measurement

To determine if the prototypes were fast enough to operate at ultrasonic frequencies, an impulse-response measurement was made by exposing a device to a 10 ns square wave optical pulse from a 780 nm wavelength diode laser. Fig. 13 shows the recorded. response.

Beyond a certain bias level, the decay time of the response drops off dramatically, presumably due to the filling of shallow traps by excess dark current. Background illumination from a secondary laser diode also reduces the decay time of the detector response, presumably due to shallow trap filling.

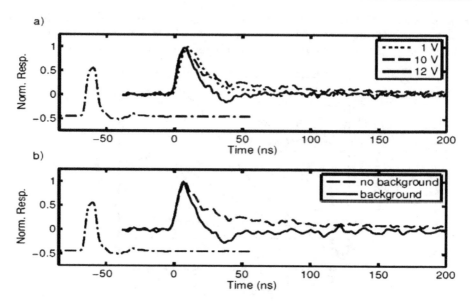

Fig. 13. Photodetector response to a 10 ns square pulse from a 780 nm diode laser under a) variable bias voltage with no background illumination and b) 10 V, bias with and without DC background illumination. Photodiode response for comparison is time-shifted to an earlier time for clarity.

Two things are immediately apparent. First, the bias voltage and background illumination greatly affect the decay time of the response. Second, even at its greatest, the detector response time is on the order of 50 ns. This indicates that the device should adequately measure signals well into the megahertz range with maximum internal gain, already measured greater than 20.

The impulse response appears to be composed of two components, a fast initial response and a slower decay. The slower component can be reduced by adding more bias voltage or by adding more background illumination. Fig. 14(a) shows the actual detector response, under minimal background illumination, modeled well by a double exponential decay with two time-constants, one at $\tau_1$ = 6 ns and one at $\tau_2$ = 90 ns. This secondary response, with the long decay rate, is responsible for the high gain at lower bandwidths up to a few MHz, as can be seen in Fig. 14(b). It is suppressed by either the excessive addition of bias voltage or by the application of strong background illumination, also illustrated in Fig. 14. Carrier lifetimes in this range, tens or hundreds of nanoseconds - much less than in bulk silicon - are often found in SOI structures (Mendicino, 1999; Rong, 2004). The lifetime shortening is due to the proximity of the active area to $Si/SiO_2$ interfaces.

We can make an order-of-magnitude estimate of the lower limit of trap carrier density by assuming that all of the high-intensity background light is absorbed by the long-lived trapping states, and furthermore that the states are evenly distributed throughout the device.

The power required to saturate those states multiplied by their effective recombination lifetime equals the total number of states in the device. We can see from Figs. 11 and 12 that the traps seem to saturate when 1 $\mu$W or more is absorbed by the device. With a response time of 90 ns and a device volume on the order of $10^3$ $\mu$m$^3$, this indicates that the trap density is at least $4 \times 10^{12}$ cm$^{-3}$. Where they exist, these traps are probably considerably more dense, since a) the background light is not exclusively absorbed by the trapping states and b) the trapping states are most likely associated with the device interfaces and not distributed evenly through the device volume.

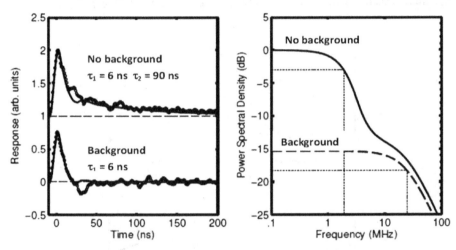

Fig. 14. a) Measured impulse response at 10 V bias (circles) overlaid with a fit of the double exponential impulse response (solid) for the detector with no background illumination and a single exponential response for the detector with significant background illumination. b) The frequency response of the fitted impulse responses. Note that detectors with sufficient (but not too much) bias voltage and low levels of background illumination show a significantly boosted response with bandwidths up to 2 MHz.

### 3.4.4 Detector noise

As already discussed, the best performance of the detector will be achieved when the SNR is maximized for a given level of illumination. This SNR value depends on both the noise levels and gain of the device. As long as the noise is dominated by the thermal noise term, which is unaffected by the bias voltage, the highest bias voltage (below saturation) and thus the highest possible gain should be used to maximize the SNR. Fig. 15(a) shows both the measured gain for the SOI device together with the combined thermal and current shot noise levels based on the dark current measurements. Assuming a 120 $\Omega$ load resistance and a 1 MHz detection bandwidth, the overall noise is dominated by the thermal noise component and is virtually unchanging until the bias voltage exceeds 10 V. All the while, the gain continues to rise with increasing bias, improving the signal-to-noise ratio.

However, as the bias voltage is increased, so is the dark current, which in turn increases the current shot noise. At some point, the dark current shot noise will become a significant portion of the overall noise. This increased noise will cancel the benefit of increased gain. The bias beyond which this happens can be determined by plotting both the gain and the noise on the same semilog plot, as in Fig. 15(b). When the slope of the noise versus voltage exceeds the slope of the gain versus voltage on a logarithmic scale, the noise is increasing faster than the gain, and no benefit is obtained by turning up the bias more. The best SNR is obtained for this device somewhere around 15 V. The benefit of using detectors with enhanced internal gain in detector regimes dominated by thermal noise is nicely illustrated by this particular measurement. If the detector gain was limited to unity, about 1/20 of the actual measured value, the best the detector could achieve in terms of SNR is less than 0.2% of the actual measured value. This is obviously a significant improvement.

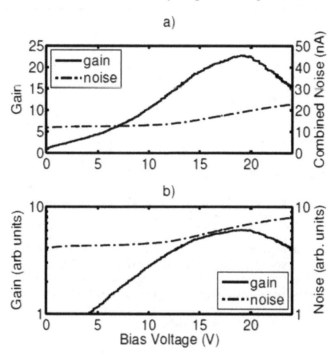

Fig. 15. a) Internal gain and detector noise as a function of bias voltage; b) The same values plotted on a semilog scale. When the slope of the noise exceeds the slope of the gain, the noise increases faster than the gain, and there is no benefit from increasing the bias. Combined detector noise is calculated assuming thermal noise of a 120 Ω load resistance and the shot noise associated with the measured dark current values. A detection bandwidth of 1 MHz is assumed.

### 3.5 Summary

Gallium aresenide photoconductors of the type discussed in (Heinz, 2004, 2005, 2007, 2008) were determined to be too slow for megahertz detection of speckle motion due to the buildup of space charge, even with the slight improvement of transparent electrical contacts and background illumination. Silicon mesa structures were developed that do not suffer from the same space-charge bandwidth limitations. The mesa architecture provides spatial isolation of each detector, and the prototypes proved to be fast enough to operate at megahertz frequency.

The current-versus-voltage curves for the prototypes show behavior that is more like MSM photodiodes than a true photoconductor, but this is acceptable since MSM devices can have greater than unity gain, which is the important characteristic for high-speed speckle-monitoring. Indeed, the internal gain was measured as a function of bias voltage and found to have a maximum for one device at about 23. The gain is greatest for biases below 20 V and at low levels of illumination due to the trap filling effects of large dark currents and photo-generated charges.

It was also shown that when operating at MHz bandwidths, the noise is indeed dominated by the thermal noise component of the load resistance even at relatively high bias voltages. The SOI device was determined to have a maximum SNR roughly 500 times greater than an equivalently sized silicon *pn* junction diode operating under the same conditions.

## 4. Optical generation and detection of acoustic waves with an experimental direct speckle monitoring system

This section discusses experimental demonstrations of speckle monitoring-based vibration measurement using the photo-conductor mesa detectors coupled with optical techniques for acoustic excitation. There are many reasons to develop a system that exploits the optical excitation of acoustic waveforms as well as their optical detection. Fully non-contact systems can be less invasive, less disruptive to the device under test, and in some cases more rapidly deployed. Extensive work on the subject is available in the literature (Scruby, 1990; Davies, 1993; Pierce, 1998) that will only be summarized briefly here. Some understanding of the different mechanisms involved will help us better understand the experimental results obtained from tests with the photo-conductive mesa detectors.

### 4.1 Optical generation of acoustic waves

Generation of ultrasound waves with laser beams in untreated surfaces can be divided into two categories (Davies, 1993):

i.    The thermoelastic regime describes acoustic waveforms generated by expansion and compression resulting from the localized heating caused by an optical pulse. Typical peak optical intensities supplied to the surface in this regime are $< 10^7$ W cm$^{-2}$ (Scruby, 1990).

In the thermoelastic regime, the acoustic source (the bit of solid material that actually moves in direct response to heating, setting the acoustic wave in motion) is roughly a disc with a diameter equal to the width of the impinging laser beam and a depth determined by the thermal conductivity of the material and the rise-time of the laser pulse (Scruby, 1990; Davies, 1993; Pierce, 1998). For metal samples, and for nanosecond pulses (from a Q-switched Nd:YAG pulse, for instance), the source depth is on the order of microns, but can grow to hundreds of microns for millisecond rise times (Scruby, 1990; Pierce, 1998). As a result of heating, the disc exhibits expansion, predominantly in the plane of the surface (Pierce, 1998).

ii.   The ablation regime pertains to acoustic waveforms generated by the momentum transfer of evaporated material ejected from a rapidly, locally heated component of the test object's surface. Typical peak optical intensities supplied to the surface in this regime are $> 10^7$ W cm$^{-2}$ (Scruby, 1990).

In the ablation regime, the acoustic source is a cloud of vaporized material forming a plasma just above the surface and moving rapidly away from it. The result is an acoustic impulse directed predominantly normal to the surface.

In addition to peak powers and the *temporal* profile of the laser beam, the *spatial* profile of the laser beam as it intersects the solid surface is crucial in determining the final acoustic waveform injected on the test object (Monchalin, 2007; Khang, 2006]. The beam shape can be tweaked to encourage or discourage coupling into one or more acoustic modes. For instance, an optical grating can be imposed on the object surface with a spacing corresponding to the wavelength of the desired surface acoustic wave. This can greatly enhance coupling into that mode. Since no effort was made to explore this aspect of optical ultrasonic generation during the course of this project, no further discussion will be given to this topic. However, this remains a rich area for study, in particular learning how to promote the in-plane acoustic modes – we have already seen how photoconductive mesas are well suited to the detection of such waves, through speckle monitoring.

## 4.2 Nd:YAG pulsed laser generation of ultrasound waves

An experiment was performed in which ultrasonic waveforms were produced in solid metal targets by surface-ablation with a Q-switched Nd:YAG laser pulse. The various samples under test were thin strips made from unpolished aluminum, copper, and stainless steel. The resulting acoustic waves were allowed to propagate through the strips and were detected by a speckle monitoring system that observed a point on the test object located some distance away from the acoustic source.

The interest in these experiments is two-fold. Firstly, they show how well the photoconductive mesas monitor direct speckle motion, providing a sensitive method for remote ultrasound detection, especially for detection of in-plane surface motion to which other detection schemes are not as selectively sensitive (Monchalin, 1986). Secondly, they provide a demonstration of the bandwidth capabilities of the SOI photoconductive detector mesa prototypes made in our laboratories.

### 4.2.1 $A_0$ antisymmetric mode Lamb waves

Acoustic waves that propagate in the plane of thin plates and membranes can be decomposed into different Lamb wave modes. The two main classes of Lamb wave are the antisymmetric modes, for which the front and back surfaces of the plate move out of phase with respect to the middle of the plate and the symmetric modes, for which the front and back surfaces move in phase. For low frequencies, below some cut-off value that depends on surface thickness and plate material, we saw that only the lowest order modes, $A_0$ and $S_0$ exist. Which of these gets excited depends on the type and geometry of the exciting pulse.

Fig 16 shows the experimental arrangement for the first test, in which the Nd:YAG laser is directed onto the back planar surface of the strip. This directs most energy into a direction perpendicular to the surface (flexing), which is the primary motion of the $A_0$ mode. The antisymmetric mode should be the dominant acoustic waveform resulting from this excitation (Pierce, 1998). The detector output confirms this hypothesis, as there is clear evidence of antisymmetric Lamb wave propagation in the strips, and very little evidence of symmetric Lamb waves. In Fig. 17, we see a comparison of the recorded signal and the theoretical calculations of the $A_0$ mode propagating through the same distance of an aluminum sheet. The calculations assume a square impulse excitation of the $A_0$ mode with the bandwidth of the recorded signal. The dispersion relation is $V_p = 2V_S(1 - V_S^2/V_L^2)\beta b/(2 \bullet 3^{1/2})$, where $V_p$ is the phase velocity of the mode, $V_S$ is the shear velocity of the material, $V_L$ is the longitudinal velocity of the material, $\beta$ is the wavenumber of the mode, and $b$ is the plate thickness, according to Chapter 9 of (Cheeke, 2002).

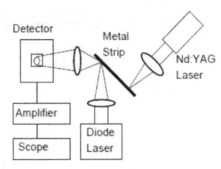

Fig. 16. Experimental arrangement for pulsed Nd:YAG laser excitation of ultrasound in metal strips. Detection of ultrasound was made with a speckle monitoring photoconductor.

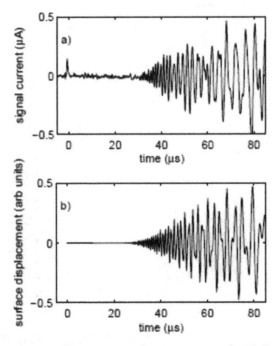

Fig. 17. a) Oscilloscope trace of an acoustic vibration generated with 7 cm of separation between the Nd:YAG pulse and the reference spot. The initial spike in photocurrent comes from the Nd:YAG pulse at time $t = 0$. b) Calculated surface displacement caused by $A_0$ Lamb waves propagating a distance of 7 cm. One reflection from the far edge of the strip (1 cm further from the detection point) is included to illustrate the effect of interference.

Figure 18a depicts the signal obtained from the copper strip, showing the arrival of multiple distinct waves. By looking the spectrogram in Fig. 18b, it is clear that each of these waves shares the same dispersion relation. Thus each wave is an instance of the same acoustic mode ($A_0$). These secondary instances are most likely reflections from one of the copper strip edges.

The antisymmetric mode is certainly useful, especially for extracting material properties from samples. For just one example, there is great interest in online monitoring of paper stiffness, consistency, and quality with laser ultrasonics exploiting antisymmetric Lamb waves (Brodeur, 1997). The thickness and shape of thin film structures, especially during fabrication (Pei, 1995), is another parameter of interest to which the propagation of such Lamb waves are sensitive. Any all-optical method for such measurements is certainly worth a consideration.

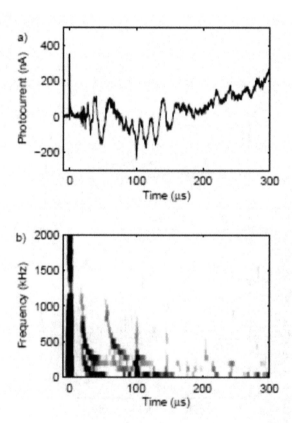

Fig. 18. (a) Oscilloscope trace of ultrasonic waves traveling in a copper strip with (b) corresponding spectrogram showing the spectral content of distinct waves as a function of time. The initial spike at time t = 0 comes from stray Nd:YAG light and serves as a marker for the onset of sonic excitation. At least the first two modes appear to have the same dispersion relation, indicating that they represent the reflection of a wave in the same mode ($A_0$).

## 4.2.2 $S_0$ symmetric mode Lamb waves

The lowest order symmetric mode ($S_0$) is of particular interest to applications that record the timing of pulse echoes. This would include parts inspections for internal cracks and discontinuities (Lowe, 2002). The reason for this is that the $S_0$ mode suffers very little dispersion, limiting pulse breakup and allowing for better event timing. Furthermore, at low frequencies, the $S_0$ mode is dominated by in-plane surface motion (Cheeke, 2002), so that is it is particularly well-suited to observation by a speckle monitoring system.

One problem with using the $S_0$ mode is that it is more difficult to excite by Nd:YAG laser pulse than the $A_0$ mode. Striking the flat surface of a plate tends to put most acoustic energy into the $A_0$ mode. The dominant motion of the $S_0$ mode is compression in the plane of the plate, so it is easier to excite this mode by striking the thin plate edge, as in the arrangement of Fig. 19. Although edge-excitation has been investigated with traditional contact transducers (Lowe, 2002), this novel arrangement is the first all-optical, end-on excitation technique specifically designed to excite only the symmetric mode. With further improvements, it could be adapted to generate in-plane vibrations for testing in any kind of plate-like structure.

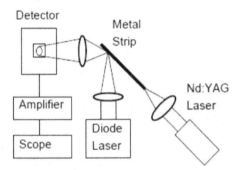

Fig. 19. Edge excitation of metal strip with Nd:YAG laser pulse.

When edge excitation is used, the coupling efficiency into the $S_0$ mode is dramatically enhanced, as can be seen in the oscilloscope trace in Fig. 20, showing the resulting vibration of such excitation in a stainless steel strip. This arrangement provides better timing information due to the non-dispersive nature of the $S_0$ waveforms. The waveform in this case is more or less a single pulse, about 10 $\mu$s wide, with a slightly exaggerated tail.

The round-trip travel of the wave is clearly discernible from the detector output in the first 300 $\mu$s of the recorded signal. The wave first arrives (after traveling just under one length of the strip) after 18 $\mu$s, and subsequently every 37 $\mu$s or so after that.

Although the Nd:YAG pulses produce ultrasonic pulses in these strips that can be used to investigate the material and geometric properties (for example, sheet thickness and points of internal reflection), this method can hardly be called "nondestructive." The material ablation that causes the acoustic impulse leaves a crater in the object surface. Repeated impulses, especially on delicate structures and thin films, require a gentler approach.

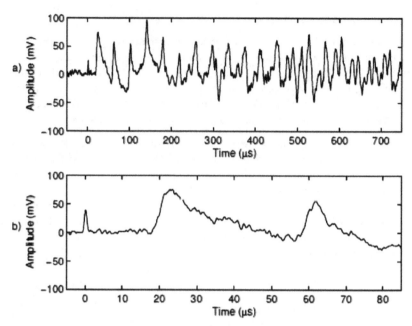

Fig. 20. Vibration signals recovered during edge excitation experiment. The spike at time $t =$ 0 is from scattered Nd:YAG laser light, marking the onset of acoustic excitation. This $S_0$ waveform is much less dispersive than the $A_0$ mode, allowing for more precise timing of discrete events, such as reflections. a) For the first 300 μs or so, the waveform is dominated by repeated reflections of the non-dispersive acoustic wave from either end of the strip as it traverses the detection point. b) At a smaller time scale, the time separations between peaks are easier to see. The time (37 μs) between the first pass of the wave and the second pass, or one round trip of the wave, is just slightly more than twice the delay (18 μs) between excitation onset at the far end of the strip and the leading edge of the first pass. This is precisely what is expected from the experimental geometry, depicted in Fig. 19. The mode velocity as measured by the temporal separation of the peaks is 5200 m/s.

### 4.3 Laser diode generation of ultrasound

To produce ultrasound in a truly nondestructive manner, one must restrict oneself to the thermoelastic regime of optical ultrasound generation. This can be achieved by lowering the peak power below the damage threshold. That approach favors the use of technical finesse over raw power, for which laser diodes are a good alternative to the Q-switched Nd:YAG laser. Consider the experiment in Fig 21, in which the Nd:YAG excitation laser has been replaced by an InGaAs current-modulated diode laser with a peak output power of approximately 50 mW in a surface-normal configuration. Figure 22 shows the detector output for an aluminum foil strip with dimensions 10 cm × 0.5 cm in response to excitation pulses of various widths. A 1 ms excitation pulse produces a large vibration that is easily measured by the speckle-monitoring detector. Even with such low coupling-efficiency, there is some diagnostic capability with this simple configuration, for instance monitoring low-frequency standing waves in small structures.

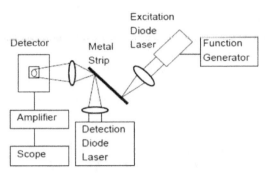

Fig. 21. Experimental setup for measuring the impulse response excited with a current-modulated laser diode. For the data presented in Figures 22 and 23, the metal targets were strips of aluminum foil.

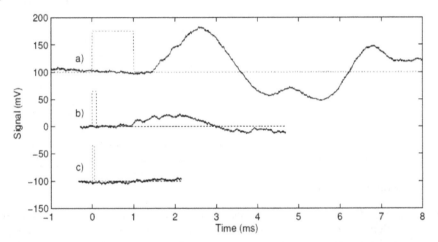

Fig. 22. Waveforms in an aluminum foil strip resulting from thermoelastic generation with an InGaAs laser diode pulse with a peak power of 50 mW and spot size of approximately 1 mm. Dotted lines indicate the period of laser diode excitation, solid lines indicate the detector signal. Three pulse lengths are shown: a) 1000 $\mu$s, b) 100 $\mu$s, and c) 50 $\mu$s. Decreasing the overall optical power delivered to the metal shows a corresponding decrease in the amplitude of the detected signal. For pulse widths less than 50 $\mu$s, the signal is no longer discernible from the noise.

Consider the data shown in Fig. 23. Three aluminum foil strips, 10 cm × 0.5 cm, were gently stretched flat and anchored on either end with approximately equal tension. Two of the three strips were left unaltered, but in the third strip a small tear was introduced into the strip's edge halfway between the excitation and detection points. Each strip was excited by a 500 $\mu$s square wave impulse from the excitation laser diode. The impulse responses of the two undamaged strips look nearly identical, a fundamental vibration with a period of about 4.7 ms ringing down at approximately the same rate. The damaged strip does not ring as cleanly. Although the fundamental frequency looks approximately the same as for the undamaged strip, the signal is disrupted by some other vibrations. Without analyzing the precise differences in the impulse response signature between the healthy and the damaged strips, it is fair enough to say that there is a clearly discernible difference. Such a distinction could be used to detect the presence of damage.

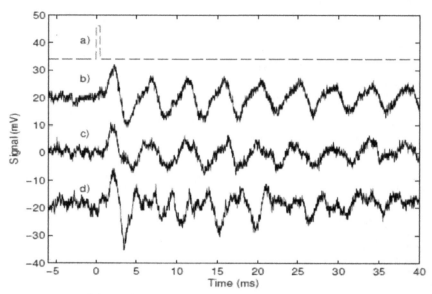

Fig. 23. Response of three identically sized aluminum foil strips under equal tension to a 500 μs rectangular pulse from a diode laser. a) Excitation diode current (arbitrary vertical scale), b) strip 1, no defect, c) strip 2, no defect, and c) strip 3, a tear introduced between points of acoustic generation and detection.

Although useful, this low-frequency, resonance technique is not perfect for every solution. Some applications are best served by a different approach to acoustic imaging that requires temporally and spectrally resolved pulse information such as time-of-flight and dispersion data. This generally precipitates the use of higher-bandwidth acoustic regimes.

The obvious drawback to operating in the thermoelastic regime is the decrease in acoustic power delivered to the object of interest. This problem can be addressed by using selective filtration methods on the detector signal, such as lock-in amplification and input-output signal correlation techniques.

### 4.4 Summary

In this section it was shown that speckle-monitoring is a viable technique for the remote detection of acoustic waves, in particular ultrasonic waveforms that are useful for material characterization and structural examination. The possibility of coupling speckle-monitoring with optical ultrasound generation is particularly attractive, as it allows for a completely non-contact system. Speckle-monitoring is especially appealing for use in measuring in-plane ultrasonic waves, such as the $S_0$ Lamb wave mode measured in the edge-excitation experiment, since it is selectively sensitive to that type of motion, unlike most other optical vibration detection schemes. There is an incentive to restrict optical excitation to the lower-power thermoelastic regime, such as by using current modulated CW laser diodes.

## 5. SOI speckle monitoring: detection limits compared to alternatives

Having laid the framework for characterizing the SOI detector prototype and conducting preliminary experiments, we are in a position to make some statements about the viability of this technique as an alternative to other optical vibration measurements. This section quantifies those comparisons and indicates certain applications to which speckle monitoring with photoconductor arrays is best suited. Numerical values are reported as rough estimates constructed from the data gathered in the previous chapters.

### 5.1 Pure in-plane-motion

After establishing that measurement with such a system is indeed possible, the most obvious question to ask about a particular measurement scheme for a given application is "how sensitive is it?" To that end, this section will attempt to answer that question for the direct speckle-monitoring method and other comparable techniques under similar circumstances.

Since we have already established that speckle-monitoring is most effective for in-plane measurement, we will focus on that application. As a convenient metric for the lower limits of sensitivity, let us define the quantity $\delta_{min}$ to be the minimum detectable displacement that produces a signal with amplitude equal to the noise for a given system output. Since we have seen that sensitivity generally scales in proportion to optical power and inversely to bandwidth, $\delta_{min}$ will be given units of m Hz$^{-1/2}$ W, where the power is the total available laser power.

### 5.2 SOI Detectors, single element detection limit

The minimum detectable photocurrent is equal to the noise current, which from experiments is roughly 20 nA at 1 MHz bandwidth with an overall gain of 20 across a load resistor $RL = 50\Omega$. Since noise scales with the square root of the bandwidth, we can estimate a noise figure of $\sigma_n = 20$ pA•Hz$^{1/2}$. A gain of 20 corresponds to a responsivity of about $\mathfrak{R} = 10$ A/W at 633 nm. This means that the minimum detectable change in collected power per element is $P_{min} = 2$ pW•Hz$^{1/2}$.

The amount of speckle translation required to achieve this change in photocurrent depends on the optical system and how much optical power is collected. For a small scattering spot whose width is $S$, illuminated with a total power $P_S$, the speckle size in the detector plane is approximately $\varepsilon = \lambda B/S$, where $B$ is the optical matrix element. Assuming the light is scattered equally in all directions and the scattering spot size $S \ll B$, the average speckle intensity is approximately $I_0 = P_S\lambda^2/(2\pi\varepsilon^2 S^2)$ as shown in (Dainty, 1975, Bessette, 2010). For a small translation $\delta_s$ of the speckle pattern, this means the detector will see on average (considering the ensemble of all possible random speckle patterns) a change in collected power given by $\Delta P = I_0 \varepsilon^2 (\delta_S/\varepsilon) = P_S\lambda^2\delta_S/(2\pi\varepsilon S^2)$, where we have assumed that the speckle width is matched to the detector width. Setting $\Delta P =$ to $\Delta P_{min}$ and solving for $\delta_s$, we find that the minimum measurable motion is

$$\delta_{S,min} = 2\pi\varepsilon S^2 \Delta P_{min} / \lambda^2 P_S$$

Using a typical scattering spot size of 0.5 mm and a detector-matched speckle size of 20 $\mu$m, this gives $\delta_{S,min}$ = 160 pm Hz$^{1/2}$ W. This is the minimum detectable translation of the speckle pattern. Typically, there will be optical magnification in this system with a value ~10, so this yields a minimum detectable displacement on the order of $\delta_{min}$ = 16 pm Hz$^{1/2}$ W.

For example, with a 1 W laser source and a bandwidth of 1 MHz and using a 50$\Omega$ load resistor, the detection limit is approximately $\delta_{min}$ = 16 nm for a single element. This value can be brought down even more by using a larger load resistor, as large as the bandwidth limitations will allow.

## 5.3 Other methods

Classical interferometry can be used to detect in-plane motion under certain conditions. For instance, rather than using one beam to illuminate the object and one reference beam, the differential interferometry approach uses two beams to illuminate the object (Monchalin, 1986). If the input aperture of the system and the illuminating spot are sufficiently small, then what forms in the detector plane is essentially a single speckle. The phase of that one speckle is sensitive to motion parallel to the difference in wave-vectors of the two illuminating beams. If the two beams illuminate the surface at equal but opposite angles with respect to the normal, the setup is sensitive purely to motion in the plane of the object. Using a 0.5 mm spot size and a HeNe laser with a wavelength of 633 nm gives a detection limit $\delta_{min}$ = 16 pm Hz$^{-1/2}$ W. With a 1 W laser and 1 MHz bandwidth, a sensor with these parameters has a detection limit on the order of $\delta_{min}$ = 16 nm. Again, this can be reduced by increasing the load resistor to the maximum the bandwidth requirements will allow. We see that the detection limit for the direct speckle monitoring method with SOI detectors is roughly equal to the interferometric method. The speckle monitoring method has the greatly added advantage that it does not require two coherent beams to be precisely overlapped at a specific angle in a small spot.

Speckle photography relies on tracking individual speckle features as they move across the detection plane with an electronic detector array. These methods are generally used in low speed applications -- CCD arrays with low read-out rates and long integration times (on the order of milliseconds). A recent report on different algorithms (Cofaru, 2010) finds that resolution is on the order of 10$^{-3}$ pixel widths, generally speaking around $\delta_{min}$ = 10 nm. Although this is on the same order as the detection limit of the SOI sensor, the SOI photoconductive mesa sensor described here does so at 1 MHz bandwidth.

Spatial filtering of moving speckles is a subcategory of speckle photography in which image processing of the speckle pattern is done in real-time by spatially filtering the speckle pattern with a periodic mask (for example a simple Ronchi grating) and integrating the filtered light (Komastu, 1976). As the speckle pattern translates across the grating, the integrated signal is periodic with a period equal to the time it takes an individual speckle to translate a distance equal to the grating spacing. The frequency of the integrated signal is proportional to velocity, so this technique is generally used to measure surface velocity. The measurement depends on determining a temporal frequency, so there must be at least one full period observed. This means that the speckle must travel at least the distance of a grating spacing. Since signal-to-noise is best when the speckle is about the same width as the grating spacing (Kamshilin, 2009), the detection limit is approximately 10 µm. This method is not nearly as sensitive to small displacements as the other methods discussed; however it works well for high in-plane velocities when the object is moving a large distance.

## 5.4 Summary

We have shown that photoconductive mesas directly monitoring speckle motion have an important place among applications requiring higher bandwidth detection of in-plane motion. The detection limit for a single element monitoring in-plane motion is on the order of $\delta_{min} = 16$ pm Hz$^{-1/2}$ W.

## 6. Arrays of photoconductive mesas

The performance of speckle monitoring systems can be dramatically increased by using multiple detectors to make simultaneous measurements of a single speckle pattern displacement. An array configuration of photodetectors can be used to enhance the signal to noise ratio of displacement measurements. In addition to a lower noise floor, uncertainty in absolute displacement measurements can be mitigated. Additionally, the adverse effects owing to harmonic distortion can be lessened by the use of arrays, all allowing for better overall system calibration.

The signals from multiple array elements can be combined in a way that reduces noise in the same spirit as simple averaging. The contribution of each element to the average, however, must be weighted according to the particular speckle that element is monitoring. The primary challenge is the ambiguity in both the sign and magnitude of the slope in photo-intensity as a function of speckle displacement. If the signals are corrected so that every element has the same phase (i.e., all elements are made to have a positive slope), then the weighted average shows an increase in signal to noise ratio as the number of elements is increased.

A demonstration of the power of arrays for improving signal-to-noise was made by comparing the signals of individual detector elements with a phase-corrected average of 18 combined elements. The 18 independent signals were obtained by using a three-element array to monitor six different random speckle patterns, each undergoing the same displacement. The SNR of the phase-corrected average was superior by approximately 10 dB over the typical single element signal. In this demonstration case, the phases were sorted manually during post processing.

## 7. Conclusions

This chapter has shown how photoconductive detectors tens of microns in size can provide more sensitivity than photo-diodes. Very small photo-diodes have very small areas, which limits their sensitivity at high bandwidth. They are useful in applications such as single-mode fiber optics telecommunications, because the light is concentrated to a very small area. When signals are as small as a nanowatts, photo-multipliers are often used. However, in applications where spatial resolution at high bandwidth is required, such as monitoring mechanical vibrations, photoconductive elements can have the best useful figure of merit. We have demonstrated a minimum detectable power of $P_{min}$ is 2 pW•Hz$^{1/2}$.

Experiments showed the most sensitivity with photoconductive GaAs mesas, but speed was limited to tens of kHz because of slow space-charge buildup in regions of sharp transition between light and dark. Reducing the shadows with transparent ITO contacts and reducing the background resistivity with continuous illumination with an LED, provided a response time of about 10 μs, three times faster than without these special measures. Nonetheless, GaAs mesas were unable to respond fast enough to measure ultrasonic waves.

The solution was to use photoconductive mesas in silicon, which does not suffer from space charge buildup. Because carriers can diffuse very far in silicon, it is important to use silicon-on-insulator (SOI) technology, to remove any diffusion into the substrate. It is also necessary to electrically isolate mesas by etching all the way through the silicon layer. With careful etching, mesas 15 μm wide were etched and contacted on either side. These photoconducting mesas were demonstrated to respond in 2 MHz with a 10V bias. Analysis of speckle motion measured with these SOI mesas indicated a sensitivity of $\delta_{min} = 16$ pm Hz$^{1/2}$ W.

Preliminary experiments were carried out to demonstrate improvements when arrays of photo-conducting mesas are used, as long as the signals are properly combined. Further work on arrays will enable system calibration as well as measurement. All results, both experimental and theoretical, indicate chapter that this technology is certainly amenable to further development in practical applications of vibration and motion sensing, particularly at high speeds and small distances.

## 8. References

Bessette, J. & Garmire, E. (2010). Silicon-on-insulator photoconducting mesas for high-speed detection of laser speckle motion. *Journal of Selected Topics in Quantum Electronics*, Vol. 16, No. 1, Jan/Feb 2010, pp. 93-99.

Bessette, J., Gogo, A. & Garmire, E. (2006). Four-point photo-conductance monitoring array for speckle and fringe motion sensing. *Proceedings of SPIE* , Vol. 6379, Oct 2006, pp. 637903(1-10).

Biswas, K., Das, S. & Kal, S. (2006). Analysis and prevention of convex corner undercutting in bulk micromachined silicon microstructures. *Microelectronics Journal*, Vol. 37, No. 8, Aug 2006, pp. 765-769.

Brodeur, P.H., Johnson, M.A., Berthelot, Y.H. & Gerhardstein, J.P. (1997). Noncontact laser generation and detection of Lamb waves in paper. *Journal of Pulp and Paper Science*, Vol. 23, No. 5, May 1997, pp. J238-J243.

Bube, R.H. (Ed) (1960). *Photoconductivity of Solids*. John Wiley & Sons, Library of Congress Control Number lc60010309, New York, 1960.

Burm, J. & Eastman, L.F. (1996). Low-frequency gain in MSM photodiodes due to charge accumulation and image force lowering. *IEEE Photonics Technology Letters*, Vol. 38, No. 1, Jan 1996, pp. 113-115.

Carrano, J.C., Li, T., Grudowski, P.A., Eiting, C.J., Dupuis, R.D. & Campbell, J.C. (1998). Comprehensive characterization of metal-semiconductor-metal ultraviolet photodetectors fabricated on a single-crystal GaN. *Journal of Applied Physics*, Vol. 83, No. 11, June 1998, pp. 6148-6160.

Celler, G.K. & Cristoloveanu, S. (2003). Frontiers of silicon on insulator. *Journal of Applied Physics*, Vol. 93, No. 9, May 2003, pp. 4955-4978.

Cheeke, J.D.N. (2002). *Fundamentals and Applications of Ultrasonic Waves*. CRC Press, ISBN 0-8493-0130-0, Boca Raton, Florida.

Cofaru, C., Philips, W. & Van Paepegem, W. (2010). Evaluation of digital image correlation techniques using realistic ground truth speckle images. *Measurement Science and Technology*, Vol. 21, No. 5, March 2010, 055102 (17pp).

Dainty, J.C. (ed)(1975). *Topics in Applied Physics. Vol. 9, Laser Speckle and Related Phenomena*, Springer-Verlag, ISBN 0387131698, 9780387131696, Berlin and New York.

Davies, S.J., Edwards, C., Taylor, G. S. & Palmer, S. B. (1993). Laser-generated ultrasound: its properties, mechanisms and multifarious applications. *Journal of Physics D: Applied Physics*, Vol. 26, No. 3, Nov 1993, pp. 329-348.

Heinz, P. (2008). *Optical Vibration Detection with a Four-Point Photoconductance-Monitoring Array*. Ph.D. thesis, Thayer School of Engineering at Dartmouth College, 2008.

Heinz, P. & Garmire, E. (2004). Optical vibration detection with a photoconductance monitoring array. *Applied Physical Letters*, Vol. 84, No. 16, April 2004, pp. 3196–3198.

Heinz, P. & Garmire, E. (2005) Low-power optical vibration detection by photoconductance-monitoring with a laser speckle pattern. *Optics Letters*, Vol. 30, No. 22, Nov 2005, pp. 3027–3029.

Heinz, P. & Garmire, E. (2007). Photoconductive arrays for monitoring motion of spatial optical intensity patterns. *Applied Optics*, Vol. 46, No. 35, Dec 2007, pp. 8515–8526.

Jhang, K.Y., Shin, M.J. & Lim, B.O. (2006). Application of the laser generated focused Lamb wave for non-contact imaging of defects in plate. *Ultrasonics*, Vol. 44, No. 1, Dec 2006, pp. e1265-e1268.

Kamshilin, A., Miridonov, S.V., Sidorov, I.S., Semenov, D.V. & Nippolainen, E. (2009). Statistics of dynamic speckles in application to distance measurements. *Optical Review*, Vol. 16, No. 2, March 2009, pp. 160-166.

Komatsu, S., Yamaguchi, I. & Saito, H. (1976). Velocity measurement using structural change of speckle. *Optics Communications*, Vol. 18, No. 3, Aug 1976, pp. 314-316.

Lee, D.B. (1969). Anisotropic etching of silicon. *Journal of Applied Physics*, Vol. 40, No. 11, Oct 1969, pp. 4569-4574.

Liu, J.M. (2005). *Photonic Devices*. Cambridge University Press, ISBN: 9780521551953, UK, 2005-04-21.

Lowe, M.J.S. & Diligent, O. (2002). Low-frequency reflection characteristics of the $S_0$ Lamb wave from a rectangular notch in a plate. *Journal of the Acoustical Society of America*, Vol. 111, No. 1, Jan 2002, pp. 64-74.

Macdonald, D., Sinton, R.A. & Cuevas, A. (2001). On the use of a bias-light correction for trapping effects in photoconductance-based lifetime measurements of silicon. *Journal of Applied Physics*, Vol. 89, No. 5, March 2001, pp. 2772-2778.

Mendicino, M.A. (1999), in *Properties of Crystalline Silicon*. Hull, R. (ed.) (1999), The Institution of Electrical Engineers, ISBN: 0852969333, London, UK, 1999.

Monchalin, J.P. (2007). Laser-Ultrasonics: Principles and Industrial Applications, chapter 4, in: *Ultrasonic and Advanced Techniques for Nondestructive Testing and Material Characterization*, C.H. Chen (Ed), pp. 79-116, World Scientific Publishing Co., ISBN: 978-981-270-409-2, Singapore.

Neamen, D. (2003). *Semiconductor Physics and Devices*, McGraw-Hill, ISBN: 0072321075, New York.

Pei, J., Degertekin, F.L., Khuri-Yakub, B.T. & Saraswat, K.C. (1995). *In situ* thin film thickness measurement with acoustic Lamb waves. *Applied Physics Letters*, Vol. 66, No. 17, April 1995, pp. 2177-2179.

Pierce, S.G., Culshaw, B. & Shan, Q. (1998). Laser generation of ultrasound using a modulated continuous wave laser diode. *Applied Physics Letters*, Vol. 72, No. 9, March 1998, pp. 1030-1032.

Rong, H., Liu, A., Nicolaescu, R. & Paniccia, M. (2004). Raman gain and nonlinear optical absorption measurements in a low-loss-silicon waveguide. *Applied Physics Letters*, Vol. 85, No. 12, Sept 2004, pp. 2196-2198.

Rose, A. (1955). Space-charge-limited currents in solids. *Physical Review*, Vol. 97, No. 6, March 1955, pp.1538-1544.

Rose, A. & Lampert, M.A. (1959). Photoconductor performance, space-charge currents, and the steady-state Fermi level. *Physical Review*, Vol. 113, No. 5, March 1959, pp. 1227-1235.

Scruby, C.B. & Drain, L.E. (1990). *Laser ultrasonics: techniques and applications*. CRC Press, ISBN 0750300507, New York.

Soares, S.F. (1992). Photoconductive gain in a Schottky barrier photodiode. *Japanese Journal of Applied Physics*, Vol. 31, No. 2A, Feb 1992, pp. 210-216.

Sze, S.M. & Ng, K.K. (2007). *Physics of Semiconductor Devices, 3rd Ed*. John Wiley & Sons, ISBN: 978-0-471-14323-9, New York, 2007.

Tu, S.L. & Baliga, B.J. (1992). On the reverse blocking characteristics of Schottky power diodes. *IEEE Tranactions on Electron Devices*, Vol. 39, No. 12, Dec 1992, pp. 2813-2814.

Wu, X.P. & Ko, W.H. (1989). Compensating corner undercutting in anisotropic etching of (100) silicon. *Sensors and Actuators*, Vol. 18, No. 2, June 1989, pp. 207-215.

# 5

# Ultrafast Imaging in Standard (Bi)CMOS Technology

Wilfried Uhring[1] and Martin Zlatanski[2]
[1]*University of Strasbourg and CNRS*
[2]*ABB Switzerland Ltd.*
[1]*France*
[2]*Switzerland*

## 1. Introduction

Around 1822, the French inventor Niépce made the first photographic image by the use of a *camera obscura*. He formed the image passing through the hole on a metal plate with a bitumen coating. After 8 hours of exposure, the bitumen on the illuminated sections of the plate was hardened. By washing the unhardened regions, a print of the observed scene appeared. After Niépce's death in 1833, Daguerre worked on the improvement of the chemical process involving interaction of the plate with light. In 1839 he announced the invention of a new process using silver on a copper plate. This invention reduced the exposure time to 30 minutes and denotes the birth of modern photography. During the following years, improvement on the photographic processes led to increased sensitivity and allowed shorter exposure times. In 1878, Muybridge gave an answer to a popular question at this time: whether all four hooves of a horse are off the ground at the same time during a gallop. By taking the first high-speed sequence of 12 pictures, each picture spaced about 400 ms from the neighbouring one with an exposure time of less than 500 µs. In 1882, George Eastman patented the roll film, which led to the acquisition of the first motion pictures. Four years later, a student of Daguerre, Le Prince, patented a *Method of, and apparatus for, producing animated pictures*. Through its *16 lens receiver*, as he called his camera, and by the use of an Eastman Kodak paper film, Le Prince filmed the first moving picture sequences known as the *Roundhay Garden Scene*, which was shot at 12 frames per second (fps) and lasted less than 2 seconds. Two years later, Edison presented the *Kinetoscope*, a motion picture device capable of acquiring sequences at up to 40 fps. It creates the illusion of movement by conveying a strip of perforated film filled with sequential images over a light source through a mechanical shutter. In 1904, the Austrian physicist Musger patented the *Kinematograph mit Optischem Ausgleich der Bildwanderung* which is capable of recording fast transients and projecting them in slow motion. In acquisition mode, the light is turned off and a rotating mirror, projecting them in slow motion. In acquisition mode, a rotating mirror mechanically coupled to the film shifting mechanism, reflects images onto the film. During the projection, the light is turned on and the same operation is carried out, but at a much slower rate. This high-speed photographing principle was used during the First World War by the German company Ernemann Werke AG to develop the *Zeitlupe*, a 500 fps camera used mainly for ballistic purposes. In 1926, Heape and Grylls constructed *Heape and*

*Grylls Machine for High Speed Photography*, in which the film shifting mechanism was replaced by a film lining the inside of a rotating drum, driven by an 8 horse power engine (Connel, 1926). The relaxed constraints on the film strength allowed to the 4 tons instrument to reach a frame rate of 5000 fps. In later realizations the drum was powered by an electric motor, leading to much more compact instruments (Lambert, 1937). Nowadays drum cameras are still in use and their fundamental principle is still the same. They employ a rotating mirror mechanically coupled to the drum and use the Miller's principle to perform the shuttering between frames. Miller's principle states that if an image is formed on the face of a mirror, then it will be almost static when relayed by lens to a film. The fastest drum cameras produce frame records at up to 200 000 fps which is limited by the mechanical constraints imposed by the need of a synchronous rotation of the mirror and the drum at very high speeds. Rotating mirror cameras have been developed toward the end of World War II to answer the need of photographing atomic explosions. The image formed by the objective lens is relayed back to a rotating mirror which sweeps the reflected image focused through an arc of relay lenses and shuttering stops on a static film (Fig. 1). The first camera of this type was built by Miller in 1939 and reached a speed of 500 000 fps. The concept was patented in 1946 (Miller, 1946) and in 1955, Berlin Brixner achieved a speed of 1 million fps using the same principle (Brixner, 1955). Cordin's Model 510 rotating mirror camera reaches 25 million fps, by driving the mirror up to 1.2 million rpm in a helium environment using a gas turbine (Cordin 2011). The speed of the drum and the rotating mirror cameras has not been improved for more than 40 years, the limits of this technology being set by the maximum rotation speed of the mirror and the materials (Frank & Bartolick, 2007).

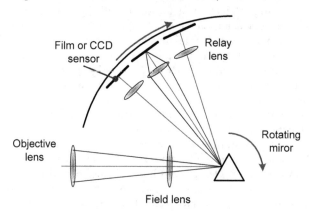

Fig. 1. Principle of the rotating mirror high speed video camera

Shortly after the first CMOS image sensors, the first Charge-Coupled Device (CCD) imagers appeared (Amelio et al., 1970). The CCD, initially developed for semiconductor bubble memory purposes, was invented in 1969 by Boyle and Smith at the Bell Laboratories. After promising initial results, this technology was rapidly adapted to be used for image sensor applications and little-by-little succeeded the photographic film (Smith, 2001). The first high-speed CCD cameras appeared during the 80's and were capable of acquiring up to 185 fps at a resolution of 512 × 512 pixels with a dynamic range of 8 bits. Their CMOS counterparts were able to go even further by integrating the signal conditioning electronics directly on-chip (Krymski et al., 1999; Kleinfelder et al. 2001). In both technologies, the shuttering is operated electronically without the use of a mechanical part and can be potentially operated in few nanoseconds. Nevertheless, in both CCD and CMOS architectures, the maximum frame rate is limited by the time interval required for an image to be read out. Actual CMOS high-speed video cameras are capable of producing about 1 Mega pixel images at a rate of approximately 1000 fps or every other combination of image resolution and frame rate, which maintains the read-out rate to several GS/s. Indeed, by reducing the image size, the frame rate can be increased. A summary of the fastest high-speed video cameras in 2011 is given in Table 1. The maximum amount of data rate at the output (about 100 Gb/s for the fastest available camera) is generally limited by the bandwidth of the output bus. Parallelization of the channels seems to be a unpromising solution, since it leads to an excessive power consumption and unreasonable chip area (Meghelli, 2004; Swahn et al. 2009).

| Manufacturer | Model | Resolution | Frame rate @ full resolution (total sampling rate) |
|---|---|---|---|
| Framos | MT9S402 (Sensor) | 512×512 | 2500 ips (655MS/s) |
| Aptina (ex Micron) | MT9M413 (Sensor) | 1280×1024 (10 bit) | 500 ips (655 MS/s) |
| Micron | (Krymski et al. 2003) | 2352×1728 (10 bits) | 240 ips (975 MS/s) |
| Optronis | CR5000x2 (Camera) | 512×512 (8 bits) | 5000 ips (1.3GS/s) |
| Cypress | LUPA 3000 (Sensor) | 1696×1710 8 bits | 485 ips (1.4 GS/s) |
| Vision research | Phantom v640 (Camera) | 2560×1600 (8 or 12 bits) | 1500 ips (6.1 GS/s) |
| Vision research | Phantom v710 (Camera) | 1280×800 (8 or 12 bits) | 7530 ips (7.7 GS/s) |
| Photron | Fastcam SA5 (Camera) | 1024×1000 (12 bits) | 7500 ips (7.7 GS/s) |
| Photron | Fastcam SA2 (Camera) | 2048×2048 (12 bits) | 1000 ips (4.2 GS/s) |
| IDT | Y4-S3 (Caméra) | 1016×1016 (10 bits) | 9800 ips (10 GS/s) |

Table 1. State-of-the-art of the fastest electronic high-speed video cameras in 2011

As seen from Table 1, today's high speed video cameras present a temporal resolution in the range of 1 ms in full resolution down to 1 μs in a reduced image size of about 1 thousand pixels. These performances are still far away from those of the rotating mirror cameras. To break the GS/s-order limit, a radically different approach than trying to operate the acquisition and the read-out progressively was required.

## 2. Ultrafast imaging concepts

### 2.1 In situ storage concept

The conventional high-speed video cameras face the input/output bus bandwidth constraint, which limits their data rate and temporal resolution to about 10 GS/s and 100 μs in full frame format, respectively. To overcome this bottleneck, the solution is simply not to extract the data from the sensor. Indeed, the fastest video sensors reported in the literature employ the in-situ concept, which consists in storing the acquired data on-chip rather than continuously extracting it and operate the readout afterwards. The use of this method in

CCD and CMOS optical sensors allowed to push the total sampling rate to about 1 TS/s, i.e. 100 times faster than the fastest high speed video cameras. The architecture of a sensor using the *in-situ* concept is shown in Fig. 2. Each pixel integrates a photodetector, generally a photodiode, and its own *in-situ* memory. Work on this concept started during the 90's and 1 million fps cameras have been reported (Elloumi et al. 1994; Lowrance & Kosonocky, 1997). These first sensors with *in-situ* frame memory, in which each pixel is composed by a photodetector and a linear CCD storage sensor were able to store 30 frames. In 1999, Professor Etoh proposed a sensor with 103-deep CCD storage (Etoh et al., 1999, 2005).

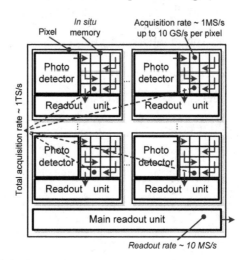

Fig. 2. Architecture of a high speed image sensor with the in-situ storage concept

A comparison between Cordin's Model 121 rotating mirror camera and Shimadzu's Model HPV-1 high-speed CCD imager with *in-situ* storage is made in (Franck and Bartolick, 2007). The test clearly shows the superiority of rotating mirror cameras in terms of spectral range, temporal and spatial resolutions. The authors conclude that at the moment a solid-state replacement for the rotating mirror cameras does not exist, but the *in-situ* based sensor architectures have the potential to reach higher performance through specifically optimized designs. Some studies forecast theoretical rates as high as 100 million fps for *in-situ* CCD detectors (Son et al. 2010), at the cost of reduced image quality.

The development of CMOS technologies allowed the integration of signal processing electronics and photodetecting site on the same substrate, which resulted in an enlargement of the spectrum of applications of solid-state imagers (Bigas et al., 2005). Since the bandwidth of a field-effect transistor in a submicron silicon technology easily exceeds the GHz, a number of these applications were naturally oriented towards high-speed imaging. In 2004, a $12 \times 12$ pixels demonstrator processed in standard 0.35 µm CMOS technology employing the *in-situ* storage concept achieved more than 10 million fps (Kleinfelder et al., 2004). The memory depth was of 64 and the pixel size of 200 µm × 200 µm. Recently, a $32 \times 32$ pixel prototype with an *in-situ* memory of 8 frames has been presented (Desouki et al. 2009). Processed in a 0.13 µm standard CMOS, it achieved a rate of 1.25 billion fps. The main drawback of this method is the limited memory depth to a maximum of about several hundred frames. Indeed, the silicon area required increases with the number of stored frames and in 2D imaging, the in-pixel embedded memory decreases drastically the fill factor, e.g. less than 10% for only 8 frames in (Desouki et al., 2009). Nevertheless this is not a real issue, since in practice at these rates, generally only a few frames are analyzed. 3D technologies or die stacking may lead to fill factors as high as 100 %. However, ultrafast optical sensors employing the in-situ concept and featuring high fill factor can be processed in a standard (Bi)CMOS technology if streak-mode images are taken.

## 2.2 Streak-mode imaging concept

A feature proper to all the devices presented until now is that they produce 2-dimensional $(x,y)$ images $I_f$ at equally spaced in time intervals $\Delta t$. Consequently, $I_f$ is a function of the two spatial dimensions $x,y$ with a constant time $t_0+n\cdot\Delta t$, where $n$ is an integer:

$$I_f = f\left(x,y,t_0+n\cdot\Delta t\right) \tag{1}$$

This is referred as framing mode photography. The advantage of a frame record is that information in two spatial dimensions is recorded, so the recorded image is an easily recognized version of the subject. However, if finer temporal information is required, another imaging method must be used. Speed and spatial information size of a camera being strongly related, a considerable gain in the frame rate can be achieved if the spatial information is reduced. If a single spatial dimension is selected, a picture containing the continuous temporal evolution of the spatial information is obtained. The recorded image $I_s$ contains the spatial information $x$ which crosses the slit, observed at different times $t$:

$$I_s = f\left(x,t\right) \tag{2}$$

This is denoted as streak-mode imaging. A streak record is made by placing a narrow mechanical slit between the event and the camera. Next, the temporal evolution of the one-dimensional spatial information crossing the slit is swept along a taping material. Thus, a continuous record containing position, time and intensity information is obtained. The recording rate, called sweep speed, translates the distance covered on the record surface for a unity of time and is measured in mm/μs or ps/pixel.

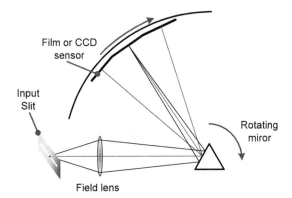

Fig. 3. Rotating mirror camera operating in streak-mode photography

To operate a rotating mirror camera in streak-mode, the relay lenses have to be removed and a slit positioned on the input optical path, (Fig. 3). With such a device, a maximal sweep speed of 490 ps/pixel has been reported (Cordin, 2011). Fig. 4 shows a frame (top) and a streak record (bottom) of a projectile fired through a near-side cut plastic chamber to impact an explosive sample. It clearly shows how a streak record provides continuous inter-frame temporal information of the slit image. The improvement in temporal resolution reaches about 2 orders of magnitude. Streak-mode imaging is ideal for studying events of uniform growth, e.g., an expanding sphere, where the rate of dimensional change is to be measured. To avoid misinterpretation and provide the maximum amount of information, often a combination of streak and framing records are carried out. Thus, framing and streak-mode cameras are not comparable, but rather complementary (Fuller, 2005).

Fig. 4. Illustration of the streak imaging concept: frame record at 100 000 fps (top) and streak record at 0.53 mm/μs (bottom) of a ballistic event acquired by a Cordin Model 330 camera (Cordin, 2011)

The rotating mirror technology is limited by the centrifugal force applied to the mirror by the very high rotation speed. Even the use of light metals as beryllium does not prevent the mirror from explosion when the rotation speed exceeds 2 millions rpm (Igel & Kristiansen, 1997). To overcome this limitation a completely different technology based on a vacuum tube has to be used. The fastest device for direct light measurement available, known as the conventional streak camera allows picosecond-order temporal resolution to be reached. The operation principle of a conventional streak camera is depicted in Fig. 5. The very core of the streak camera is a modified first-generation sealed vacuum image converter tube, known as a streak tube, comprising four main sections: a photon-to-electron converter, an electron bunch focusing stage, an electrostatic sweep unit, and an electron-to-photon conversion stage. On some streak tubes, an internal Micro Channel Plate (MCP) is added in front of the phosphor screen for signal amplification. A mechanical slit is illuminated by the time varying luminous event to be measured and is focalized on the photocathode of the streak tube. The incident photons on the photocathode are converted to photoelectrons with a quantum efficiency depending on the type of the photocathode. A mesh is placed in the proximity of the photocathode and a high static voltage is applied between these two components in order to generate a high electrical field, which extracts the photoelectrons from the photocathode, makes their velocities uniform, and accelerates the pulse of photoelectrons along the tube. At this stage, the photogenerated electrons represent a direct image of the optical pulse, which reached the photocathode surface. When the photoelectrons approach the sweep electrodes, a very fast voltage ramp $V(t)$ of several hundred Volts per nanosecond is applied at a timing synchronized with the incident light. During the voltage sweep, the electrons which arrive at slightly different times, are deflected in different angles in the vertical direction. As far, first a photo-electrical conversion is carried out by the photocathode, and then a translation from time to space is operated through the sweep electrodes. After being deflected, the electrons enter the MCP, where they are multiplied several thousands of times by bouncing on the internal channel walls acting as continuous dynodes and are extracted by the high electrical field applied on both sides of the MCP. At the end of the streak tube the photoelectrons impact against a screen which emits a number of photons proportional to the incident electron density. To prevent dispersion, an objective or a tapper made of a fine optical fiber grid is positioned in-between the phosphor screen and the read-out CCD or CMOS camera to guide the emitted photons.

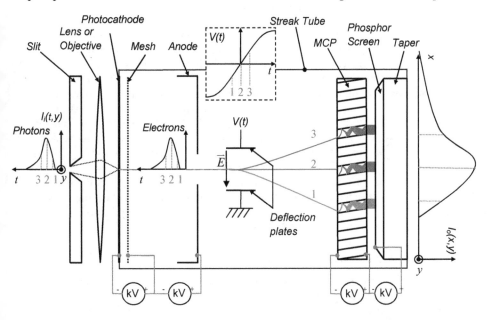

Fig. 5. Operation principle of a conventional streak camera

As the time-varying electric field caused by the voltage ramp between the electrodes is assumed to be spatially uniform, the spatial distribution of the light is directly obtained on the phosphorus screen without temporal modification. Finally, the temporal axis is found along the vertical direction of the screen, the position in the horizontal direction corresponds to the location of the incident light, and the brightness is proportional to the intensity of te respective optical pulses. The relationship between a given vertical position $x$ and the time $t$ depends on the slew rate $S_R$ [V/s] of the sweep voltage $V(t)$ and the deflection sensitivity $D_S$ [V/mm] of the streak tube is:

$$t = t_0 + \frac{D_s}{S_R} \cdot (x - x_0),$$
(3)

where $x_0$ is the position in mm of the slot image on the phosphorus screen when $V(t_0)=0$. As for the rotating mirror camera operating in streak-mode, the sweep speed of the camera is found as the ratio $D_S/S_R$, generally expressed in ps/mm.

| Spectral range | [41 pm, 10 μm] depending on the photocathode type |
|---|---|
| Sensitivity | Single photon detection capability (with MCP or image intensifier) |
| Repetition rate | From single shot up to 250 MHz (synchroscan) |
| Spatial resolution | From 25 μm (ns resolution) to 100 μm (ps resolution) |
| Temporal resolution | From 200 fs (single shot) or 2 ps (synchroscan) to 300 μs |
| Sweep speed | From 10 ps/mm (fast sweep unit) up to 5 ms/mm (slow sweep unit) |
| Observation time | From 60 ps up to 175 ms (phosphorus screen size ∈ [9–35 mm]) |

Table 2. Performance summary of a conventional streak camera

The conventional streak cameras are very versatile devices with high-end performances (Table 2). The wavelength detection spectrum ranges from the X ray (Scheidt & Naylor, 1999) up to the far infrared (Jiang et al. 1999). Their very high sensitivity permits the detection of a single photon event and the repetition rates extend from single shot operation up to several hundred of MHz. The typical spatial resolution of a conventional streak camera is between 25 μm and 100 μm. Specific tube designs using magnetic solenoid lens can reach a spatial resolution of 10 μm (Feng et al., 2007). Finally their temporal resolutions are very close to the physical theoretic limitation, about 100 fs (Zavoisky & Fanchenko, 1965) but with a poor signal to noise ratio whereas 1 ps can be reached with a high signal to noise ratio (Uhring et al., 2004a). Besides their extreme performances, conventional streak cameras and rotating mirror cameras have drawbacks: they are bulky, fragile, delicate to manufacture, and cost around 100 k€. Moreover there are many applications in which a temporal resolution of about 1 ns is sufficient. For these applications, solid-state technologies can offer interesting alternatives.

## 2.3 Streak camera alternatives

A high-speed video camera can be used as a streak-mode device by activating several rows on the sensor array and focusing the spatial information of interest on them. This technique is demonstrated in (Parker et al. 2010), where a temporal resolution of 1 μs and a sweep speed of 7.5 mm/μs have been reached, obviously limited by the output data rate.

In 1992, Lai from the Lawrence Livermore National Laboratory (USA) proposed a streak-mode camera in which the sweep is carried out by an externally triggerable galvanometer (Lai et al., 1992, 2003). The signal beam reaches the surface of the moving wedge-gap shaped deflector. It is subjected to multiple reflections between the surfaces of the stationary and deflecting mirror which results in an effective increase of the optical sweep speed. $N$ reflections on the moving mirror result in an output beam angular speed equal to $2N$ times the rotating speed of the deflector. The deflected beam is focused and projected on a CCD sensor. At least 2 ms of trigger delay are required to allow the mirror to reach the targeted deflection speed. A temporal resolution of 30 ns has been reported.

Nowadays there isn't any solid-state streak-mode imaging device with performances situated in-between high-speed video cameras and conventional streak and rotating mirror cameras and in the same time there is a highly potential market for such a device. Indeed, many customers cannot afford the acquisition of a streak camera because of its price and, in the same time, they do not always need picosecond temporal resolution. In many streak-mode photography applications a temporal resolution in the order of several hundreds of picoseconds is often sufficient. The following section presents a solid-state alternative to the conventional streak camera, which should give birth to new commercial and scientific activities, answers existing but unresponded demands and opens new fields of application. The prototypes are designed in a standard (Bi)CMOS technology which ensures quick prototyping and very low production cost estimated to about 10 % of the price of a traditional streak-mode imaging devices.

## 3. Integrated streak camera

Following the example of high-speed video cameras, the integration possibilities offered by CMOS technologies allow the use of the *in-situ* storage approach and its application to the first streak-mode CMOS imagers. In 1987, Professor Kleinfelder from the University of California proposed a 16 channel electrical transient waveform recording integrated circuit with 128 Sample and Hold (S/H) cells per channel operating in parallel at 100 MHz sampling frequency (Kleinfelder, 1987, 1990). In 2003, he demonstrated a 4 electrical channel 128-deep sampling circuit with up to 3 GHz operation using sub-micron silicon technology (Kleinfelder, 2003). One year later he presented a multi-channel sampling circuit in which the inputs of every channel are connected to an optical front-end (Kleinfelder et al. 2003, 2004). This was the first single-column streak-mode optical sensor. However, the very first integrated streak camera was developed in 2001 by the University of Strasbourg and was based on a pixel-array sensor architecture driven by an electronic temporal sweep unit (Casadei et al. 2003). In the following pages, both pixel array Integrated Streak Camera (MISC) and single-column Integrated Streak Camera (VISC) architectures are discussed.

### 3.1 Pixel array based integrated streak camera (MISC)

The optical setup required for a correct operation of a MISC is schematized in Fig. 6. The image of a light source illuminating a mechanical slit is uniformly spread along the temporal axis (the rows) of the 2D CMOS sensor through a cylindrical lens. Thus, each pixel of the same row is subjected to the same optical event. The distinctiveness of the integrated streak camera sensor compared to a conventional framing-mode sensors, consists in the way the pixels are operated. Indeed, a sweep unit is used to operate the time-to-space conversion carried out by the deflection plates in conventional streak cameras. The pixels of the sensor operate in photon flux integration mode. The circuit temporally sweeps the columns by shifting the beginning of their integrating phase every $\Delta T$. Consequently, each pixel of a given row is subjected to the same illumination, which is measured at different moments.

The first generation MISC uses a 3T pixel that does not allow the integration to stop (Casadei et al. 2003). In practice, the latter ends with the read-out of the pixels. Consequently, attention must be paid to ensure that no light reaches the sensor's surface until the end of the read-out. The output $\Delta V_{ij}$ of the pixel $p_{ij}$ at row $i$, column $j$ is given by:

$$\Delta V_{ij} = \frac{1}{m} G_c \int_{j \cdot \Delta T}^{\infty} E_i(t) \cdot dt, \tag{4}$$

where $G_c$ [V/$n_{ph}$] is the global conversion gain including the fill factor and the quantum efficiency, $E_i$ [$n_{ph}$/s] is the optical power received by row $i$ and $m$ the number of pixels along the temporal axis, i.e., the number of columns. It is assumed that the time origin ($t=0$) corresponds to the first column ($j=0$). Processing the differential of the pixels along the temporal axis, i.e., $S_{ij} = (\Delta V_{i(j+1)} - \Delta V_{ij})$ allows the reconstruction of the luminous event. Nevertheless, during this process, the high frequency noise is amplified and the signal obtained has a poor signal to noise ratio. Moreover, the dynamic range of the sensor is reduced since the first pixel integrates the entire signal and, at the same time, saturation must be avoided. The repetition rate of the acquisitions is limited by the read-out time of the array which is about 20 ms.

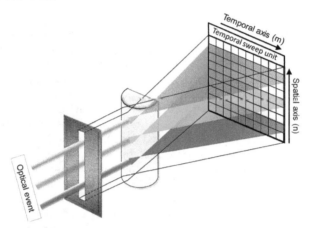

Fig. 6. Optical setup of a MISC

The second generation MISC employs a 6T pixel architecture in order to eliminate the faults of the first generation MISC by featuring a shutter capability. The detailed block diagram of a second generation pixel array-based sensor is shown in Fig. 7 (Morel et al. 2006). The circuit features an array of $n \times m$ active pixels, an $m$-stage temporal axis generator for the sweep unit, and two selection circuits and an output amplifier for serial data read-out. It temporally sweeps the columns by shifting the beginning and the end of their integrating phase, i.e., by delaying the arrival of the $RPD$ and $SH$ signals from one column to another. Thus, the integrating procedure starts with the turn off of transistor $M_{RPD}$ and ends with the turn off of the transistor $M_{SH}$ for all the pixels of column $i$ and is delayed from column $i$-1 by $\Delta T$.

The 6T active pixel is also featuring an analog accumulation capability (Morel et al. 2006). Before acquisition, the potential on the cathode of the photodiode is initialized to $V_{RPD}$ and the potential on the read-out node $RN$ to VRR. After the integrating phase, i.e., the switching off of transistor $M_{SH}$, the signal charges are trapped in node $IN$. Next, they are conveyed to the read-out node through $M_{TX}$ acting as a transfer gate, allowing the on-chip accumulation of several repetitive low-light events through successive charge transfers from node $IN$ to node $RN$. Anti-blooming is performed by keeping $M_{RPD}$ in sub-threshold conduction after reset. In this circuit, the temporal sweep unit is composed of two identical delay lines with a single cell delay value of $\Delta T$. The first delay line is used to control the reset transistor $M_{RPD}$ whereas the second controls the shutter transistor $M_{SH}$. The beginning of the acquisition procedure, which consists of feeding the Trigger signal in the temporal sweep unit, is synchronized with the arrival of the optical event, through an external photodiode and a delay box. An externally tunable delay of $T_{int}$ is added between the delay lines and allows the tuning of the integrating time. For this architecture, the output $\Delta V_{ij}$ of the pixel $p_{ij}$ is given by:

$$\Delta V_{ij} = \sum_{k=1}^{N} \left( \frac{1}{m} G_c \int_{j \cdot \Delta T}^{j \cdot \Delta T + T_{int}} E_{ki}(t) \cdot dt \right),$$  (5)

where $N$ is the number of accumulations, $G_c$ is the global conversion gain which includes the charge transfer efficiency from the photodiode to node $IN$ and from node $IN$ to node $RN$ and $E_{ki}$ is the $k_{nth}$ optical signal received by row $i$. The second generation MISC features much higher dynamic range and tunable sweep speed from 140 ps up to 1 ns per pixel. The on-chip accumulation allows the observation of low intensity and repetitive optical signals, comparably to the synchroscan mode of a conventional streak camera. High repetition rates can be reached, since the duration of the charge transfer can be reduced down to 1 ns by appropriately adjusting the $V_{RPD}$ and $V_{RR}$ voltages (Uhring et al. 2011).

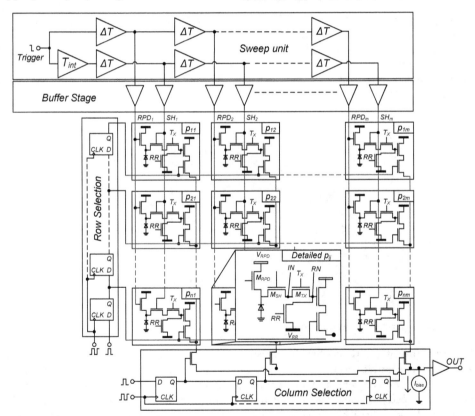

Fig. 7. Block diagram of the second generation MISC and detailed view of a pixel

Through the *in situ* concept and the parallelized operation, the MISC offers a very high speed as the only electronic limitation is the $M_{RPD}$ and $M_{SH}$ transistors bandwidth and turn off transient. The latest prototype with 93 rows channel exhibits a total sampling rate in excess of 650 GS/s (Zlatanski et al. 2010a). The acquisition of a femtosecond laser pulse at 400 nm with an integration duration $T_{int}$ of 400 ps indicates that the temporal resolution is 1.1 ns FWHM (Uhring et al., 2011). Preliminary results obtained with the picoseconds laser diode generator described in (Uhring et al. 2004b) show a temporal resolution very close to the nanosecond at 650 nm whereas the FWHM of the output pulse is increased to more than 3 ns at 808 nm. In 0.35 μm (Bi)CMOS technologies the bandwidth of a MOST switch can easily reach several GHz. The overall temporal response of a MISC in standard CMOS technology is then limited by the photodetector bandwidth. Unfortunately, the MISC suffers from a lack of sensitivity. Indeed, equations 4 and 5 clearly state that the output signal is divided by $m$, i.e. the memory depth (Zlatanski et al., 2010c). To avoid this, instead of spreading uniformly the light over the whole surface of the sensor, it should be focused to only one column. This is the basis of the vector based integrated streak cameras.

### 3.2 Vector based integrated streak camera (VISC)

The operation principle of a VISC is schematized in Fig. 8. In this case, the image of the slit is focused on the vector of photodetectors of the sensor through a spherical lens. Each one of the $n$ photodetectors is connected to an amplifier and an $m$-deep sampling and storage unit. During illumination, the incident light is transformed to an electrical signal, which is amplified by the wideband front-end electronics. The outputs of the front-ends are sampled on the *in-situ* frame storage.

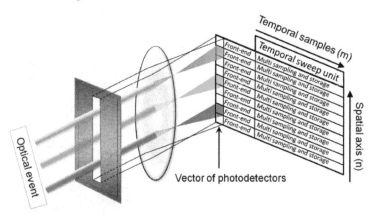

Fig. 8. Optical setup and architecture of a VISC

The different configurations of such a sensor are shown in Fig. 9. One approach is to use a transimpedance amplifier (TIA) coupled to a photodiode as a VISC front-end (Fig. 9 (a)). In this case, the photocurrent provided by the photodetector is converted to voltage with a given transimpedance gain, and is thus a direct image of the light event. At this level the circuit can be seen as a multichannel optical receiver circuit with an analog output. The sampling and storage unit is subjected to less design constraints as the signal to be sampled is in the voltage domain. The output $\Delta V$ of a pixel $p_{ij}$ is simply given by:

$$\Delta V_{ij} = G_c \cdot E_i (j \cdot \Delta T),$$ (6)

where $G_c$ is the global conversion gain of the sensor in [V·s/$n_{ph}$]. The frequency response of the photodiode is only slightly influenced, since the input impedance of a TIA is very low, typically ~100 Ω, and the bias voltage across the photodiode terminals is kept constant. However, a major drawback is the high power consumption as each front-end stage requires about 10 mW to reach a GHz-order bandwidth (Razavi, 2003).

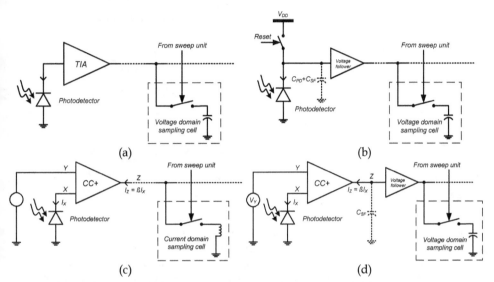

Fig. 9. Different front-end configurations: transimpedance approach (a), integrating approach (b), current to current approach (c) and hybrid integrating approach (d)

Another direct conversion could be obtained by using a Current Conveyor (CC) coupled to a line of $m$ current-mode sampling cells (Toumazou et al. 1993; Mehr & Sculley, 1998; Rajaee & Bakthiar, 2005), as shown in Fig. 9 (c). The CC ensures that the voltage applied at $Y$ is copied to $X$, keeping the photodetector under constant reverse bias. Next the current is conveyed to the output $Z$ with a gain of ß, set by the design. Thus, the CC acts as current buffer between points $X$ and $Z$. The value of pixel $p_{ij}$ is given by equation (6), in which $G_c$, now given in [A·s/$n_{ph}$], includes the current conveyor and sampling cell gain.

The voltage follower front-end configuration (Fig. 9 (b)) has been used in the first VISC (Kleinfelder, 2004). The photocurrent is integrated on the photodetector and the voltage follower input capacitances and a standard voltage-domain sampling and storage block. The output voltage $\Delta V$ of pixel $p_{ij}$ is given by:

$$\Delta V_{ij} = G_c \int_{j \cdot \Delta T}^{(j+\alpha) \cdot \Delta T} E_i(t) \cdot dt \qquad (7)$$

where $G_c$ is the global conversion gain and $\alpha = T_{int}/\Delta T$ is the ratio of the integrating time to the sample period with $\alpha \in [0\text{-}1[$ in the structure proposed in (Kleinfelder, 2004). The hybrid integrating approach (Fig. 9 (d)) uses a CC to isolate the photodiode from the voltage follower. Thus, the photodiode biasing is maintained constant and the conversion gain is increased as the integrating capacitance is reduced to the one at the input of the voltage follower.

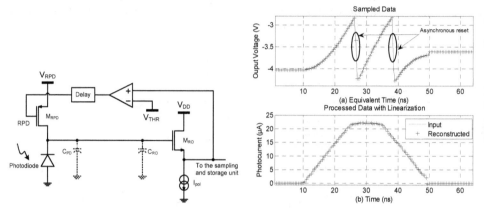

Fig. 10. Architecture of the asynchronous photodiode reset integrating VISC front-end (left) and a simulation result showing the output response to a pulse stimulus (right)

In the latter designs, the synchronous operation of the start and stop integrating phase severely limits the sampling rate to about 100 MS/s. In order to overcome this limitation, a novel asynchronous reset front-end architecture has been developed (Fig. 10-left). It includes a reset transistor $M_{RPD}$, a read-out transistor $M_{RO}$, and a feedback circuit which continuously senses the output of the front-end and resets the photodiode when a threshold voltage $V_{THR}$ is attained indicating a soon run out of dynamic range. At circuit power-on, the $RPD$ signal is externally triggered in order to provide a correct initial condition. For the fastest operation, the used comparator is an unmatched CMOS inverter. The simulated with Spectre® extracted view of this circuit showed a reset time as short as 600 ps. The photodiode reset phase being performed only when necessary together with its asynchronous with respect to the delay generator operation enable GS/s-order sampling performance associated to the high conversion gain of the integration. Data lost during the reset phase are detected and reconstructed (Zlatanski et al. 2010a).

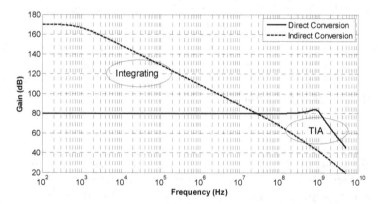

Fig. 11. Frequency responses of direct and indirect conversion architecture

The frequency response of the integrated streak camera depends on the front-end (Fig. 11). The response of the direct conversion front-end is given for a TIA with a gain of 80 dBΩ and a bandwidth of 1.3 GHz. The response of the indirect conversion front-end is given for a total capacitance of 100 fF including the photodiode and the front-end input parasitic capacitance. The cross point of these two responses is in the range of 10 MHz to 100 MHz depending on the integrating capacitance and the TIA. It is seen that the indirect conversion is appropriate for high gain measurements at low frequencies whereas the direct conversion seems to be the right choice when high bandwidth is required.

### 3.3 Temporal sweep unit

A common and important building block of the ISCs is the temporal sweep unit. The temporal sweep unit depicted in Fig. 12 allows the generation of sweep speeds in the range from 125 ps/pixel up to DC. It consists of a fast and a slow sweep unit. The Fast Sweep Unit (FSU) is useful for sub-nanosecond sampling period operation. It is composed of a Delay-Locked Loop (DLL) and two Voltage-Controlled Delay Lines (VCDLs). The observation time of the camera is set by the DLL reference clock period. The mirror VCDLs are driven by the same control voltages $V_{hl}$ and $V_{lh}$ as the master DLL allowing for asynchronous delay generation with respect to the reference clock. This makes it possible for the proposed generator to be launched from zero-tap position upon external triggering signal, ensuring the synchronization between the beginning of the acquisition procedure and the luminous phenomenon to be captured. The output taps of the VCDLs are delayed by an adjustable delay $\Delta T$ given by the ratio of the DLL Clock to the number of column $m$. The voltage controlled cell is a current starved double inverter with NMOS control transistors. It presents a large delay range and has been optimized for the shortest delay achievable in the CMOS technology employed (Mahapatra et al., 2002). The master DLL ensures the stability over temperature and the absolute precision of the delay. It is important that the VCDLs are

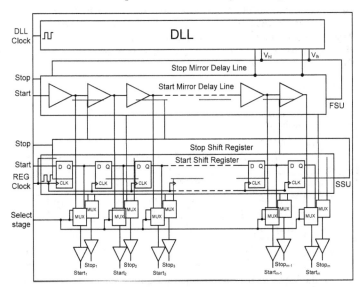

Fig. 12. Versatile sweep unit architecture including a Fast and a Slow Sweep Units

matched to ensure the same temporal shifting. For the MISC architecture, the *Start* and *Stop* signals, which control the $M_{RPD}$ and $M_{SH}$ transistors, respectively, are generated externally. Using two independent control signals allows the reduction the integration time down to 1 ns or less. The VISC architecture requires only one sampling signal, thus the mirror line generating the *Stop* signal is removed. Results from a similar FSU have been reported in (Zlatanski et al., 2011). Sweep speed of 125 ps/pixel up 1 ns/pixel with a drift of less than 0.05%/°C and temporal axis linearity better than 1% have been reported. The measured timing jitter in a completely asynchronous operation mode is less than 70 ps p-p@150ps/pixel and less than 700 ps p-p@770ps/pixel.

The Slow Sweep Unit (SSU) is useful for sweep speed of 1 ns/pixel up to DC. It uses a fast D flip-flop shift register. The sample period $\Delta T$ is equal to the period of the register clock *REG Clock*. The linearity of such a delay generator is very high as it is linked to the clock stability. Nevertheless, in an asynchronous operation, the timing jitter of the generated $start_j$ and $stop_j$ signals with respect to the trigger signal *Start* and *Stop* is equal to the sampling period peak-to-peak, i.e. $\Delta T$ p-p.

## 4. Direct optical conversion VISC

### 4.1 Front-end and sampling cell

A direct optical conversion VISC have been designed with the Austria Micro Systems 0.35μm BiCMOS process. One acquisition channel of the sensor is composed of a transimpedance amplifier as a front-end and a 128-deep memory (Fig. 13, left). The prototype features 64 channels. The temporal resolution of an integrated streak camera is determined by the overall bandwidth of the photodetector and the subsequent electronics. The target is to extend the bandwidth of the front-end beyond 1 GHz, which corresponds to a temporal resolution of less than 1 ns. The challenge consists in keeping the overall sensor bandwidth equal to that of the photodetector alone, while keeping a high gain. Moreover, the power consumption and silicon area should be kept low because of the multichannel architecture. The VISC makes use of a compact and fast voltage-domain sampling scheme shown in Fig. 13-left. An NMOS voltage sample and hold unit is appropriate for the operation as it offers a dynamic range of more than 2 volts and a bandwidth of more than 6 GHz.

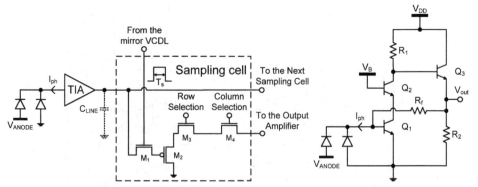

Fig. 13. A channel of the VISC with a double photodiode, a high speed TIA and the detail of a sampling cell with its read-out electronics (left). The transimpedance amplifier schematic (right).

The choice of the TIA topology is strongly restricted to low power and low area consumption architectures as some hundreds of channels are targeted in the final VISC. The designed shunt-shunt feedback common-emitter TIA, shown in Fig. 13-right, features a Gain-Bandwidth product of 17 THzΩ while consuming less than 7 mW. The simulated with extracted parasitics closed-loop gain and bandwidth are 80 dBΩ and 1.75 GHz, respectively. The circuit occupies only $26 \times 116 \, \mu m^2$ and shows a good trade-off between gain, bandwidth, and power consumption (Table 3).

| Reference | Technology | Gain | Bandwidth | GBZ | Power consumption | Notes |
|---|---|---|---|---|---|---|
| (Oh & Lee 2004) | 0.35 µm CMOS | 68 dBΩ | 1.73 GHz | 4.35 THzΩ | 50 mW | CE with cascode, shunt peaking |
| (Huang, & Chen, 2009) | 0.18 µm CMOS | 66 dBΩ | 7 GHz | 14 THzΩ | - | Differential, inductive peaking |
| (Kao et al. 2010) | 0.18 µm CMOS | 74.6 dBΩ | 2.9 GHz | 15.6 THzΩ | 31.7 mW | Differential, two stage, negative Miller capacitance |
| (Aflatouni et al. 2009) | 0.13 µm CMOS | 57 dBΩ | 6 GHz | 4.25 THzΩ | 1.8 mW | RGC-input, series and shunt peaking |
| This work | 0.35 µm BiCMOS | 80 dBΩ | 1.75 GHz | 17.5 THzΩ | 7 mW | SiGe Bipolar transistor |

Table 3. Broadband transimpedance amplifiers in standard (Bi)CMOS 2011

## 4.2 Photodiode

The most versatile photodiode in bulk CMOS processes is the $N_{WELL}$-$P_{SUB}$ photodiode since it features a wide and well-positioned depletion region, allowing the detection of long wavelengths with good responsivity. Because of the low overall speed of this diode, it is often used in a spatially modulated light detector (SMLD) configuration which consists in a finger topology of alternatively illuminated and shielded photodiodes (Genoe, 1998, Huang & Chen, 2009). However, the differential diode arrangement has several important drawbacks. The first is the large occupation area that arises from the number of alternating $N_{WELL}$ fingers. Another drawback, directly originating from the finger topology, is the increased parasitic capacitance. To limit its impact on the electrical bandwidth of the subsequent electronics, the input impedance of the amplifier must be very low. The sensitivity is also severely decreased and more than a factor of 5 in output signal is lost at 800 nm, because of the shielded regions (Tavernier et al., 2006). Finally, a SMLD requires a differential processing to carry out the signal of the illuminated and the shielded detector. This configuration is thus not adapted for use within a VISC, where the spatial resolution must be kept around 20 µm and the number of channels must exceed 100.

It has been shown in (Radovanovic et al. 2005; Hermans & Steyaert 2006) that a $P_{DIFF}$-$N_{WELL}$ photodiode with a screening $N_{WELL}$-$P_{SUB}$ photodiode features the fastest overall bandwidth among all basic photodiode structures in bulk CMOS technologies, which can exceed 1 GHz at a wavelength of 850 nm. Unfortunately, the responsivity of the screened photodiodes is 10 times lower than that of $P_{DIFF}$-$N_{WELL}$-$P_{SUB}$ double photodiodes at 850 nm and 6 times lower at 650 nm. The practical operation range of this topology is restricted to the short wavelengths, where the photogenerated carriers are efficiently collected by the surface photodiode. Employing a screened $P_{DIFF}$–$N_{WELL}$ photodiode would lead to an improvement of the temporal resolution, but at the cost of a severe loss in sensitivity.

Since the SMLD arrangement and the screened $P_{DIFF}$–$N_{WELL}$ photodiode are not adapted for use in an integrated streak camera design, a single-ended photodiode coupled to an equalizer could be employed. However, the roll-off of the $N_{WELL}$–$P_{SUB}$ photodiode at long wavelengths starts at about 1 MHz, which requires a 3rd order equalizer to achieve an effective compensation up to 1 GHz (Radovanovic et al. 2005). A single-stage, 3rd order equalizer requires not less than 2 transistors, 3 capacitors and 5 resistors. Thus, its implementation may become problematic with respect to the small implementation area available and the high power consumption constraint. Moreover, equalization should better be adaptive in order to cover multiple wavelengths and lower the impact of eventual component parameters spread. Adaptive equalization requires adjustable passive component values and thus additional control signals (Chen & Huang, 2007).

In an integrated streak camera, the raw data at the output of the sensor can be processed off-chip, allowing various post-processing such as filtering, FPN reduction through dark image subtraction, and recurrent light event accumulation. In the same vein, equalization can be carried out off-chip through software processing, provided the transfer function of the photodetector is known. If the roll-off in the frequency response can be compensated, it is better to use a photodiode with a higher responsivity in order to widen the spectral range and increase the signal to noise ratio. For this reason the photodiode with the widest spectral range and the highest responsivity in bulk CMOS technologies, i.e., the double $P_{DIFF}$–$N_{WELL}$–$P_{SUB}$ photodiode has been chosen (Radovanovic et al., 2006). The equalization process in explained is Section 6.

## 5. Characterization of the temporal resolution of the VISC

The temporal resolution of the VISC has been measured by using a femtosecond laser source at 400 nm and 800 nm. The test bench, shown in Fig. 14, operates as follow: a femtosecond laser generates a stream of 100 fs FWHM pulses at 800 nm and 80 MHz. One of the pulses is selected and amplified by a oscillator/amplifier at a repetition rate of 5 kHz. This amplified pulse can then be doubled by an optional second harmonic generator (SHG) in order to generate a wavelength of 400 nm. This laser pulse is fed into an optical fiber which is focused on to the slit of the VISC. The measurement of the laser pulse at the output of the optical fiber with a synchroscan conventional streak camera shows that the pulse width is less than 7 ps FWHM, showing that it can be considered as a Dirac impulse by the system. A set of fixed attenuators have been placed in the optical path to avoid saturation of the sensor. A trigger signal generated by the amplifier ensures the synchronization of the streak camera with the laser pulse. A delay and pulse generator allows the precise adjustment of the trigger with respect to the laser pulse.

The response of the camera at a wavelength of 800 nm is shown in Fig. 15. The laser pulses followed by a secondary trace due to reflection in the optical fiber are clearly observed. As the laser pulse is very short, the obtained response represents the impulse response of the system, which in turn stands for the temporal resolution of the ISC. Thus, the temporal resolution is 710 ps FWHM at 800 nm and 490 ps FWHM at 400 nm. To compensate for the poor performance of photodiodes in standard technologies, the frequency response of camera has been equalized.

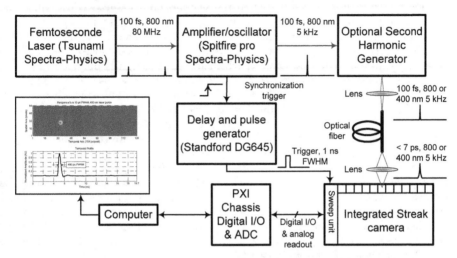

Fig. 14. Experimental setup for impulse response measurement of the VISC

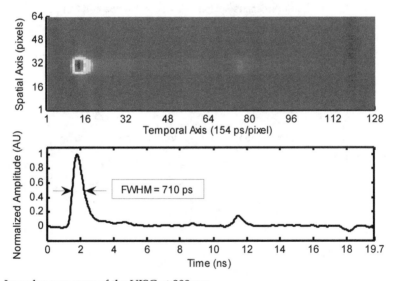

Fig. 15. Impulse response of the VISC at 800 nm

## 6. Improvement of the temporal resolution

The frequency response of the integrated photodiode has been corrected by using software equalization. The measured frequency response $H_M(z)$ of the TIA VISC, (Fig. 16) is obtained by computing the Fourier transform of the measured impulse response. It is seen that even if the -3 dB bandwidth of the photodiode is as low as 130 MHz at 800 nm, the high frequency attenuation are limited to only 15 dB at 1 GHz. The weak attenuation at high frequency allows the compensation to be efficiently carried out. Assuming that no aliasing occurs in the baseband, if the inverse of the measured transfer function is employed as equalizing filter, a flat frequency response should be obtained. In practice, doing so, the high frequency noise is amplified and becomes dominant over frequencies in which meaningful information is contained. Finally, in the time domain, the event ends completely shaded by noise. To avoid this scenario, the frequency band in which the equalization is operated has to be limited. To do so, we synthesize a filter $H_I(z)$, which exhibits flat response up to a frequency $f_{max}$ beyond which the SNR of the signal becomes unacceptably low, and next cuts off sharply any frequency situated above this bound. Indeed, this filter represents the desired global transfer function of the integrated streak camera. By computing the ratio between $H_I(z)$ and the measured transfer functions $H_M(z)$, we obtain the correcting function $H_C(z)$. Pushing the cut-off frequency of $H_I(z)$ close to $f_{max}$ requires a high-order filter to efficiently attenuate noise. However, an abrupt cut-off approaches the shape of a brick-wall filter, which causes the time response to exhibit non-causal behavior, i.e. a *sinc* function impulse response. To prevent this scenario, the order of the filter must be lowered, which implies a lower cut-off frequency to efficiently attenuate noise. In the time domain, the compromise between filter order and cut-off frequency corresponds to a trade-off between pulse shape and signal to noise ratio. Thus, for given pulse shape and signal to noise ratio, a limit in the maximal achievable temporal resolution exists.

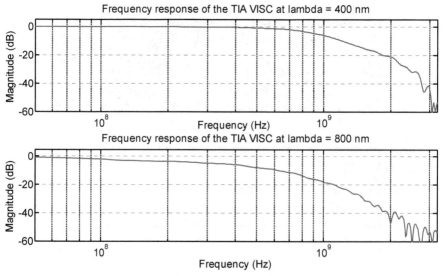

Fig. 16. Frequency response of the TIA VISC at 400 nm and 800 nm

Gaussian-type window has been chosen for the filter, since it matches best the expected and desired pulse shape at the output of a streak camera, excited by a delta function. Application of this global equalization at a wavelength of 800 nm using a 10th-order Gaussian filter $H_I(z)$ with a cut-off frequency of 1 GHz as desired frequency response yields the spectrum of the temporal response $H_E(z)$ shown on Fig. 17. Comparing the measured $H_M(z)$ and equalized $H_E(z)$ responses we notice a significant improvement up to ~2 GHz. Indeed, the low-frequency roll-off is removed, resulting in the suppression of any slow variations in the observation window. The high frequencies have been amplified with a gain of about 15 dB, boosting the high speed signal variations. Finally, the attenuation of $H_C(z)$ above 2.5 GHz improves the signal to noise ratio of the equalized signal.

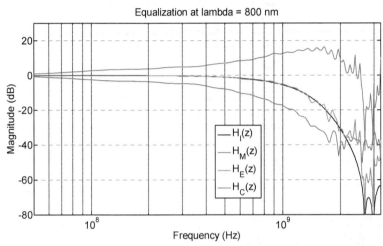

Fig. 17. Spectrum of the measured impulse response $H_M$ (blue), targeted frequency response $H_I$ (black) equalization filter $H_C$ (red) and obtained frequency response after equalization $H_E$ (green)

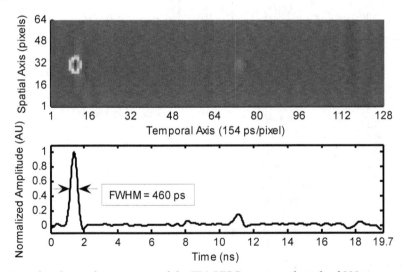

Fig. 18. Equalized impulse response of the TIA VISC at a wavelength of 800 nm

Result of the equalized impulse response of the TIA VISC at a wavelength of 800 nm is shown on Fig. 18. The FWHM is of 460 ps, which is more than 35 % lower than the measured value shown in Fig. 15. Interestingly, after equalization, the camera exhibits similar temporal resolution and pulse shape at both wavelengths of interest. With such a temporal resolution, sub-nanosecond FWHM events can be detected and nanosecond-order phenomena measured with good accuracy. Equalization is a powerful technique, which allows the frequency response of an integrated streak camera to be made uniform over its entire spectral range. This is an important feature, since in streak-mode imaging the incident wavelength is often monochromatic or well known. For example, in one of the most common streak camera applications, the time-resolved spectroscopy, the streak image can be dynamically equalized, such as no distortion is introduced by the degraded behavior of the photodetectors with increasing wavelength.

| Sensor | Sampling rate Temporal resolution | Number of channel | Memory depth per channel | Total sampling rate | Notes |
|---|---|---|---|---|---|
| UHFR – (Lowrance & Kosonocky, 1997) | 1 MHz 1 µs | 360 × 360 | 30 | 130 GS/s | First ultrafast CCD |
| ISIS-V4 (Etoh et al. 2006) | 1 MHz 1 µs | 420 × 720 | 144 | 300 GS/s | CCD technology. Colour version exists |
| 3T MISC (UDS/CNRS (Casadei et al. 2003) | 1,25 GHz <6 ns | 64 | 64 | 80 GS/s | First integrated streak camera |
| 6T MISC$_a$ (UDS/CNRS) (Morel et al. 2006) | 1,5 GHz <6 ns | 64 | 64 | 96 GS/s | On chip analog accumulation feature |
| 6T MISC$_c$ (UDS/CNRS) (Zlatanski et al. 2010a) | 7 GHz <1.1 ns@400 nm | 93 | 64 | 650 GS/s | Adjustable sweep speed |
| VISC$_b$ TIA (UDS/CNRS) (Zlatanski & Uhring 2011) | 7 GHz 500 ps@400~800 nm | 64 | 128 | 450 GS/s | TIA Gain 10000 V/A. First sub 500 ps VISC |
| VISC Source Follower (Zlatanski et al. 2010a)] | 7 GHz < 1 ns | 64 | 128 | 450 GS/s | Asynchronous reset |
| Kleinfelder SC (Kleinfelder et al, 2009) | 100 MHz 10 ns | 150 | 150 | 15 GS/s | Synchronous reset |
| J. Deen (Desouki et al., 2009) | 1,25 GHz <1 ns | 32 × 32 | 8 | 1,28 TS/s | 0,130 µm, 1st video camera > 1 Gfps |
| Near future of integrated streak camera | 10 GHz 100 ps | 1024 | 1024 | 10 TS/s | Do not exist for the moment |
| Phantom v12.1 in streak-mode (Parker et al. 2010) | 1 MHz 1 µs | 128 × 8 | 2 s, i.e. 2 GS | 1 GS/s | High speed video camera in streak-mode |
| Optronis SC10 streak camera | ~5 THz 2 ps | ~1024 | ~1024 | ~5 PS/s | Sweep speed 200 fs/pixel |

Table 4. State of the art of ultrafast imaging solid-state sensors, the grey cells refer to other technologies than solid-state sensor with *in-situ* feature

# 7. Conclusion and perspectives

The concepts of ultrafast imaging and their application in ultrafast BiCMOS optical sensors have been presented. The TIA VISC prototype processed in standard BiCMOS technology reached an overall bandwidth of 1 GHz and a global sampling rate of more than 400 GS/s.

This work demonstrates that sub-nanosecond imaging with an equivalent frame rate of several billion fps is possible using standard (Bi)CMOS technologies. Table 4 summarizes the performances of the state of the art ultrahigh speed imaging solid-state sensors.

Phototransistors or avalanche photodiodes reported in standard technologies could be an interesting alternative to the standard photodiode structures as they can reach a photoelectron amplification factor of 100. The difference in sensitivity between the integrated and the conventional streak camera can thus be reduced down to one order of magnitude. Consequently, the sensitivity of the integrated streak camera has a high potential of evolution. In the near future, integrated streak cameras could be a competitive alternative to conventional streak-mode imaging instruments for applications in which a temporal resolution of about several hundreds of picoseconds is required.

# 8. References

Aflatouni F., & Hashemi H. (2009). A 1.8mW Wideband 57dBΩ Transimpedance Amplifier in 0.13μm *CMOS, IEEE Radio Frequency Integrated Circuits Symposium*, 2009, pp. 57

Amelio G. F., Tompsett M. F., & Smith G. E. (1970). Experimental Verification of the Charge-Coupled Semiconductor Device Concept, *The Bell System Technical Journal*, vol. 49, 1970, pp. 593–600

Bigas M., Cabruja E., Forest J. & Salvi J. (2006). Review of CMOS Image Sensors, *Microelectronics Journal*, vol. 37, 2006, pp. 433–451

Brixner B. (1955). One Million Frame per Second Camera, *Journal of the Optical Society of America*, vol. 45, no. 10, 1955, pp. 876–880

Casadei B., Le Normand J. - P, Hu Y., & Cunin B. (2003). Design and Characterization of a Fast CMOS Multiple Linear Array Imager for Nanosecond Light Pulse Detections, *IEEE Trans. on Instrumentation and Measurement.*, vol. 52, no. 6, 2003, pp. 1892–1897

Chen W. -Z, & Huang S. -Hao. (2007). A 2.5 Gbps CMOS Fully Integrated Optical Receiver with Lateral PIN Detector, *IEEE Custom Integrated Circuits Conference*, 2007

Connell H. W. (1926). The Heape and Grylls Machine for High-speed Photography, *Journal of Scientific Instruments*, Vol. 4, 1926, pp. 82–87

Cordin. (2011). *Rotating mirror camera specifications*, 10/05/2011, Available from: http://www.cordin.com/

Desouki M., Deen M. J., Fang Q., Liu L., Tse F., & Armstrong D. (2009). CMOS Image Sensors for High Speed Applications, *Sensors*, vol. 9, 2009, pp. 430–444

Elloumi M., Fauvet E., Goujou E., & Gorria P. (2004) The Study of a Photosite for Snapshot Video, *SPIE High Speed Imaging and Photonics*, vol. 2513, 2004, pp. 259–267

Etoh T. G., & Mutoh H. (2005). An Image Sensor of 1 Mfps with Photon Counting Sensitivity, *Proceedings of SPIE*, vol. 5580, 2005, pp. 301–307

Etoh T. G., Mutoh H., Takehara K., & Okinaka T. (1999). An Improved Design of an ISIS for a Video Camera of 1000000 fps, *Proc. of SPIE*, vol. 3642, 1999, pp. 127–132

Feng J. et al. (2007). An X-ray Streak Camera with High Spatio-temporal Resolution, *Applied Physics Letters*, vol. 91, 2007, no. 13

Frank A. M., & Bartolick J. M. (2007). Solid-state Replacement of Rotating Mirror Cameras, *Proc of SPIE*. vol. 6279, 2007, 62791U

Fuller P. (2005), Some Highlights in the History of High-Speed Photography and Photonics as Applied to Ballistics, *High-Pressure Shock Compression of Solids VIII Shock Wave and High Pressure Phenomena*, 2005, pp. 251-298

Genoe J., Coppe D., Stiens J.H., Vonekx R.A., Kuijk M. (2001).Calculation of the current response of the spatially modulated light CMOS detector, *IEEE Transactions on Electron Devices*, Vol. 48, 2001, pp. 1892

Hermans C., & Steyaert M. S. J. (2006). A High-speed 850-nm Optical Receiver Front-end in 0.18-μm CMOS, *IEEE Solid-State Circuits*, vol. 41, no. 7, 2006, pp. 1606–1614

Huang S. H., & Chen W. Z. (2009). A 10 Gb/s CMOS Single Chip Optical Receiver with 2-D Meshed Spatially-Modulated Light Detector, *IEEE Custom Integrated Circuit Conference*, 2009, pp. 129

Igel, E. A. & Kristiansen M. (1997). *Rotating mirror streak and framing cameras*, SPIE, ISBN 0-8194-2461-7

Jiang Z., Sun G. G., & Zhang X. C. (1999). Terahertz Pulse Measurement with an Optical Streak Camera, *Optics Letters*, vol. 24, no. 17, 1999, pp. 1245–1247

Kao T. S.-C., Musa F. A., Carusone A. C. (2010). A 5-Gbit/s CMOS Optical Receiver With Integrated Spatially Modulated Light Detector and Equalization, *IEEE Transactions on Circuits and Systems – I: Regular Papers*, vol. 57, no. 11, 2010, pp. 2844–2857

Kleinfelder S. (1987). Development of a Switched Capacitor based Multi-channel Transient Waveform Recording Integrated Circuit, *IEEE Transactions on Nuclear Science*, vol. 35, no. 1, 1987, pp. 151–154

Kleinfelder S. (1990). A 4096 Cell Switched Capacitor Analog Waveform Storage Integrated Circuit, *IEEE Transactions on Nuclear Science*, vol. 37, no. 3, 1990, pp. 1230–1236

Kleinfelder S. (2003). Gigahertz Waveform Sampling and Digitization Circuit Design and Implementation, *IEEE Trans. on Nuclear Science*, vol. 50, no. 4, 2003, pp. 955–962

Kleinfelder S., & Kwiatowski K. (2003). Multi-Million Frames/s Sensor Circuits for Pulsed-Source Imaging, *IEEE Nuclear Science Symposium*, vol. 3, 2003, pp. 1504–1508

Kleinfelder S., Chen Y., Kwiatkowski K., & Shah A. (2004). High-speed CMOS Image Sensor Circuits with In Situ Frame Storage, *IEEE Transactions on Nuclear Science*, vol. 51, no. 4, 2004, pp. 1648–1656

Kleinfelder S., Chen Y., Kwiatowski K., & A. Shah (2004). Four Million Frames/s CMOS Image Sensor Prototype with on focal Plane 64 Frame Storage. *Proc. of SPIE*, vol. 5210, 2004, pp. 76-83

Kleinfelder S., Sukhwan L., Xianqiao L., & El Gamal E. A. (2001). A 10 000 Frames/s CMOS Digital Pixel Sensor, *IEEE Journal of Solid-state Circuits*, vol. 36, 2001, pp. 2049–2059

Kleinfelder S., Wood Chiang S. -H., Huang W., Shah A., & Kwiatkowski K. (2009). High-Speed, High Dynamic-Range Optical Sensor Arrays, *IEEE Transactions on Nuclear Science*, vol. 56, no. 3, 2009, pp. 1069–1075

Krymski A. et al. (1999). A High-speed, 500 Frames/s, 1024 × 1024 CMOS Active Pixel Sensor, *Symposium on VLSI Circuits*, 1999, pp. 137–138

Krymski A.I., Bock N.E., Nianrong T., Van Blerkom D., & Fossum E.R. (2003). A High-speed, 240-frames/s, 4.1-Mpixel CMOS Sensor, *IEEE Transactions on Electron Devices*, vol. 50, no. 1, 2003, pp. 130–135

Lai C. C. (1992). A New Tubeless Nanosecond Streak Camera Based on Optical Deflection and Direct CCD Imaging, *Proceedings of SPIE*, vol. 1801, 1992, pp. 454–468

Lai C. C., Goosman D. R., Wade J. T., & Avara R. (2003). Design and Field Test of a Galvanometer Deflected Streak Camera, *Proc. of SPIE*, vol. 4948, 2003, pp. 330–335

Lambert R. (1937). A Rotating Drum Camera for Photographing Transient Phenomena, *Review of Scientific Instruments*, vol. 8, 1937, pp. 13–15

Le Normand J-P, Zint V., & Uhring W. (2011). High Repetition Rate Integrated Streak Camera in Standard CMOS Technology, *Sensorcomm 2011*, 2011, pp. 322-327

Lowrance J. L., & Kosonocky W. F. (1997). Million Frame per Second CCD Camera System, *SPIE*, vol. 2869, 1997, pp. 405–408

Mahapatra N. R., Garimella S. V., Tareen A. (2000). An Empirical and Analysis Comparison of Delay Elements and a New Delay Element Design, *IEEE Computer Society Workshop on VLSI*, 2000, pp. 81–86

Meghelli M. (2004). A 132-Gb/s 4:1 Multiplexer in 0.13-Mm SiGe Bipolar Technology, *IEEE Journal of Solid-state Circuits*, vol. 39, no. 12, 2004, pp. 2403–2407

Mehr I. & Sculley T. L. (1998). Oversampling Current Sample/Hold Structures for Digital CMOS Process Implementation, *IEEE Trans. on Circuits and Systems–II: Analog and Digital Signal Processing*, vol. 45, no. 2, 1998, pp. 196–203

Miller C. D. (1946). U. S. Patent 2400887, 1946.

Morel F., Le Normand J.-P., Zint C.-V., Uhring W., Hu Y., & Mathiot D. (2006). A New Spatiotemporal CMOS Imager With Analog Accumulation for Nanosecond Low-Power Pulse Detections, *IEEE Sensors Journal*, vol. 6, 2006, pp. 11-20

Oh Y. - H., & Lee S. - G. (2004). An Inductance Enhancement Technique and its Application to a Shunt-peaked 2.5 Gb/s Transimpedance Amplifier Design, *IEEE Transactions on Circuits and Systems–II: Express Briefs*, vol. 51, no. 11, 2004, pp. 624–628

Parker G. R., Asay B. W., & Dickson P. M. (2010). A Technique to Capture and Compose Streak Images of Explosive Events with Unpredictable Timing, *Review of Scientific Instruments*, vol. 81, 2010, no. 016109

Radovanović S., Annema A. - J., & Nauta B. (2006). High-Speed Photodiodes in Standard CMOS Technology, *Springer*, 2006

Radovanovic S., Annema A.-J., & Nauta B. (2005). A 3-Gb/s Optical Detector in Standard CMOS for 850-nm Optical Communication, *IEEE Journal of Solid-State Circuits*, vol. 40, no. 8, 2005, pp. 1706–1717

Rajaee O. & Bakhtiar M. S. (2005). A High Speed, High Resolution; Low Voltage Current Mode Sample and Hold, *IEEE Symp. Circuits and Syst.*, vol. 2, 2005, pp. 1417–1420

Razavi B. (2003). Design of Integrated Circuits for Optical Communications, *McGraw-Hill*, 2003.

Scheidt K.,& Naylor G.(1999). 500 fs Streak Camera for UV-hard X-rays in 1 kHz Accumulating Mode with Optical-jitter Free-synchronisation, *4th Workshop on Beam Diagnostics and Instrumentation for Particle Accelerators*, 1999, pp. 54–58

Smith G. E. (2001). The Invention of the CCD, *Nuclear Instruments and Methods in Physics Research, A*, Vol 471, 2001, pp. 1–5

Son D. V. et al. (2010). Toward 100 Mega-Frames per Second: Design of an Ultimate Ultra-High-Speed Image Sensor, *Sensors*, vol. 10, 2010, pp. 16–35

Swahn T., Baeyens Y., & Meghelli M. (2009). ICs for 100-Gb/s Serial Operation, *IEEE Microwave Magazine*, vol. 10, no. 2, 2009, pp. 58–67

Tavernier F., Hermans C., & Steyaert M. (2006). Optimised equaliser for differential CMOS photodiode, Electronic Letters, vol. 42, no. 17, 2006, pp. 1002–1003

Toumazou C., Lidgey F. J., & Haigh D. G. (1993). Analogue IC Design: The Current Mode Approach, *Institution of Engineering and Technology*, 1993

Uhring W. Zint C.V. Bartringer J. (2004b). A low-cost high-repetition-rate picosecond laser diode pulse generator, *Proc. of SPIE*, Vol 5452, 2004, pp. 583-590

Uhring W., Jung M., & Summ P. (2004a). Image Processing Provides Low-frequency Jitter Correction for Synchroscan Streak Camera Temporal Resolution Enhancement, *Proc. of SPIE*, vol. 5457, 2004, pp. 245 252

Zavoisky E. K., & Fanchenko S. D., Image Converter High-speed Photography with 10⁻⁹–10⁻¹⁴ sec Time Resolution, *Applied Optics*, vol. 4, no. 9, 1965, pp. 1155-1167.

Zlatanski M. Uhring W. Zint C-V. Le Normand J-P., Mathiot D. (2010a). Architectures and Signal Reconstruction Methods for Nanosecond Resolution Integrated Streak Camera in Standard CMOS Technology, *DASIP*, 2010, pp. 1-8

Zlatanski M., & Uhring W. (2011). Streak-mode Optical Sensor in Standard BiCMOS Technology, *IEEE sensor conference*, 2011, In Press

Zlatanski M., Uhring W. Le Normand J-P. Zint C-V. Mathiot D. (2010b). 12 × 7.14 Gs/s rate Time-resolved BiCMOS Imager, *8th IEEE International NEWCAS*, 2010, pp. 97-100

Zlatanski M., Uhring W., Le Normand J.P., & Mathiot D. (2011). A Fully Characterizable Asynchronous Multiphase Delay Generator, *IEEE Transactions on Nuclear Science*, Vol. 58, 2011, pp. 418-425

Zlatanski M., Uhring W. Le Normand J-P. Zint C-V. Mathiot D. (2010c). Streak camera in standard (Bi)CMOS (bipolar complementary metal-oxide-semiconductor) technology, *Measurement science & technology*, Vol. 21, 2010, pp. 115203

# 6

# Atmospheric Clock Transfer Based on Femtosecond Frequency Combs

Lingze Duan and Ravi P. Gollapalli
*The University of Alabama in Huntsville*
*USA*

## 1. Introduction

Precise timing and frequency synchronization has become ubiquitously needed as the world enters the age of global connectivity. Today's communication and computer networks are running under the regulation of synchronized time bases to ensure efficient data routing and information transfer. As the carrier frequencies of these networks continue to rise to accommodate the ever-increasing data traffic, the need for high-fidelity clock distribution becomes imperative (Prucnal et al., 1986). Within the scope of basic sciences, ultra-stable timing dissemination also becomes increasingly important as more and more highly sophisticated instrument and research facilities, such as space-borne atomic clocks (Chan, 2006), long-baseline radio telescope arrays (Cliché & Shillue, 2006), and particle accelerator-based X-ray pulse sources (Altarelli, 2007), require unprecedented level of synchronization in order to explore the uncharted physical parameter regimes.

Recently, there has been a growing interest in the research of high-fidelity, remote transfer of reference clock signals using femtosecond frequency combs (Foreman et al., 2007; Kim et al., 2008). The rationale behind the idea is that frequency combs can conveniently acquire their timing stability from optical atomic clocks, the next generation timing standard, and transfer the stability of an atomic clock at a fixed frequency onto a number of spectral lines (as many as $\sim 10^5$-$10^6$) over a broad spectrum in *both* optical and microwave frequency regions. These features give frequency combs enormous advantage as a carrier of timing and frequency references over conventional single-frequency lasers, which can only transfer a single channel of clock signal in *either* the optical or the microwave frequency region.

Over the last two years, we have proposed and experimentally demonstrated atmospheric remote clock transfer based on a femtosecond frequency comb (Alatawi et al. 2009; Gollapalli & Duan, 2010). It is a complementary scheme to the previously studied fiber-optic remote delivery of frequency combs (Holman et al., 2004). Compared to fiber-optic delivery, transferring frequency combs through the atmosphere does not rely on the existence of a wired system and hence reduces the cost while offering a much better flexibility. It is particularly desirable for timing synchronization between moving transmitters and receivers, such as motor vehicles, ships, airplanes and satellites, and it is the preferred scheme when a close-range, short-term transfer link is needed. Moreover, future space-terrestrial network will need high-fidelity optical up-and-down links to synchronize space-borne atomic clocks with ground-based atomic clocks (Chan, 2006). Pulse laser-based timing synchronization offers a viable solution to the establishment of such links.

Photodetection plays a key role in laser-based synchronization systems because it converts the timing signals stored in the photons into electronic form so the information can be read by the instruments under synchronization. Photodetectors are essential components in the study of laser remote clock transfer. Their unique impacts on the characterization of atmospheric frequency comb transmission are two-fold. First, the "square-law" nature of photodetectors leads to two completely different system configurations for the transfer of timing signals in the radio/microwave frequency (RF) range and in the optical frequency range, i.e. RF heterodyne and optical heterodyne. Second, atmospheric optical effects such as beam wander and speckle induce strong amplitude noise in the heterodyne signal, which enters the phase noise characterization as a limiting noise source due to the power-to-phase coupling effect in photodetectors.

In this chapter, we will give an overview of the recent progress in our study of atmospheric clock transfer based on femtosecond frequency combs. We shall first review the concept and the basic characteristics of femtosecond frequency combs. Then we will discuss the technical specifics associated with the atmospheric clock transfer experiment, especially the roles of photodetectors in different heterodyne schemes and the impact of power-to-phase coupling in the phase noise measurement. The two sections following that will be devoted to the experimental results, first with the clock signal in the RF range ($\sim 10^9$ Hz) and then with the clock in the optical frequency range ($\sim 10^{14}$ Hz). We shall compare our results with similar experiments based on cw (continuous-wave) lasers and summarize at the end.

## 2. Femtosecond frequency combs

### 2.1 The concept of frequency combs

The concept of optical frequency combs was initially introduced as a way to bring frequency metrology into the optical region (Udem et al., 2002). The advance in precision spectroscopy and optical sensing requires that optical frequencies, which fall in the range of 100–1000 terahertz (THz) for the IR-Visible-UV band, be measured with accuracies comparable to microwaves. Such a level of accuracy can only be achieved with a metrological link that can directly reference optical frequencies to Cs atomic clocks, which operate at about 9.2 GHz (Hall & Ye, 2003). The emergence of solid-state femtosecond lasers in the late 1990s allowed this concept to be experimentally realized (Jones et al., 2000). In essence, a femtosecond frequency comb is simply a frequency-stabilized ultrafast laser (Holzwarth et al., 2000). Such lasers are able to produce optical pulses as short as a few femtoseconds ($10^{-15}$ s) at very fast pulse repetition rates ($\sim 0.1$–10 GHz). The result is a pulse train with a constant pulse interval, as shown in Fig. 1 (a). The spectrum of such a pulse train consists of a series of equally spaced spectral lines (often referred to as comb lines), spanning across an extremely broad spectral range owing to the ultra-short temporal profiles of the femtosecond pulses.

A conceptual illustration of such a supercontinuum spectrum is depicted in Fig. 1 (b). With proper electronic control, the pulse repetition rate (i.e. the spacing of the comb lines) and the carrier-envelope offset (CEO) frequency (i.e. the offset frequency of the comb when it is extended to zero), both in the microwave frequency range, can be stabilized against an atomic reference, such as a Cs clock (Udem et al., 2001). This establishes a grid of well-defined spectral markers through the simple relation of $f_m = f_{CEO} + m f_R$, where $f_R$ and

$f_{CEO}$ are the repetition rate and the CEO frequency, respectively, and $m$ is the index number of the comb line in concern. Unknown spectral features within the span of the supercontinuum can be compared with these comb lines through interference and their exact frequencies can be determined with kilohertz-level resolutions or even better (Fortier et al., 2006).

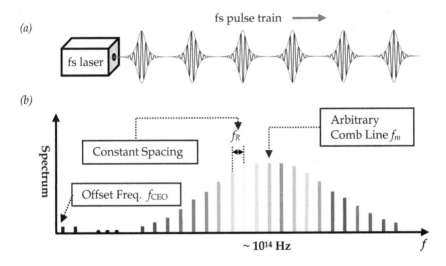

Fig. 1. Concept of femtosecond frequency combs. (a) time-domain picture – a femtosecond pulse train with a fixed pulse interval; (b) frequency-domain picture – a supercontinuum that consists of a series of equally-spaced spectral lines. The frequency of an arbitrary comb line has two degrees of freedom defined by $f_m = f_{CEO} + mf_R$. Locking $f_R$ and $f_{CEO}$, or any two independent combinations of them to atomic clocks will transfer the frequency accuracy of the clocks onto the entire spectrum.

## 2.2 Key characteristics of frequency combs

Once stabilized, a femtosecond frequency comb offers some unique features unattainable with other types of laser systems. For example, it combines a supercontinuum spectrum, often spanning across more than an octave, with ultrahigh spectral precision for individual comb lines, which in the past could only be achieved with cw lasers operating at a single frequency. Sub-hertz linewidth across the entire comb spectrum (e.g. 580–1080 nm) has been reported (Bartels et al., 2004). This feature allows a frequency comb to combine the accuracy of a single-frequency laser and the flexibility of a supercontinuum source, and hence leads to a wide range of applications such as trace-molecule detection (Thorpe et al., 2006), precise ranging (Ye, 2004) and extrasolar planet search (Li et al., 2008).

Another interesting feature for a femtosecond frequency comb is that, besides the precisely defined grid of *optical* frequencies, the comb also carries a grid of highly stable *microwave* frequencies stemming from the constant pulse repetition rate. These frequency references can be easily accessed via photodetection. A photodetector essentially converts the ultrastable timing of the pulses into a series of discrete spectral lines consisting of the harmonics of the repetition rate. A femtosecond frequency comb hence can be treated as a carrier of timing/frequency references, or "clocks". But unlike conventional clock carriers, such as cw lasers or RF communication channels, which can only carry one single frequency at a time, a frequency comb simultaneously carries multiple (~ $10^5$–$10^6$ in some cases) clock signals in both microwave and optical frequency ranges. Such a capability offers femtosecond frequency combs unprecedented versatility in remote clock transfer (Foreman et al., 2007), which we shall focus on in the rest of the chapter.

## 2.3 Frequency comb-based clock transfer

One important recent development in the application of femtosecond frequency combs is the demonstration of comb-based high-fidelity remote clock transfer through optical fibres (Holman et al., 2004; Holman et al., 2005; Foreman et al., 2007; Kim et al., 2008). In the microwave frequency range, a fractional frequency instability of <9×$10^{-15}$ for a 1-s averaging time has been achieved through a 6.9-km installed fibre link with the help of active noise cancellation (Holman et al., 2005). In the optical frequency range, two lasers connected by a 300-m fibre link have been synchronized for 12 hours with a relative timing instability of 9×$10^{-21}$ (Kim et al., 2008). These results are comparable to or exceeding the performance of the conventional techniques, which use a modulated cw laser to transfer a single microwave clock and rely on an ultra-stable single-frequency laser to deliver an optical frequency reference (Foreman et al., 2007). Thus, femtosecond frequency combs are not only efficient and versatile, but effective as well in terms of remote clock delivery via fibre networks.

The success of comb-based clock transfer in fibre logically leads to the question about free-space clock distribution using femtosecond lasers. This will not be a trivial extension from the work done in fibre because the transmission medium now becomes air, which is much more dynamic than glass. New phenomena are expected to dominate and new strategies are needed. In fact, the research on optical pulse propagation in the air dates back to the 1970s. In the context of optical communications, various models of pulse propagation in turbulent random media have been studied (Su & Plonus, 1971; Liu et al., 1974; Hong et al., 1977; Young et al., 1998). Today, femtosecond transmission in the atmosphere continues to attract much attention because of its potential applications in remote sensing and Light Detection and Ranging (LIDAR). Topics in this area that attract major research interest include nonlinear effects, ionization, atmosphere turbulence, and their impact on pulse propagation (Akozbek et al., 2000; Mlejnek et al., 1999; Sprangle et al., 2002; Rodriguez et al., 2004). However, most of the previous work on atmospheric propagation of femtosecond pulses focuses on the interactions between air and individual pulses. Little attention has been given to the propagation of a pulse train as a whole in the atmosphere, especially in terms of the precision degradation of its Fourier frequencies. The work described in the following aims to address this important aspect through experimental characterizations.

## 3. Atmospheric remote transfer of frequency combs

Before we begin the discussion about specific experiments, it is necessary to review several general technical aspects associated with these experiments so that readers can gain a better understanding of the physical background behind some of the techniques involved.

## 3.1 Noise sources in atmospheric transmission

The key question we seek to answer through the current research is: How much extra noise will be added to the clock signals as the femtosecond pulse train propagates through the air? To answer the question, we must first take a look at the potential noise sources in the atmosphere. Two major mechanisms can cause the degradation of clock precision in the air - temporal fluctuation of the refractive index and scattering (Ishimaru, 1978). Air turbulence creates random density fluctuations, resulting in a time-dependent average refractive index over the path of the laser beam, which leads to variation of pulse arrival time. On the other hand, density inhomogeneities in the air, along with scattering particles such as aerosols, may cause pulses being randomly scattered, as shown in Fig. 2. This may lead to pulse distortion and dephasing, which degrade the timing precision of the original pulse train and broaden the comb lines.

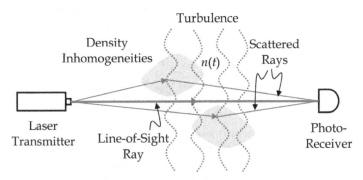

Fig. 2. The effects of index fluctuation and scattering on atmospheric optical transmission.

For ground-level short-distance (e.g. < 1 km) transmission links, one can assume clear air and weak turbulence (Andrews & Phillips, 2005). Under such conditions, it has been shown that scattering plays a far less important role in affecting the pulse arrival time compared to pulse wandering due to index fluctuation (Liu & Yeh, 1980). The transmission distance in our experiments is about 60 m, and we usually avoid taking measurement under extreme weather conditions such as rainy, foggy or highly windy days. Therefore, it is justified to ignore scattering in our research and focus only on the index fluctuation in line-of-sight beam propagation. In addition, it should be pointed out here that turbulence and scattering also cause beam pointing fluctuation and speckle, which affect the received pulse amplitude as well. Special care must be taken in the design of the optical system to minimize the impact of these effects. Specific discussions about some of these measures will be made in Section 3.4.

## 3.2 Excess phase noise and frequency instability

For an infinite periodical pulse train, the temporal variation of the pulse arrival time due to atmospheric transmission broadens the spectral lines in the Fourier domain, introducing excess phase noise to the initially highly precise grid of Fourier frequencies. The phase noise spectrum can be determined via a heterodyne process, which compares the transmitted pulse train with its copy from a reference path that induces negligible phase noise. By spectrally analyzing the heterodyne signal, we can measure the power-spectral density (PSD) of the phase-noise generated sidebands, $S\varphi(f)$, representing the mean-squared phase fluctuation at a Fourier frequency $f$ away from the centre frequency with a 1-Hz bandwidth of measurement (Foreman et al., 2007),

$$S_\varphi(f) = [\delta\varphi(f)]^2 \quad (\text{rad}^2/\text{Hz})$$

(1)

Sometimes it is more convenient to characterize phase noise in terms of pulse timing jitter, which is defined as

$$\delta T(f) = \frac{\delta\varphi(f)}{2\pi\nu_0} \quad (\text{s}/\sqrt{\text{Hz}}),$$

(2)

where $\nu_0$ is the centre frequency. Integrating the spectral density of the timing jitter over the measurement bandwidth, from $f_l$ to $f_h$, leads to the total rms timing jitter of the pulses

$$T_{rms} = \sqrt{\int_{f_l}^{f_h} [\delta T(f)]^2 \, df} \quad (\text{s}).$$

(3)

The spectrum of the excess phase noise or timing jitter essentially characterizes the short-term (usually < 1 s) stability of the clock transmission link.

The second aspect in the characterization of clock stability is how much the time-averaged frequency varies over different lengths of the averaging period. It is usually used to describe the stability of a frequency signal over relatively longer term, e.g. > 1 s. Experimentally, this is done by measuring the Allan deviation of the frequency signal over an averaging time $\tau$, which is defined as

$$\sigma_y(\tau) = \left\langle \frac{1}{2} [\bar{y}(t + \tau) - \bar{y}(t)]^2 \right\rangle^{1/2},$$

(4)

where $\bar{y}(t)$ is the time average of the instantaneous fractional frequency deviation from the centre frequency (Allan, 1966).

The phase noise/timing jitter spectrum and the Allan deviation characterize clock instability from different perspectives. They are the main quantities we seek to determine through our experiments.

### 3.3 RF heterodyne vs. optical heterodyne

The heterodyne technique has long been used in RF technologies to shift information signals from one frequency channel to another (Nahin, 2001). In the current research, in order to study the phase noise sidebands due to pulse propagating in the atmosphere, we need to shift the noise sidebands to lower frequencies where a fast Fourier transform (FFT) analyzer can effectively resolve the noise spectrum. Conceptually, this can be easily done by beating the noisy signal with a noise-free local reference, which has a frequency identical or close to the centre frequency of the noisy signal. At the output of the heterodyne mixer, the noise sidebands are shifted to both the sum and the difference frequencies. With a low-pass filter, we can select the downshifted bands at the difference frequency. In the case the difference frequency is zero, i.e., the noisy signal and the reference have the same centre frequency, the noise sidebands are shifted to the base band, and the lower sideband is folded into the positive side by the FFT. Fig. 3 (a) & (b) give a conceptual sketch of this heterodyne process. The reference signal can be obtained by splitting a portion of the original clock at the output of the source. This also greatly reduces the impact of the source noise to the measurement as the source noise is common-mode and hence is cancelled at the difference frequency.

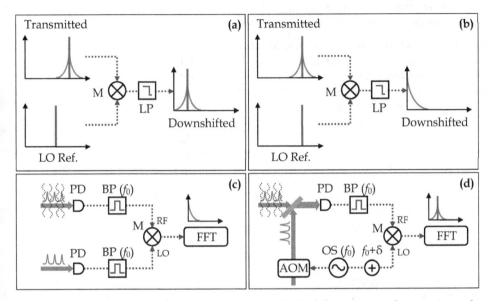

Fig. 3. Concept of heterodyne noise measurement with (a) different centre frequencies and (b) same centre frequency. Configurations of (c) RF heterodyne and (d) optical heterodyne. BP, bandpass filter; LP, lowpass filter; M, mixer; OS, oscillator; PD, photodiode.

In practice, two different heterodyne configurations are used for the two clock frequency ranges, namely RF heterodyne and optical heterodyne. The necessity originates from the use of square-law detectors. Photodetectors, such as photodiodes, are known to be square-law detectors. They cannot directly respond to the optical field due to the short optical cycles (~ fs). Instead, they respond to the quantized excitations of the optical field, i.e. photons. This is equivalent to finding the time-averaged intensity, which is proportional to the "square" of the optical field (Davis, 1996). As a result, the optical cycle is averaged out and the detector can only probe the slow variation of the field envelope. However, a square-law detector can respond to the beat frequency of two optical fields with a small frequency difference when they are incident on the detector simultaneously. To understand this effect, we shall assume optical waves $A$ and $B$ with a slight frequency offset $\Delta\omega$ superpose on a photodetector. The total field can be written as

$$\vec{E} = \vec{A}\cos\omega t + \vec{B}\cos(\omega + \Delta\omega)t .$$ (5)

The detector responds to the time-averaged intensity $I$, which can be expressed as

$$I \equiv \varepsilon_0 c < \vec{E} \cdot \vec{E} > \ = \varepsilon_0 c \left( \frac{A^2}{2} + \frac{B^2}{2} + \vec{A} \cdot \vec{B} \cos \Delta \omega t \right), \tag{6}$$

where <> indicates time-average, and $\varepsilon_0$ and $c$ are free-space permittivity and speed of light, respectively. The last term on the right hand side is the interference term, which oscillates at the beat frequency. It can be detected by the photodetector as long as it falls within the detector's bandwidth. From this point of view, a photodetector can be used as a heterodyne mixer to generate an electric oscillation at the difference frequency of the two optical fields. This feature of photodetectors becomes the basis of optical heterodyne.

In Fig. 3 (c) and (d), we use our experiments as examples to illustrate the concepts of RF heterodyne and optical heterodyne. In the RF heterodyne, two identical photodetectors convert the femtosecond pulse trains from both the transmitted arm and the reference arm into a series of electric oscillations at the harmonics of the pulse repetition rate. One of these frequency components is selected by bandpass filers in both arms as the clock under study. The transmitted clock with excess phase noise is then mixed with the reference clock at a double-balanced mixer to generate the heterodyne beat signal, which is spectrally analyzed by an FFT analyzer to give the noise spectrum. In the optical heterodyne, the pulse trains from the two arms are first mixed on a photodetector. The reference beam gains a slight frequency shift $f_0$ at an acousto-optical modulator (AOM). As a result, the photodetector produces a beat signal at $f_0$. This beat signal is further downshifted via RF heterodyne to make it accessible by the FFT analyzer. As we shall discuss in Section 5, for optical clock transfer, it is more convenient to keep the final beat signal away from the base band so that the full line shape can be seen.

One unique feature of optical heterodyne with femtosecond frequency combs is the process of multiheterodyne. When two frequency combs are mixed on a photodetector, each pair of corresponding comb lines from the two arms produces a beat signal. The total heterodyne signal is a coherent superposition of all these beat notes from individual comb lines. The analysis of multiheterodyne of frequency combs is out of the scope of this chapter. In-depth discussions can be found elsewhere (Gollapalli & Duan, 2011).

## 3.4 Power-to-phase coupling in photodiodes

When a photodiode is directly used to extract microwave signals from a modulated optical signal, additional phase noise is generated in the spectra of the microwave signals due to power-to-phase conversion in photodiodes (Tulchinsky & Williams, 2005). When the optical signal is an ultrafast pulse train, this excess phase noise can be found at the harmonics of the pulse repetition rate, causing extra fluctuations to the microwave clock signals extracted from the pulse train (Ivanov et al., 2003). The exact physical mechanisms of the power-to-phase conversion in photodiodes are not completely clear. One likely cause is the saturation of photodetectors (Liu et al., 1999). When the average power of the optical signal reaches certain level, space-charge build-up in the depletion region of the photodiode reduces the electric field in the depletion region, leading to a reduction of the velocity of photogenerated carriers and hence a phase delay in the photocurrent. Clear correlations between the arrival times of the photocurrent pulses and the incident optical power have been observed experimentally (Bartels et al., 2005).

The presence of power-to-phase coupling in photodetectors results in a detector-induced excess phase noise, which, in the case of microwave clock transfer, could mask the excess phase noise due to transmission. Two unique factors affecting the quality of atmospheric laser communication are beam wander and speckle (Andrews & Phillips, 2005). Both effects have been visually observed in our experiment. One of their consequences is optical power fluctuation on the photodetector. The excess phase noise due to power-to-phase coupling can make a significant contribution to the total measured phase noise. In order to evaluate the impact of this detector-induced phase noise, we use a microwave power detector to monitor the power of the extracted clock signal and measure the correlation between the detected phase noise and the clock power (see Section 4.2 and Fig. 5 for description of the experimental setup). Under linear operation, the power of the recovered clock scales as the square of the optical power received by the photodiode. However, the fluctuation of the clock power can be treated as linearly proportional to the fluctuation of the optical power near the average power level. Therefore, the correlation between the phase noise and the fluctuation of the clock power actually reflects the correlation between the phase noise and the optical power fluctuation. Correlation is determined by recording the coherence function between the measured phase noise and clock power during each data run. Here "coherence" is defined as

$$C_{yx}(f) = \frac{\Phi_{yx}(f)}{\sqrt{\Phi_{yy}(f) \cdot \Phi_{xx}(f)}}, \tag{7}$$

where $\Phi_{xx}(f)$ and $\Phi_{yy}(f)$ are the power spectral densities of time series $x$ and $y$ (as functions of Fourier frequency $f$), and $\Phi_{yx}(f)$ is the cross-power spectral density between the two series. By definition, the coherence function ranges between 0 and 1.

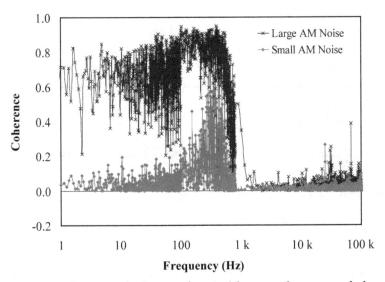

Fig. 4. The correlation (in terms of coherence function) between the measured phase and clock power indicates the impact of power-to-phase coupling in the photodiode.

Fig. 4 shows the measured phase-power coherence under two experimental configurations. The blue cross trace is obtained with a fibre-coupled photodiode and a fibre collimator receiving the transmitted laser beam. Since the coupling efficiency of the collimator critically depends on beam pointing and transverse intensity distribution of the incident beam, the power coupled to the detector is very sensitive to beam wander and speckle. As a result, a significant amount of phase noise is generated by the photodiode due to power-to-phase coupling. This is indicated by the coherence values close to 1 below 1 kHz. Apparently, such a system configuration would not allow us to effectively evaluate the transmission-induced phase noise because of the overwhelming contribution from the detector. Therefore, in the second configuration, we use large-diameter optics (e.g. 2 inch) in the receiving system and focus the beam directly onto the photodiode with the size of the focus much smaller than the active area of the detector. Such a configuration keeps the receiving system insensitive to the beam pointing drift and the transverse beam-profile variation. As the red diamond trace shows, the coherence function is close to zero over a wide frequency range, indicating little phase-power correlation at these frequencies. A moderate peak of coherence (~ 0.5) at a few hundred hertz can be attributed to temporal fluctuation of the irradiance due to scintillation of the laser beam (Andrews & Phillips, 2005). But its impact to the phase noise measurement is very limited because the noise spectrum rolls off quickly at such high frequencies.

## 4. Atmospheric transfer of microwave timing signals

As pointed out in Section 2, the ultrastable pulsing rate of a femtosecond frequency comb can be used as a clock signal. The repetition rates of most common femtosecond lasers range from tens of MHz to 1 GHz or so. Their harmonics, however, can be much higher. A femtosecond pulse train inherently carries all these harmonics simultaneously, and their retrieval as timing references is only limited by the bandwidths of the photodetectors used.

### 4.1 A rooftop transmission link

In order to study frequency comb-based atmospheric clock transfer, we established an outdoor laser transmission link. The link is located on the roof of our laboratory building on

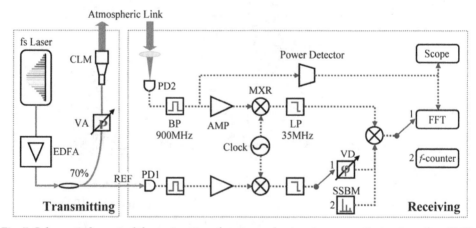

Fig. 5. Schematic layout of the test system for atmospheric microwave timing transfer. AMP, RF amplifiers; BP, band-pass filters; LP, low-pass filters; MXR, double-balanced mixers, PD, photodiodes; VA, variable attenuator; and VD, variable delay.

the campus of the University of Alabama in Huntsville. The four-floor building has an observation platform about 20 meters above the ground. There is no high-rise building or other tall structure nearby to create local turbulence. A sturdy steel tripod is anchored on the platform to house a metal beam reflector (a 2-inch gold mirror). The laser beam is launched from a nearby astronomical observatory into the open air and the reflector sends the beam back to the observatory, where all the signal processing and measurement takes place. The round-trip propagation distance is about 60 m.

## 4.2 Experimental setup

Fig. 5 shows the schematic of the experimental setup for microwave timing transfer. The system is divided into transmitting and receiving subsystems. In the transmitting part, a femtosecond fiber laser (Precision Photonics FFL-1560) generates a train of 150-fs pulses centered at 1560 nm, with a 90-MHz repetition rate and a 4-mW average power. An erbium doped fiber amplifier (EDFA) boosts the average power to 100 mW. A 70:30 broadband fiber coupler directs the majority of this power into a fiber collimator, which launches a 7 mm-diameter beam into the atmospheric transmission link. A fiber-coupled variable attenuator is inserted before the collimator as a power regulator to provide precise control over the total optical power reaching the receiving photodiode.

In the receiving subsystem, the transmitted beam is tightly focused onto a high-speed photodiode, which recovers the repetition frequency of the femtosecond laser as well as its harmonics. The 10th harmonic at 900 MHz is chosen as the microwave clock under test and is selected by a bandpass filter. To further reject the side modes, a local clock is used to beat the 900 MHz signal down to 35 MHz, where sharp low-pass filters can effectively remove all the remaining harmonics. Meanwhile, a reference clock signal is obtained by coupling a small portion of the EDFA output directly into the receiving subsystem via optical fiber (the REF path in Fig. 1) and then using a microwave circuit similar to the transmitted clock. The resulted frequency signal serves two sets of measurement. In the phase noise measurement, the reference clock passes through an adjustable delay line to gain a proper phase before it beats the transmitted clock in quadrature at a double-balanced mixer. The generated phase signal is frequency analyzed by a Fast Fourier Transform (FFT) analyzer (SRS SR785) and timing jitter spectral density can be calculated from the phase noise spectrum. In frequency stability measurement, the reference clock is frequency-shifted by 500 kHz through a single-sideband modulator (SSBM) and then mixed with the transmitted clock. This leads to a 500-kHz beat note, which is then measured with a frequency counter (SRS SR620) to determine its stability. It should be noted here that, although the repetition rate of the femtosecond laser is not stabilized in the test, the noise from the laser and the EDFA does not affect the measurement because it is common mode in the above heterodyne scheme. This ensures correct characterization of the excess clock instability due to the atmospheric propagation.

## 4.3 Laboratory test

In order to verify that the sensitivity of our experimental system is sufficient to measure the excess phase noise due to small index fluctuation in the air, we have first conducted a proof-of-principle test inside the lab, where the airflow can be better controlled. The main source of index fluctuation in this case is the air-conditioning (AC) airflow, predominantly moving downward across the laser beam. To verify the coincidence between the AC airflow and the phase noise, we fix the total length of the transmission link (~10 m) but change the effective beam path exposed in the AC airflow by inserting rigid plastic tubes (2.5 cm inner diameter) in the beam path. We find that the tubes can effectively control the measured phase noise. In addition, we have also made measurement with the AC system off and found the noise to be much lower. These facts confirm the effectiveness of our measurement system (Alatawi et al. 2009). Fig. 6 (a) shows the spectra of the transmission-induced timing jitter at four effective beam path lengths over a Fourier frequency range of $10^{-2}$ – $10^5$ Hz, along with the system noise (measured with a negligible beam path in the air). The system noise, mainly attributed to the RF amplifiers and the mixers, has a $1/f$ power frequency dependence below 1 kHz,

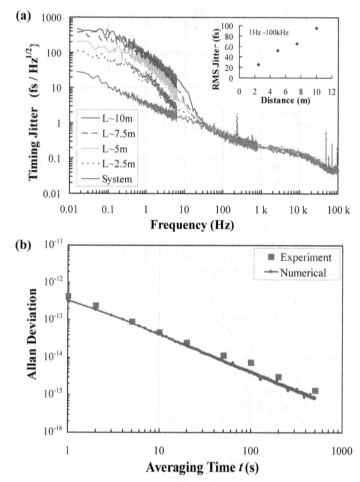

Fig. 6. (a) Spectral density of the excess timing jitter for four effective transmission distances, along with the system noise floor; inset: the total rms jitter integrated from 1 Hz to 100 kHz vs. effective beam path length in AC airflow. (b) The measured Allan deviation with a 10-m effective beam path and the numerical estimation based on the phase noise measurement.

indicating a Flicker phase noise that commonly exists in RF amplifiers within this frequency range (Halford et al., 1968). As the frequency increases, the noise floor gradually shifts toward white noise. The transmission-induced phase noise only becomes appreciably above the system noise below a few tens of hertz. As the Fourier frequency goes down, the power law of the noise gradually changes and eventually shows a trend toward $f^0$ below 0.1 Hz. Such behaviour is markedly different from what has been observed in fibre (Holman et al., 2005). To further verify the dependence of the excess phase noise on the AC airflow, we have calculated the rms timing jitter at different effective beam path lengths by integrating the timing jitter spectra in Fig. 6 (a) from 1 Hz to 100 kHz. For effective path lengths of 2.5, 5, 7.5, and 10 m, the total rms jitters are 37.5, 58.9, 71.2, and 99.0 fs, respectively, and the total rms jitter caused by the system noise is 27.5 fs. Fig. 6 (a) inset shows a near linear relation between the total *actual* rms timing jitter (system noise excluded) and the beam path length exposed in the airflow, confirming a direct correlation between airflow and phase noise.

The long-term transfer instability for 10-m effective beam path is evaluated by measuring the Allan deviation $\sigma_y$ of the beat note frequency at a set of different averaging times. Fig. 6 (b) shows a typical set of experimental data, along with a set of numerical data based on the timing jitter measurement shown in Fig. 6 (a). The calculation uses the relation between the spectral density of phase noise and the Allan deviation (Rutman & Walls, 1991)

$$\sigma_y^2(\tau) = 2 \int_0^{f_h} S_y(f) \frac{\sin^4(\pi\tau f)}{(\pi\tau f)^2} df , \qquad (8)$$

where $f_h$ is the bandwidth of the system low-pass filter. The excellent agreement between the two sets of data demonstrates the effectiveness of our measurement in assessing the excess clock fluctuation caused by atmospheric transmission.

### 4.4 Outdoor test

Once the measurement scheme passes the lab test, we apply it to the rooftop transmission link. Excess phase noise has been measured at different times of a day and under various weather conditions (except rainy days). The corresponding timing jitter is then calculated from the phase noise spectrum using (2). Fig. 7 (a) shows several typical traces of the jitter spectral density, along with the system noise baseline. The excess timing jitter is above the baseline only at frequencies below several hundred hertz. The magnitude and frequency dependence of the jitter spectra are strongly affected by the weather conditions, especially the wind speed, leading to a group of different spectral traces under otherwise similar conditions. The system baseline is mainly attributed to the RF amplifiers in the receiving system, and the noise spikes around 10-200 Hz are believed to be due to electric interference caused by the utility circuitry in the observatory. Fig. 7 (b) shows the rms jitter integrated from 1 Hz to the frequency in concern for all five traces in (a). Clearly, most contributions to the rms jitter come from noise below 100 Hz, indicating the dominance of slow phase modulations. The total rms jitters integrated from 1 Hz to 100 kHz range from several hundred femtoseconds to about two picoseconds. The system noise proves to have a negligible effect in the measurement as evident from its sub-100-fs effective rms jitter.

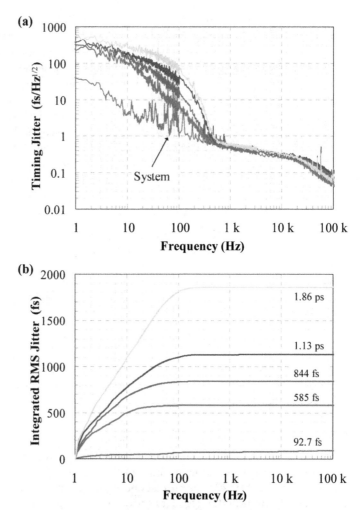

Fig. 7. Transmission-induced pulse timing jitter after a femtosecond frequency comb propagates 60 m through open atmosphere under various weather conditions. (a) Typical timing jitter spectra along with the system noise. (b) The integrated rms jitters for the five spectra in (a). The numbers are the RMS jitters integrated from 1 Hz to 100 kHz.

The result of the Allan deviation measurement is shown in Fig. 8, which also includes the trace from the 10-m indoor test for the purpose of reference. The fractional frequency stability after 60-m atmospheric transmission is in the order of a few parts per trillion at a 1-s averaging time $\tau$. This is comparable to the frequency stability of most commercial atomic timing references, such as Cs or Rb clocks. At longer averaging times, the Allan deviation falls at a rate close to $\tau^{-1}$, indicating white phase fluctuations (IEEE, 1983). This is different from fiber optic transmission, where the Allan deviation falls at $\tau^{-1/2}$ (Holman et al., 2004). It should also be noted that most atomic timing references also have a $\tau^{-1/2}$ behaviour. As a result, transferring a master microwave timing reference through free space becomes more advantageous than maintaining a less-precise local clock when the clock signal is averaged over longer time.

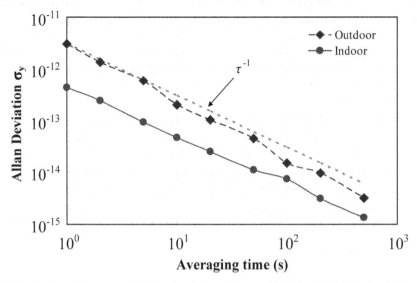

Fig. 8. A typical set of measured Allan deviation (diamond) of the 900 MHz clock signal after 60-m atmospheric propagation shows an approximate $\tau^{-1}$ dependence over the averaging time $\tau$. The baseline data (circle) are obtained under indoor conditions with a much lower wind speed.

### 4.5 Discussion

The propagation of an optical pulse train through an atmospheric communication channel is susceptible to the refractive-index fluctuation caused by clear-air turbulence (Su & Plonus, 1971). From the rms timing jitter, we can derive the rms fluctuation of the group index $n_g$ by using the relation $\Delta n_g = (c / L)\Delta T$, where $c$ is the speed of light in vacuum, $L$ is the total propagation distance, and $\Delta T$ represents the rms timing jitter. By using $\Delta T = 2$ ps and $L = 60$ m, we find the value of $\Delta n_g$ to be $1 \times 10^{-5}$. Meanwhile, it has been shown that $\Delta n_g \approx a \cdot \Delta n$, where $\Delta n$ is the fluctuation of the phase index and the proportional constant $a$ is approximately equal to 3 in the visible and near infrared wavelength range (Ciddor, 1996; Ciddor & Hill, 1999). This leads to an estimated rms phase index fluctuation of several parts per million, which agrees with the well-known scale of such fluctuation due to clear-air turbulence (Shaik, 1988).

The measured timing transfer stability is compared with a similar rooftop experiment over a 100-m open link using the conventional carrier-modulation scheme (Sprenger et al., 2009). The fractional instability at 1 s measured in our test is several times smaller than the result presented in the reference. Moreover, their Allan deviations appear to have an averaging-time dependence close to $\tau^{-1/2}$ below 100 s, while our result is close to $\tau^{-1}$. Such a difference in Allan deviation behaviour indicates possible difference in the underlying mechanism of instability. In fact, as pointed out by the authors, the earlier experiment is likely limited by the stability of the frequency synthesizers, and therefore only offers an upper bound of the propagation-induced instability. This seems to be supported by the fact that the electronic timing instrument usually shows a $\tau^{-1/2}$ characteristic. Such a system-noise limitation is partly because of the low clock frequency (80 MHz) used in the earlier work. In comparison, our experiment is free from such a restriction, as shown by the baseline data, because the wide bandwidth of the femtosecond pulses allows the use of a higher harmonic of the repetition rate as the clock signal.

## 5. Atmospheric transfer of optical frequencies

While clock signals in the microwave frequency range can be delivered by a femtosecond frequency comb and conveniently extracted with a photodetector, it is the ability to carry many optical frequency references that makes frequency combs very attractive as a clock transfer means. This is not only because the much higher frequencies of optical clocks lead to better fractional stability, but because optical frequency references can directly benefit from the next-generation atomic clocks, which will work in the optical frequency range.

### 5.1 Experimental setup

In order to demonstrate the capability of a femtosecond frequency comb to transfer optical frequency references through free space, we convert our rooftop experimental setup shown in Fig. 5 into an optical heterodyne configuration. Fig. 9 shows the new schematic layout. The system is again divided into transmitting and receiving subsystems. The transmitting subsystem is similar to the one used in the microwave clock transfer tests. The receiving

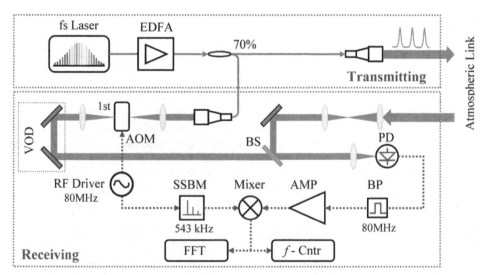

Fig. 9. Schematic layout of the experimental system for atmospheric transfer of optical frequency references. AMP, microwave amplifiers; BS, beam splitter; PD, photodiode; SSBM, single-side band modulator; and VOD, variable optical delay.

subsystem features an optical heterodyne configuration as discussed in Section 3.3. The reference beam passes through an AOM, which is driven by a RF driver at 80 MHz, and the first-order deflection is selected. Meanwhile, the beam transmitted through the atmospheric link is coupled into the receiving subsystem by a lens pair, which converts the expanded beam into a comparable size to the reference beam. The two beams are then combined into collinear propagation at a broadband beam splitter and focused onto a photodiode. A variable optical delay line in the reference arm ensures temporal overlapping of the transmitted pulses and the reference pulses. The photodiode produces an 80-MHz beat note from the two beams. The beat note is then phase-compared with a portion of the original 80-MHz driving signal at a double-balanced mixer. In order to measure the linewidth of the beat note, a small frequency shift (about 543 kHz) is added to the original 80-MHz signal through a single-sideband modulator (SSBM). The resulted 543-kHz beat signal is analyzed by an FFT analyzer for direct spectral width measurement and a frequency counter for frequency stability measurement. It should be noted here that the design of the receiving optical system has been made with the consideration of minimizing the impact of speckle and beam wander as discussed in Section 3.4.

## 5.2 Experimental results

The frequency transfer test was performed over the course of three months in the summer of 2010. Data have been taken under low-wind conditions with the wind speed below 2 m/s. Strong phase fluctuations in the multiheterodyne signal have been observed as pictured in Fig. 10 (a) inset. Time-domain phase traces like Fig. 10 (a) inset are taken by mixing the 80-MHz beat note from the photodiode directly with the original 80-MHz AOM driving signal without the SSBM. The fast oscillation of the beat note phase indicates a phase modulation index much greater than $2\pi$. In order to examine the Fourier spectrum of the beat note, we add a small frequency offset of 543 kHz using the SSBM onto the 80-MHz reference from the AOM driver so that the beat signals from the mixer are shifted out of the base band. The resulted beat note spectrum is shown in Fig. 10 (a), plotted against the offset frequency from 543 kHz. Also plotted is the same beat signal without atmospheric transmission, realized by diverting the pulse train in the transmission arm through a fiber link of about 40 m long. The fiber is cut in such a way that it produces the same amount of time delay as the 60-m atmospheric link. Such an arrangement allows us to assess the contribution of the laser and the amplifier in our phase noise measurement.

Due to the imbalance of the interferometer, amplitude and phase noise from the laser and the EDFA could enter the heterodyne signal. One particular concern is that, because of the free-running laser, the drift of the laser repetition rate and carrier-envelope offset frequency may also broaden the beat note. With the fiber link, transmission-induced excess phase noise is much smaller (Holman et al., 2004), allowing us to assess the upper limit of the noise contribution from the laser and the amplifier. In Fig. 10 (a), both traces have a resolution bandwidth of 16 Hz, and are averaged over 1 s. The resolution limited sharp peak in the spectrum from fiber transmission indicates that the femtosecond laser and the EDFA has a negligible effect in the current excess phase noise characterization. In comparison, the atmospheric transmission causes significant spectral broadening. The broadened beat note spectrum has a full width of approximately 1 kHz.

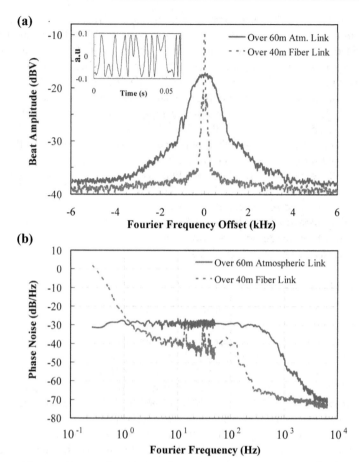

Fig. 10. (a) Fourier spectra of the 543-kHz heterodyne beat note for 60-m atmospheric transfer (solid) and 40-m fiber optic transfer (dashed). Significant linewidth broadening (~ kHz) is caused by the atmospheric propagation in comparison with fiber transmission. Inset: A time-domain trace of the homodyne beat note for atmospheric transfer demonstrates the large phase fluctuation. (b) Phase noise spectra of the 543-kHz beat note with the above two transfer cases.

A close look at the noise characteristics of the two transmission cases near the nominal frequency can be gained through their phase noise spectra, which are assembled from data over two frequency spans and are shown in Fig. 10 (b). It has been shown that such heterodyne phase noise can be understood as the zeroth-order excess phase noise of all the comb lines (Gollapalli & Duan, 2011). Since the beat note from the atmospheric transmission has no coherent spike as the carrier, the two curves in Fig. 10 (b) are normalized to their total signal powers (i.e., the integration of the beat note spectra) instead of the carrier power. A frequency-independent noise spectrum is seen below 300 Hz for the beat note from the atmospheric transmission. It is followed by a quick roll-off above 500 Hz. Such characteristics seem to agree with the spectral features for a slow phase modulation with a very large modulation index. In comparison, transmission in fibre with the same time delay results in a much smaller phase noise. The spikes between 10–100 Hz are believed to be the result of acoustic noise and electronic interference from the power circuits inside the lab. The above comparison shows that atmospheric fluctuations add a significant amount of phase noise to the optical frequency references transmitted through the air. As a result, atmospheric transfer of optical frequencies in general suffers much larger short-term fluctuations when compared with fibre-optic transfer.

For many applications, however, a more relevant parameter is the long-term stability of the frequency transfer scheme, which can be evaluated by measuring the Allan deviation of the 543-kHz beat note. Shown in Fig. 11 (a) are several sets of measured data of the Allan deviation versus the averaging time through the 60-m transmission link under various

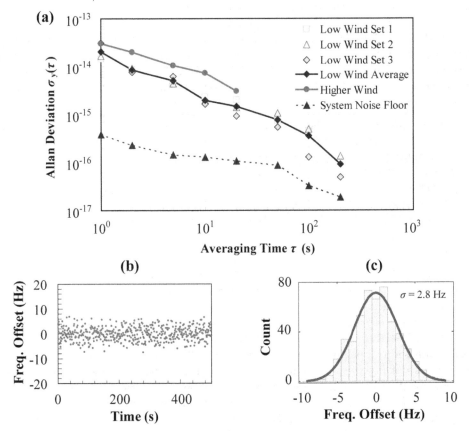

Fig. 11. (a) Allan deviations of the 543-kHz heterodyne beat note for 60-m atmospheric transfer under various weather conditions and the system noise floor. (b) A data set of 500 consecutive frequency measurements at 1-s gate time, plotted relative to the nominal frequency. (c) The histogram of (b), overlaid with a fitting to the normal distribution, leads to a standard deviation of 2.8 Hz for the frequency references transferred across the atmospheric link.

weather conditions. Among them, three sets were taken under low-wind conditions, with the wind speed generally staying below 5 mph (miles-per-hour), or 2.2 m/s, throughout the data collection periods (except for Set #1 at 200 s, which is not included here). The average of these data shows an overall behaviour of the Allan deviation close to $\tau^{-1}$ within the measurement range. The fractional frequency stability at 1 s is about 2×10⁻¹⁴. When averaged over 100 s, the stability improves to 10⁻¹⁶ level. Over even longer time, however, the beat note often develops large power fluctuation, which prevents a reliable measurement of the Allan deviation with an averaging time longer than 200 s. The fluctuation of the beat note power is believed to be caused by the transmitted and the reference pulses drifting away from their optimum overlapping position due to the change of the effective path length of the transmission link. To demonstrate the effect of wind, we also include here a set of data taken under stronger wind, with wind speed ranging between 7–14 mph over the data collection period. Below 20 s, the higher-wind data stay roughly 3 dB above the average low-wind data and display a similar $\tau^{-1}$ dependence. No Allan deviation was successfully measured above 20 s due to the power fluctuation of the beat note. In addition, a set of baseline data is also included in Fig. 11 (a) to mark the system-limited measurement resolution. It is obtained by reflecting the transmission beam immediately into the receiving subsystem so the length of atmospheric propagation is negligible.

To further verify the scale of the frequency fluctuation, we take a close look at the measured beat note frequencies. Fig. 11 (b) shows consecutive readings of the frequency counter over several minutes with a 1-s gate time, taken under the low-wind condition. The frequencies have been offset to the nominal value of 543 kHz. A histogram analysis shows that the frequencies have a normal distribution around the nominal frequency, as shown in Fig. 11 (c), and a Gaussian fitting results in a standard deviation of 2.8 Hz. This frequency fluctuation level is in agreement with the Allan deviation measurement for low wind.

### 5.3 Discussion

To put our work into context, it is interesting to compare the above results with some of the previous reports on optical frequency transfer in the atmosphere. Sprenger et al. made a similar rooftop demonstration using a cw diode laser over a 100-m atmospheric link at the ground level (Sprenger et al., 2009). The group achieved an Allan deviation of 2×10⁻¹³ at 1 s. In addition, a statistics of the beat note frequency measured with a 1-s gate time shows a 70.5-Hz full width at half maximum. Compared with this work, the current study reports a frequency instability one order of magnitude lower at 1 s. Apart from the apparent shorter transmission distance, the smaller frequency fluctuation in the current work is likely due to the low-wind condition (< 2 m/s), which means lower index fluctuation due to air turbulence. Such a fact is evident in Fig. 11 (b), where continuous frequency measurement over several minutes shows a consistently low frequency fluctuation. More recently, Djerroud et al. have reported a coherent optical link across 5 km of turbulent atmosphere based on a cw Nd:YAG laser (Djerroud et al., 2010). A remarkably low Allan deviation of 1.3×10⁻¹⁴ is achieved at 1 s, and the stability further improves to 2×10⁻¹⁵ at 100 s. The authors, however, did point out that the experiment took place at an observatory 1323 m above the sea level, which could account, at least in part, for the high degree of frequency stability over a long distance. Based on the standard Kolmogorov model of atmospheric turbulence, the power spectral density of index fluctuation, $\Phi_n$, is given by $\Phi_n(\kappa) = 0.033 C_n^2 \kappa^{-11/3}$, where $C_n$ is the index structure constant and $\kappa$ is the spatial wave number. Experimentally, it has been shown that $C_n$ strongly depends on altitude (Ishimaru, 1978). For example, the altitude of Huntsville is 200 m, which corresponds to a $C_n$ of approximately $10^{-7}$ m$^{-1/3}$ for sunny days. At 1323 m, however, $C_n$ is about $10^{-8}$ m$^{-1/3}$ under similar conditions, which indicates a refractive index fluctuation roughly two orders of magnitude smaller than it at 200 m altitude.

## 6. Summary

In summary, we have experimentally demonstrated the transfer of clock signals in both microwave and optical frequency ranges over an open atmospheric transmission link using a femtosecond frequency comb. The excess phase noise due to the atmospheric propagation is successfully characterized in both cases, and the both results show large, slow-frequency phase modulations due to the index fluctuation of air. The fractional frequency stability for a 60-m transmission under typical calm weather conditions is in the order of a few parts per $10^{12}$ for microwave clock transfer and a few parts per $10^{14}$ for optical clock transfer, both with a 1-s averaging time. The much better fractional stability for optical clock transfer is mainly attributed to the much higher clock frequencies, which demonstrates the advantage of shifting reference clocks toward higher frequencies. Our measurement also shows an approximate $\tau^{-1}$ dependence of the Allan deviation on the averaging time in both frequency ranges up to 200 s. This is an encouraging finding for atmospheric clock transfer because it means that, over longer time, transferring a highly stable master clock via free space can be more advantageous than maintaining a less-precise local clock, which normally has a $\tau^{-1/2}$ dependence. Further study under more diverse weather conditions and over longer transmission distances is needed to confirm this results.

Two key properties of photodetectors play critical roles in the current research. The square-law detector allows us to perform optical heterodyne to characterize the excess phase noise in optical frequency transfer. Meanwhile, power-to-phase coupling in photodetectors puts a requirement on the receiving optics to keep the impact of beam wander and speckle low.

Further improvement of the clock transfer stability would require the use of active noise cancellation. Such a scheme has been demonstrated to be able to lower the transfer-induced instability by a factor of 10 in fiber-optic systems (Holman et al., 2005). To apply a similar technique in free space, retroreflection is needed to allow the returning beam to travel exactly the same path back in order to avoid differences in phase noise caused by different paths. In addition, the locking system likely needs to have a much wider bandwidth compared to the fiber-optic scheme even for a moderate transmission distance (e.g. 100 m) because of the relatively fast turbulence-induced index fluctuation in the air.

## 7. Acknowledgment

We would like to acknowledge Mr. Ted Rogers and Dr. Robert Lindquist of the Center for Applied Optics, the University of Alabama in Huntsville, for their assistance in the setup of the atmospherics transmission link. Ms Ayshah Alatawi made key contributions to the lab test of the microwave clock transfer. The work was supported in part by grants from UAH.

# 8. References

Akozbek, N.; Bowden, C. M.; Talebpour, A. & Chin, S. L. (2000). Femtosecond pulse propagation in air: Variational analysis, *Physics Review E*, Vol.61, pp. 4540-4549

Alatawi, A.; Gollapalli, R. P. & Duan, L. (2009). Radio frequency clock delivery via free-space frequency comb transmission, *Opt. Lett.*, Vol.34, pp. 3346-3348

Allan, D. W. (1966). Statistics of atomic frequency standards, *Proc. IEEE*, Vol.54, pp. 221-230

Altarelli, M (eds) et al. (2007). XFEL: The European X-Ray Free Electron Laser, *Technical Design Report DESY* 2006-097 (DESY, Hamburg, 2007), at http//xfel.desy.de

Andrews, L. C. & Phillips, R. L. (2005). *Laser beam propagation through random media*, SPIE Press, 2nd Ed.

Bartels, A.; Oates, C. W.; Hollberg, L. & Diddams, S. A. (2004). Stabilization of femtosecond laser frequency combs with subhertz residual linewidths, *Optics Letters*, Vol.29, pp. 1081-1083

Bartels, A.; Diddams, S. A.; Oates, C. W.; Wilpers, G.; Bergquist, J. C.; Oskay, W. H. & Hollberg, L. (2005). Femtosecond-laser-based synthesis of ultrastable microwave signals from optical frequency references, *Optics Letters*, Vol.30, pp. 667-669

Brookner, E. (1970). Atmosphere propagation and communication channel model for laser wavelengths, *IEEE Trans. Commun. Technol.*, COM-18, pp. 396-416

Chan, V. W. S. (2006). Free-space optical communications, *J. Lightwave. Technol.* Vol.24, pp. 4750-4762

Ciddor, P. E. (1996). Refractive index of air: new equations for the visible and near infrared, *Appl. Opt.*, Vol. 35, pp. 1566-1573

Ciddor P. E. and Hill, R. J. (1999). Refractive index of air. 2. group index, *Appl. Opt.*, Vol.38, pp. 1663-1667

Cliche, J. F. & Shillue, B. (2006). Precision timing control for radioastronomy, *IEEE Control Sys. Mag.* Vol.26, pp. 19-26

Davis, C. C. 1996. *Lasers and Electro-Optics Fundamentals and Engineering*, Cambridge University Press

Djerroud, K.; Acef, O.; Clairon, A.; Lemonde, P.; Man, C. N.; Samain, E. & Wolf, P. (2010). Coherent optical link through the turbulent atmosphere, *Opt. Lett.*, Vol.35, pp. 1479-1481

Foreman, S. M.; Holman, K. W.; Hudson, D. D.; Jones, D. J. & Ye, J. (2007). Remote transfer of ultrastable frequency references via fibre networks, *Review of Scientific Instruments*, Vol.78, pp. 021101

Fortier, T. M.; Le Coq, Y.; Stalnaker, J. E.; Ortega, D.; Diddams, S. A.; Oates, C. W. & Hollberg, L. (2006). Kilohertz-resolution spectroscopy of cold atoms with an optical frequency comb, *Physical Review Letters*, Vol.97, pp. 163905

Gollapalli, R. P. & Duan, L. (2010). Atmospheric timing transfer using a femtosecond frequency comb, *IEEE Photon. Journal*, Vol.2, pp. 904-910

Gollapalli, R. P. & Duan, L. (2011). Multiheterodyne Characterization of Excess Phase Noise in Atmospheric Transfer of a Femtosecond-Laser Frequency Comb, *Journal of Lightwave technology*, Vol.29, pp. 3401-3407.

Halford, D.; Wainwright, A. E. &Barnes, J. A. (1968). Flicker noise of phase in RF amplifiers and frequency multipliers: characterization, cause, and cure, *22nd Annual Symposium on Frequency Control*, pp. 340.

Hall, J. L. & Ye, J. (2003). Optical frequency standards and measurement, *IEEE Trans. Instrum. Meas.*, Vol.52, pp. 227-230

Holman, K. W.; Jones, D. J.; Hudson, D. D. & Ye, J. (2004). Precise frequency transfer through a fiber network by use of 1.5µm mode-locked sources, *Optics Letters*, Vol.29, pp. 1554-1556

Holman, K. W.; Hudson, D. D., Ye, J. & Jones, D. J. (2005). Remote transfer of a high-stability and ultralow-jitter timing signal, *Optics Letters*, Vol.30, pp. 1225-1227

Holzwarth, R.; Udem, T.; Hansch, T. W.; Knight, J. C.; Wadsworth, W. J. & Russell, P. St. J. (2000). Optical frequency synthesizer for precision spectroscopy, *Physical Review Letters*, Vol.85, pp. 2264-2267

Hong, S. T.; Sreenivasiah, I. & Ishimaru, A. (1977). Plane wave pulse propagation through random media, *IEEE Transactions on Antennas and Propagation*, AP-25, pp. 822-828

IEEE Std. 1139-1988, IEEE standard definitions of physical quantities for fundamental frequency and time metrology (IEEE, 1983).

Ishimaru, A. 1978. *Wave Propagation and Scattering in Random Media*, Academic, New York

Ivanov, E. N.; Diddams, S. A. & Hollberg, L. (2003). Analysis of noise mechanisms limiting the frequency stability of microwave signals generated with a femtosecond laser, *IEEE J. Select. Topics Quantum Electron*, Vol.9, pp. 1059-1065

Jones, D. J.; Diddams, S. A.; Ranka, J. K.; Stentz, A.; Windeler, R. S.; Hall, J. L. & Cundiff, S. T. (2000). Carrier-envelope phase control of femtosecond mode-locked lasers and direct optical frequency synthesis, *Science*, Vol.288, pp. 635-639

Kim, J.; Cox, J. A.; Chen, J. & Kaertner, F. X. (2008). Drift-free femtosecond timing synchronization of remote optical and microwave sources, *Nature Photonics*, Vol.2, pp. 733-736

Li, C. H. et al. (2008). A laser frequency comb that enables radial velocity measurements with a precision of 1 cm/s$^{-1}$, *Nature*, Vol.452, pp. 610-612,

Liu, C. H.; Wernik, A. W. & Yeh, K. C. (1974). Propagation of pulse trains through a random medium, *IEEE Transactions on Antennas and Propagation*, Vol.22, pp. 624-627

Liu, P. L.; Williams, K. J.; Frankel, M. Y. & Esman, R. D. (1999). Saturation characteristics of fast photodetectors, *IEEE Trans. Microwave Theory Tech*, Vol.47, pp. 1297-1303

Mlejnek, M.; Kolesik, M.; Moloney, J. V. & Wright, E. M. (1999). Optically turbulent femtosecond light guide in air, *Physics Review Letters*, Vol.83, pp. 2938-2941

Nahin, P. J. 2001. *The Science of Radio*, AIP Press, New York: Springer-Verlag

Prucnal, P.; Santoro, M. & Sehgal, S. (1986). Ultrafast All-Optical Synchronous Multiple Access Fiber Networks, *Selected Areas in Communications*, Vol.4, pp. 1484-1493

Ricklin, J. C. & Davidson, F. M. (2003). Atmospheric optical communication with a Gaussian Shell beam. *J. Opt. Soc. Am. A*, Vol.20, pp. 856-866

Rodriguez, M.; Bourayou, R.; Mejean, G.; Kasparian, J.; Yu, J.; Salmon, E.; Scholz, A.; Stecklum, B.; Eisloffel, J.; Laux, U.; Hatzes, A. P.; Sauerbrey, R.; Woste, L. & Wolf, J. (2004). Kilometer-range nonlinear propagation of femtosecond laser pulses, *Physics Review E*, Vol.69, pp. 036607

Rutman, J. & Walls, F. (1991). Characterization of frequency stability in precision frequency sources, *Proc. IEEE*, Vol.79, pp. 952-960

Shaik, K. S. (1988). Atmospheric propagation effects relevant to optical communications, *TDA Progress Report*, pp. 42–94

Sprangle, P.; Penano, J. R. & Hafizi, B. (2002). Propagation of intense short laser pulses in the atmosphere, *Physics Review E*, Vol.66, pp. 046418

Sprenger, B. ; Zhang, J.; Lu, Z. H. & Wang, L. J. (2009). Atmospheric transfer of optical and radio frequency clock signals, *Opt. Lett.*, Vol.34, pp. 965-967

Su, H. H. & Plonus, M. A. (1971). Optical-pulse propagation in a turbulent medium, *J. Opt. Soc. Am.* Vol.61, pp. 256-260

Thorpe, M. J.; Moll, K. D.; Jones, R. J.; Safdi, B. & Ye, J. (2006). Broadband cavity ringdown spectroscopy for sensitive and rapid molecular detection, *Science*, Vol.311, pp. 1595-1599

Tulchinsky, D. A. & Williams, K. J. (2005). Excess amplitude and excess phase noise of RF photodiodes operated in compression, *IEEE Photonics Technology Letters*, Vol 17, pp. 654-656

Udem, Th.; Diddams, S. A.; Vogel, K. R.; Oates, C.W.; Curtis, E. A.; Lee, W. D.; Itano, W. M.;Drullinger, R. E.; Bergquist, J.C. & Hollberg, L. (2001). Absolute frequency measurements of the Hg and Ca optical clock transitions with a femtosecond laser, *Physical Review Letters*, Vol.86, pp. 4996-4999

Udem, T.; Holzwarth, R. & Hansch, T. W. (2002). Optical frequency metrology, *Nature*, Vol.416, pp. 233-237

Ye, J. (2004). Absolute measurement of a long, arbitrary distance to less than an optical fringe, *Optics Letters*, Vol.29, pp. 1153-1155

Young, C. Y.; Andrews, L. C. & Ishimaru, A. (1998). Time-of-arrival fluctuations of a space-time Gaussian pulse in weak optical turbulence: An analytic solution, *Applied Optics*, Vol.37, pp. 7655-7660

# 7

# Measurement of the Polarization State of a Weak Signal Field by Homodyne Detection

Sun-Hyun Youn

*Department of Physics, Chonnam National University, Gwangju*
*Korea*

## 1. Introduction

Information carried by an optical beam of light can usually be conveyed in the form of a temporal modulation of the intensity, phase, frequency or the polarization of the constituent mode(s). In that regard, homodyne detection is one of the most popular and standard method to measure the quantum mechanical properties of light Yuen & Shapiro (1978); Yuen et al. (1979); Yuen & Shapiro (1980); Yurke (1985). The quantum theory of homodyne detection originated principally from the works of Yuen and Shapiro et al. Yuen & Shapiro (1978); Yuen et al. (1979). In this method, a weak signal field is combined with a strong local oscillator field at a beam splitter, and the resulting signal is measured as a photocurrent. With the homodyne detection method, the quantum state of the signal field, such as the quadrature amplitude of the squeezed state, is easily measured Slusher et al. (1985); Wu et al. (1986 ); Polzik et al. (1992), and the quasiprobability distribution function can be measured by using the so-called optical homodyne tomography Smithey et al. (1993); Banaszek & W'odkiewicz (1996); Wallentowitz & Vogel (1996); Youn et al. (2001). Furthermore, optical homodyne detection is used to eavesdrop on the quantum key in quantum cryptography Hirano et al. (2000).

In quantum information science, the photon is a useful source for manipulating quantum information. To now, it has not been easy to make a consistent single-photon source, so the signal photon state is usually a weak coherent state. Therefore, it is very important to obtain the quantum state of an unknown signal field. The quantum mechanical properties of an unknown signal field, such as the amplitude squeezed state and the quadrature squeezed state, can be characterized by using the quasiprobability distribution (Wigner distribution) or the density matrix. The quasi probability distribution defines the statistical characteristics of the signal field . The well-known method to obtain quasi probability distributions (Wigner distributions) or density matrices is optical homodyne tomography Leonhardt (1997); Schiller et al. (1996).

Any kind of state reconstruction technique, however, in optical homodyne tomography requires repeated measurements of an ensemble of equally prepared signals. Therefore, this method is not adequate for finding the polarization state of a signal field that is changing pulse by pulse. Our novel scheme of polarization-modulated homodyne detection can obtain the polarization state of the signal field in a single-shot scheme. Even if the quasi probability of the signal field is not known in detail for a given single pulse, the varying polarization state of the signal field can be obtained pulse by pulse. In this respect, our novel scheme has an advantage: It can determine the varying polarization state of the signal against the usual optical homodyne tomography.

In our earlier work Youn & Noh (2003), we proposed a polarization-controllable homodyne detection scheme for a local oscillator field whose polarization state and global phase delay were changed by using an electro-optic modulator and a piezo- electric device, respectively. In that scheme, the polarization angle and the global phase of the local oscillator field must be scanned to obtain information on the polarization state of the signal field. Therefore, the scan time is non-zero, which limits the amount of polarization information that can be obtained from the pulsed signal field. The ordinary polarization modulated homodyne detection method, which requires no scan time, can obtain polarization information on a one-time pulsed signal field Youn (2005). Polarization-modulated ordinary homodyne detection does have an advantage in that it needs only one charged coupled device, but it also has a disadvantage in that it is not free from the noise of the local oscillator field.

In section 2, we propose a polarization-modulated balanced homodyne detection method. By inserting a set of wedged wave plates in the local oscillator port, we can modulate the relative phase of the two orthogonal components as well as the overall phase retardation of the local oscillator electric field Youn & Bae (2006). Using this spatially modulated local oscillator field in the homodyne detection scheme, we can obtain the amplitudes and the relative phase of the two orthogonal components of the signal electric field. We propose a practical method to measure the polarization state of the weak signal in a single-shot scheme without any scanning time. Our proposal will be one of the essential techniques in quantum information science for reading the polarization state of an unknown signal field in a single-shot scheme.

In section 3, a polarization-modulated homodyne detection scheme using photodetectors that measure four (temporally) simultaneous photocurrent signals corresponding to four (spatially) different quadrant-shaped combinations of wave plates is proposed Youn & Jain (2009). As an extension of our previous work, we essentially do a spatial phase modulation of the two orthogonal polarization components of the local oscillator (LO) electric field by inserting a system of waveplates in the path. Information about the Jones vector associated with the signal electric field can then be obtained, and this characterization can be performed in real-time, i.e., on a single-shot basis. In particular, we also articulate an analysis to discriminate between some typical polarization states, which has implications for the security offered by standard quantum cryptographic systems.

In our current work on polarization-modulated balanced homodyne detection, we are able to determine the polarization state of the signal field on a single-shot basis. This method could be described as a hybrid of ordinary homodyne detection and quantum polarization tomography, such as the one in Ref. James et al. (2001), employing an ensemble measurement of Stokes parameters for characterizing single qubits. However, while our scheme doesn't disseminate information about the quasi-probability distribution of the signal field for a given single pulse, it's usefulness comes into play in the determination of the varying polarization state of an arbitrary and unknown signal field in real-time.

In section 4, we discuss the applicability of the two methods for measuring the polarization state of a weak signal field.

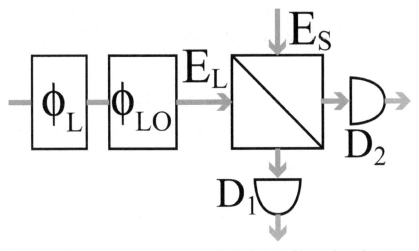

Fig. 1. Schematic diagram of polarization-controllable balanced homodyne detection. $\phi_L$ represents the polarization state of the local oscillator. $\phi_{LO}$ is the phase difference between the signal field and the local oscillator field ( BS: Beam splitter, D: detector, E: Electric field).

## 2. Novel scheme of polarization-modulated balanced homodyne detection for measuring the polarization state of a weak field

### 2.1 Polarization-modulated Homodyne detection

In general, the polarizations of the signal field and the local oscillator are assumed to be linear and equal to each other. We propose a scheme for polarization-controlled balanced homodyne detection Youn & Noh (2003). When the polarizations of the signal and the local oscillator are different from each other, even the ordinary homodyne detection scheme can obtain various information about the polarization states of the signal field by varying the polarization of the local oscillator. In this paper, we propose an experimental scheme that can determine the polarization state of the input signal field by using polarization-modulated balanced homodyne detection.

The scheme of polarization-modulated homodyne detection is shown in Fig. 1. The signal and the local oscillator fields, both with arbitrary polarizations, are combined at a beam splitter (BS). The electric field in the local oscillator is given as Yariv (1989)

$$\vec{\mathcal{E}}_{LO}(\vec{r}, t) = \left( \frac{2\hbar\omega}{\epsilon} \right)^{1/2} \left[ \vec{E}_{LO}(\vec{r}) \cos(\omega t + \phi_{LO}) + \frac{1}{2} \Delta \vec{E}_{LO}(\vec{r}, t) \right] , \tag{1}$$

where $\epsilon$ is the dielectric constant and $\omega$ is the frequency of the local oscillator field. In Eq. (1), $\Delta \vec{E}_{LO}(\vec{r}, t)$ is a fluctuation term, with the time average of $\Delta \vec{E}_{LO}(\vec{r}, t)$ vanishing. The monochromatic term can be decomposed into two frequency parts as follows:

$$\vec{E}_{LO}(\vec{r}, t) \equiv \left( \frac{2\hbar\omega}{\epsilon} \right)^{1/2} \vec{E}_{LO}(\vec{r}) \cos(\omega t + \phi_{LO})$$

$$= \vec{E}_{LO}^{(+)}(\vec{r}, t) + \vec{E}_{LO}^{(-)}(\vec{r}, t) , \tag{2}$$

where $\phi_{LO}$ is the overall phase of the local oscillator field relative to the signal field, and the positive $(\vec{E}_{LO}^{(+)})$ and the negative $(\vec{E}_{LO}^{(-)})$ frequency component are given by

$$\vec{E}_{LO}^{(+)}(\vec{r},t) = \left(\vec{E}_{LO}^{(-)}(\vec{r},t)\right)^*$$
$$= \left(\frac{\hbar\omega}{2\epsilon}\right)^{1/2} \vec{E}_{LO}(\vec{r})e^{-i\omega t}e^{-i\phi_{LO}} . \tag{3}$$

In addition, the fluctuating term can be decomposed into positive and negative frequency parts as

$$\Delta\vec{E}_{LO}(\vec{r},t) = \Delta\vec{E}_{LO}^{(+)}(\vec{r},t) + \Delta\vec{E}_{LO}^{(-)}(\vec{r},t) . \tag{4}$$

We also put the signal field in the signal port as

$$\vec{\mathcal{E}}_s(\vec{r},t) = \vec{\mathcal{E}}_s^{(+)}(\vec{r},t) + \vec{\mathcal{E}}_s^{(-)}(\vec{r},t) , \tag{5}$$

where

$$\vec{\mathcal{E}}_s^{(+)}(\vec{r},t) = \left(\vec{\mathcal{E}}_s^{(-)}(\vec{r},t)\right)^* = \vec{E}_s^{(+)}(\vec{r},t) ,$$
$$\vec{E}_s^{(+)}(\vec{r},t) = \left(\frac{\hbar\omega}{2\epsilon}\right)^{1/2} \vec{E}_s(\vec{r})e^{-i\omega t} . \tag{6}$$

The electric fields of the local oscillator and the signal are combined at a 50-50 BS, and the resultant electric fields at ports 1 and 2 are given by

$$\vec{E}_1(t) = \frac{1}{\sqrt{2}}\left[\vec{\mathcal{E}}_{LO}(\vec{r},t) - \vec{\mathcal{E}}_s(\vec{r},t)\right]$$
$$= \vec{E}_1^{(+)}(t) + \vec{E}_1^{(-)}(t) ,$$
$$\vec{E}_2(t) = \frac{1}{\sqrt{2}}\left[\vec{\mathcal{E}}_{LO}(\vec{r},t) + \vec{\mathcal{E}}_s(\vec{r},t)\right]$$
$$= \vec{E}_2^{(+)}(t) + \vec{E}_2^{(-)}(t) , \tag{7}$$

respectively, where $\vec{E}_{1,2}^{(\pm)}(t)$ can be written as

$$\vec{E}_1^{(+)}(t) = \left(\vec{E}_1^{(-)}(t)\right)^*$$
$$= \frac{1}{\sqrt{2}}\left[\vec{E}_{LO}^{(+)}(\vec{r},t) + \left(\frac{\hbar\omega}{2\epsilon}\right)^{1/2}\Delta\vec{E}_{LO}^{(+)}(\vec{r},t) - \vec{E}_s^{(+)}(\vec{r},t)\right] ,$$
$$\vec{E}_2^{(+)}(t) = \left(\vec{E}_2^{(-)}(t)\right)^*$$
$$= \frac{1}{\sqrt{2}}\left[\vec{E}_{LO}^{(+)}(\vec{r},t) + \left(\frac{\hbar\omega}{2\epsilon}\right)^{1/2}\Delta\vec{E}_{LO}^{(+)}(\vec{r},t) + \vec{E}_s^{(+)}(\vec{r},t)\right] . \tag{8}$$

Since the incident photon flux operator is proportional to the product of electric field operators which are normally ordered and since one photoelectron is generated from an incident photon, based on the assumption of ideal photodetectors, the current measured from port 1 after integration over the detection area becomes

$$I_1(t) = \frac{2e\sigma_{det}}{\hbar\omega}\sqrt{\frac{\epsilon}{\mu}}\vec{E}_1^{(-)}(t) \cdot \vec{E}_1^{(+)}(t)$$

$$= \frac{ec\sigma_{det}}{2}\left\{\vec{E}_{LO}^*(\vec{r})e^{i\phi_{LO}+i\omega t} + \left[\Delta\vec{E}_{LO}^*(\vec{r})e^{i\omega t} - \vec{E}_s^*(\vec{r})e^{i\omega t}\right]\right\}$$

$$\cdot\left\{\vec{E}_{LO}(\vec{r})e^{-i\phi_{LO}-i\omega t} + \left[\Delta\vec{E}_{LO}(\vec{r})e^{-i\omega t} - \vec{E}_s(\vec{r})e^{-i\omega t}\right]\right\} , \quad (9)$$

where $e$ is the electron charge, $c$ is the speed of light in vacuum, and $\sigma_{det}$ is the area of the detector. The current at port 2, $(I_2(t))$, is obtained similarly as above. Under the assumption that the fluctuation of the local oscillator field and the intensity of the signal field is much smaller than the mean intensity of the local oscillator field, the current difference between $I_1$ and $I_2$ can be given by

$$I_1(t) - I_2(t) =$$

$$-ec\sigma_{det}\left\{\vec{E}_{LO}(\vec{r}) \cdot \vec{E}_s^{(+)}(\vec{r},t)e^{-i\phi_{LO}-i\omega t} + \vec{E}_{LO}^*(\vec{r}) \cdot \vec{E}_s^{(-)}(\vec{r},t)e^{+i\phi_{LO}+i\omega t}\right\} . \quad (10)$$

If we set $\vec{E}_{LO}(\vec{r}) = A\hat{e}_L/\sqrt{V}$, where $\hat{e}_L$ is the Jones polarization vector, $V$ is the mode volume, and $A$ is a constant related to the intensity of the local oscillator field, the current becomes

$$I(t) \equiv I_1(t) - I_2(t) =$$

$$-\frac{ecA\sigma_{det}}{\sqrt{V}}\left[\vec{E}_s^{(-)}(\vec{r},t) \cdot \hat{e}_L e^{-i(\phi_{LO}+\omega t)} + \vec{E}_s^{(+)}(\vec{r},t) \cdot \hat{e}_L^* e^{+i(\phi_{LO}+\omega t)}\right] . \quad (11)$$

In the homodyne detection scheme, the frequency of the signal field is the same as that of the local oscillator field, as in Eq. (6). Therefore, the current can be expressed by

$$I(t) = -\frac{ecA\sigma_{det}}{\sqrt{V}}\left[E_s^*(\vec{r})e^{-i\phi_{LO}}\hat{e}_s^* \cdot \hat{e}_L + E_s(\vec{r})e^{+i\phi_{LO}}\hat{e}_s \cdot \hat{e}_L^*\right] , \quad (12)$$

where $\hat{e}_s$ is the Jones vector associated with the polarization of the signal field. Equation (12) is the final result for the measured current difference. In the general case, when both electric fields are arbitrarily polarized,

$$\hat{e}_s = a_1 e^{i\delta_1}\hat{i} + a_2 e^{i\delta_2}\hat{j}, \quad (13)$$

$$\hat{e}_L = \cos\theta_L\hat{i} + \sin\theta_L e^{i\phi_L}\hat{j} , \quad (14)$$

we have the current as

$$I = -\frac{2ecA\sigma_{det}}{\sqrt{V}}|E_s|\left[a_1\cos\theta_L\cos(\delta_1 + \phi_{LO}) + a_2\sin\theta_L\cos(\delta_2 + \phi_{LO} - \phi_L)\right] , \quad (15)$$

where the amplitudes $a_1$ and $a_2$ are non-negative real numbers which satisfy the normalization condition $a_1^2 + a_2^2 = 1$, and the phase factor $\delta_1$ and $\delta_2$ are real numbers.

## 2.2 Method to find the polarization state of the signal field

Information on the input polarization is obtained as follows: For a given signal field defined by $(a_1, a_2, \delta_1 - \delta_2)$, the intensity distribution $I$ in Eq. (15) depends on three controllable parameters, $\theta_L, \phi_L$, and $\phi_{LO}$, that are related to the local oscillator field. If we scan over $(\phi_L, \phi_{LO})$, we can obtain full information on the polarization state of the signal field Youn & Noh (2003). Usually, the parameters $\phi_L$, and $\phi_{LO}$ are controlled by using an electro-optic modulator or a piezo electric material. An electro-optic modulator changes the polarization state of the local oscillator field, and a piezo electric material changes the path length associated with the phase delay of the local oscillator field relative to the signal field. However, it is difficult to get a full scan for a single-pulse signal because the scan needs non-zero time. In this work, we propose a new scheme which does not require any scan time. We made a spatially modulated local oscillator field to perform the scan for the ( $\phi_L(t), \phi_{LO}(t)$ ) space in real spatial coordinates ( $\phi_L(x,y), \phi_{LO}(x,y)$ ). To get spatially dependent photocurrents for a single shot-scheme, we only have to insert wave plates in the optical path of the local oscillator field, as shown in Fig. 2. Fig. 2 shows a device consisting of three wave plates and gives $\phi_L$ and $\phi_{LO}$ at once. The first two wave plates consist of two isotropic wedges whose refractive indices are $n_L$ and $n_R$, respectively. The first and the second wedges are sliced by the plane $z = x \tan \alpha$ , and the length from the first wedge and to the second surface at $y = 0$ is $d_1$, as in Fig. 2.

On the other hand, the third wave plate is a uniaxial crystal, such as calcite, whose refractive indices can be $n_o$ and $n_e$. The adjacent surfaces of the second and the third wedges are sliced by the plane $z = y \tan \beta$ , and at that position, the length from the second surface at $y = 0$ to the output surface of the third wedge is $d_2$, as in Fig. 2. The refractive indices of the third wedge are $n_x = n_o$ and $n_y = n_e$.

A ray passing horizontally to the right, $+z$, through the device at some arbitrary point $(x,y)$ will traverse a thickness of $x \tan \alpha$ in the first wedge, $d_1 + y \tan \beta - x \tan \alpha$ in the second one, and $d_2 - y \tan \beta$ in the third one . The beam path delay imparted to the wave by the first wedge is $2\pi n_L x \tan \alpha / \lambda$, and that by the second wedge is $2\pi n_R (d + y \tan \beta - x \tan \alpha)/\lambda$. On the other hand, the refractive index of the third wedge depends on the polarization axis, so we have to calculate the wave retardation for two polarization axes. Let $\Gamma^o$ be the wave retardation due to the three wave plates for the $x - axis$ linearly polarized light; then $\Gamma^o$ becomes

$$\Gamma^o(x,y) = \frac{2\pi}{\lambda} \{ n_L x \tan \alpha + n_R(d_1 + y \tan \beta - x \tan \alpha) + n_o(d_2 - y \tan \beta) \} . \quad (16)$$

and $\Gamma^e$, the wave retardation caused by the three wave plates for $y - axis$ linearly polarized light becomes

$$\Gamma^e(x,y) = \frac{2\pi}{\lambda} \{ n_L x \tan \alpha + n_R(d_1 + y \tan \beta - x \tan \alpha) + n_e(d_2 - y \tan \beta) \} . \quad (17)$$

After passing through the three wedges, the incident ray falls on $(x,y)$ as the initial Jones vector $\frac{1}{\sqrt{2}}(\hat{x} + \hat{y})$ changes into

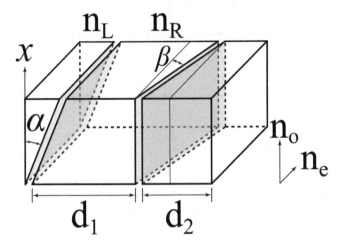

Fig. 2. Schematic diagram of two sets of wave plates with two isotropic wedges and one uniaxial wedge. The first and the second adjacent surfaces are sliced by the plane $z = x \tan \alpha$, and the refractive indices of the first and the second wedges are $n_L$ and $n_R$, respectively. The second and the third adjacent surfaces are sliced by the plane $z = y \tan \beta$, and the refractive indices of the third wedge are $n_x = n_o$ and $n_y = n_e$.

$$\begin{pmatrix} \hat{e}_L^x(x,y) \\ \hat{e}_L^y(x,y) \end{pmatrix} = \begin{pmatrix} e^{i\Gamma^o(x,y)} & 0 \\ 0 & e^{i\Gamma^e(x,y)} \end{pmatrix} \begin{pmatrix} \frac{1}{\sqrt{2}} \\ \frac{1}{\sqrt{2}} \end{pmatrix}$$

$$\equiv e^{i\Phi_{LO}(x,y)} \begin{pmatrix} 1 & 0 \\ 0 & e^{i\Phi_L(x,y)} \end{pmatrix} \begin{pmatrix} \frac{1}{\sqrt{2}} \\ \frac{1}{\sqrt{2}} \end{pmatrix}, \tag{18}$$

where the phase factors are defined as

$$\Phi_{LO}(x,y) = \Gamma^o(x,y)$$

$$= \frac{2\pi}{\lambda} \{ x(n_L - n_R) \tan \alpha + y(n_R - n_o) \tan \beta + n_R d_1 + n_o d_2 \}, \tag{19}$$

$$\Phi_L(y) = \Gamma^e(x,y) - \Gamma^o(x,y)$$

$$= \frac{2\pi}{\lambda} \{ y(n_o - n_e) \tan \beta + (n_e - n_o)n_o \}. \tag{20}$$

The overall phase delay between the signal field and the local oscillator field, and the relative phase difference between the two polarization directions $(\hat{x}, \hat{y})$ are spatially modulated by one set of wave plates. In other words, by inserting one set of wave plates, we can obtain the intensity distribution of the photocurrent difference over the entire range of the two phases

$\phi_L$ and $\phi_{LO}$. Comparing Eq. (18) and Eq. (14), the spatially dependent phtocurrent becomes

$$I(x,y) = -\frac{\sqrt{2}ecA\sigma_{det}}{\sqrt{V}}|E_s|\,[a_1\cos(\delta_1 + \phi_{LO}(x,y)) + a_2\cos(\delta_2 + \phi_{LO}(x,y) - \phi_L(x,y))]$$

$$= -\frac{\sqrt{2}ecA\sigma_{det}}{\sqrt{V}}|E_s|[a_1\cos(\delta_1 + \frac{2\pi}{\lambda}\{(n_L - n_R)x\tan\alpha + (n_R - n_o)y\tan\beta + n_Rd_1 + n_od_2\})$$

$$+a_2\cos(\delta_2 + \frac{2\pi}{\lambda}\{(n_L - n_R)x\tan\alpha$$

$$+ (n_R - 2n_o + n_e)y\tan\beta + n_Rd_1 - (2n_o + n_e)d_2\})], \tag{21}$$

where we put $\theta_L = \pi/4$ because the initial Jones vector of the local oscillator field is $\frac{1}{\sqrt{2}}(\hat{x} + \hat{y})$. Although we can find three parameters $(a_1, a_2, \delta_1 - \delta_2)$ from the modulated intensity distribution in Eq. (21), it is better to match the refractive index of the second wave plate $(n_R)$ with the refractive index of the third wave plate, $n_R = n_o$; then, the intensity distribution in Eq. (21) becomes

$$I(x,y) = -\frac{\sqrt{2}ecA\sigma_{det}}{\sqrt{V}}|E_s|[a_1\cos(\delta_1 + \frac{2\pi}{\lambda}\{(n_L - n_o)x\tan\alpha\} + \Delta_1)$$

$$+a_2\cos(\delta_2 + \frac{2\pi}{\lambda}\{(n_L - n_o)x\tan\alpha + (n_e - n_o)y\tan\beta\} + \Delta_2)], \tag{22}$$

where,

$$\Delta_1 = \frac{2\pi n_o(d_1 + d_2)}{\lambda}, \tag{23}$$

$$\Delta_2 = \frac{2\pi(n_od_1 - (2n_o + n_e)d_2)}{\lambda}. \tag{24}$$

Since the intensity distribution $I(x,y)$ depends on $x$ and $y$ independently , we can find the maximum value of the current $I_{max}$ in Eq. (22) and let the values $x_{max}$ and $y_{max}$ be the $x$ and $y$ values that will yield the maximum current. When the measured values of $x_{max}$ and $y_{max}$ are used, the difference in the phase factor of the input polarization can be expressed as

$$\delta_2 - \delta_1 = \frac{2\pi(n_e - n_o)y_{max}\tan\beta}{\lambda} + \Delta_2 - \Delta_1 + 2m\pi, \tag{25}$$

where $m$ is an integer. We can also obtain the magnitude of the polarization component of the input beam as

$$a_1 = \frac{I_0 + I_\pi}{\sqrt{2\left(I_0^2 + I_\pi^2\right)}},$$

$$a_2 = \frac{I_0 - I_\pi}{\sqrt{2\left(I_0^2 + I_\pi^2\right)}}, \tag{26}$$

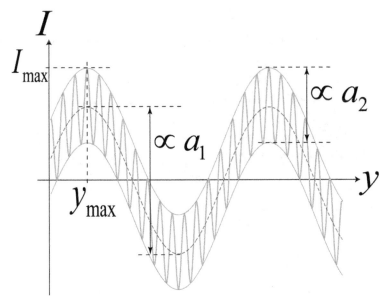

Fig. 3. Intensity modulation dependence on $y$. There are long- and short-term modulations whose modulation amplitudes depend on $a_1$ and $a_2$, respectively.

where

$$I_\eta \equiv I(x_{max}, y_{max} + \frac{\lambda}{2\pi(n_e - n_o)\tan\beta}\eta).$$ (27)

Thus, we can obtain full information on the polarization state of the input signal field, as shown in Eq. (25) and Eq. (26). Furthermore, since our results are obtained after scanning the entire space of the phase angles $\phi_L$ and $\phi_{LO}$, we do not have to fix the relative phase angle $\phi_{LO}$ between the signal field and the local oscillator field.

Our three-wave-plate system may be simplified if we let the refractive index of the first wedge ($n_L$) be the same as that of the second wedge, $n_R = n_L$. Then, the modulated intensity distribution becomes

$$I(y) = -\frac{\sqrt{2}ecA\sigma_{det}}{\sqrt{V}}|E_s|[a_1\cos(\delta_1 + \frac{2\pi}{\lambda}\{(n_R - n_o)y\tan\beta)\} + \Delta_1)$$

$$+ a_2\cos(\delta_2 + \frac{2\pi}{\lambda}\{(n_R - 2n_o + n_e)y\tan\beta\} + \Delta_2)].$$ (28)

When $|n_R - n_o| \ll |n_R - 2n_o + n_e|$, there are fast and slow modulation frequencies in Eq. (28), as shown in Fig. 3. From the modulated data, it becomes simple to find the values of the unknown parameters ($a_1, a_2,$ and $\delta_1 - \delta_2$) by using a fast Fourier transform or least square fitting method W. H. Press (1993).

Besides the complicated algorithm, a simple method is used to find the values of four parameters. From the intensity modulation data, we can roughly find the parameter $a_2$,

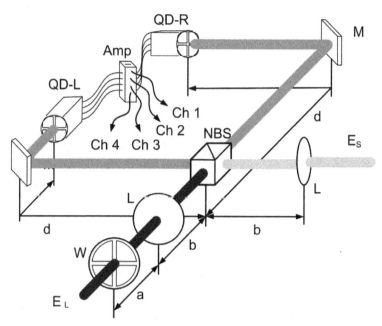

Fig. 4. Polarization-modulated homodyne detection. The organization of the waveplate quadrants for modulating the LO is explained in Fig. 5.

which is related to the short-term modulation amplitude. Furthermore, the total modulation amplitude is proportional to the sum of the two modulation amplitudes $(a_1 + a_2)$ as in Fig. 3. On the other hand, the phase factor $\delta_1$ is also approximately determined at the interpolated intensity modulation peak $I_{max}$ at $y_{max}$ in Fig. 3. Then, the phase factor $\delta_1$ becomes

$$\delta_1 = \frac{2\pi}{\lambda} \tan \beta (n_o - n_R) y_{max} - \Delta_1 + 2q\pi, \tag{29}$$

where $q$ is an integer.

With the three parameters $a_1, a_2$, and $\delta_1$, we can decide the value of the final phase factor $\delta_2$ from the intensity modulation equation, Eq. (28). This experimental setup is very practical and easy for determining the polarization of a weak signal field.

## 3. Polarization-modulated quadrant homodyne detector for single-shot measurement of the polarization state of a weak signal field

### 3.1 Spatial modulation of the local oscillator field

Our scheme of polarization-modulated balanced homodyne detection (BHD) is shown in Fig. 4. A signal having an unknown polarization state and a local oscillator with its polarization modulated deterministically, as per the scheme explained in the next few paragraphs, impinge on a 50:50 non-polarizing beam splitter. The spatial modulation of the LO polarization is carried out by using a four-quadrant double wave plate system.

Fig. 5. Quadrant wave plate assembly: Quadrants cut from different HWPs (or QWPs) with their optical axis pre-rotated as per the requirement are glued again together to form a composite HWP (or QWP).

A de-constructed and simplified view of one of such waveplates is shown in Fig. 5. Basically, birefringent half and quarter wave plates (HWP and QWP - and at most four of each) are all cut into quadrants first. Prior to this operation, they may have to be rotated as per the polarization change that is to be induced onto the incoming field by the participating quadrant. A composite HWP (or QWP) is, thus, prepared with the four quadrants glued back together - this is similar to the idea used to make a custom-made phase modulator in the experiment by Bachor Bachor (2006). Fig. 5 shows four quadrants A-D assembled from three different (say Half) wave plates. The optical axis of the quadrants are a priori rotated by $\pi/4$, $\pi/2$, $-\pi/4$ and $\pi/2$ with respect to the horizontal. Of course, these rotation angles could really be arbitrary, but we refer to these specific values because they are employed in our computational analysis, presented in the next section. Also, it is easily observed that B and D can be cutout from one single HWP with the optical axis rotated by $\pi/2$, thus the need for three waveplates, in all.

With the transfer function $\Gamma[\psi]$ of a waveplate, which induces a phase shift of $2\psi$ between the two orthogonal polarization components, being given by

$$\Gamma[\psi] = \begin{pmatrix} e^{-i\psi} & 0 \\ 0 & e^{i\psi} \end{pmatrix}$$

and a general rotation matrix given by

$$\mathbf{R}[\theta] = \begin{pmatrix} \cos\theta & -\sin\theta \\ \sin\theta & \cos\theta \end{pmatrix}$$

the output polarization state of the LO field (with input being denoted by $\hat{e}_L$), after having passed through a combination of a HWP and a QWP rotated at angles $\alpha$ and $\beta$, respectively, is given by

$$\hat{e}_L^{out} = \mathbf{R}[-\alpha]\mathbf{\Gamma}[\pi/2]\mathbf{R}[\alpha]\mathbf{R}[-\beta]\mathbf{\Gamma}[\pi/4]\mathbf{R}[\beta]\hat{e}_L. \tag{30}$$

### 3.2 Homodyne detection & polarization state measurement

As can be seen from Fig. 4, the lenses positioned in the Signal and the LO arm have a focal length $f$ such that

$$\frac{1}{a} + \frac{1}{(b+d)} = \frac{1}{f}. \tag{31}$$

Careful alignment of these lenses serves to put one image plane each at the quadrant-based detectors QD-L and QD-R, as shown in the Fig. 4. The interference between the four spatial polarization-modulated modes of LO and the signal field (assumed to be uniform spatially) at the beam splitter is thus captured just like in an ordinary homodyne detection experiment, and taking into the account the inversion (A ↔ C and B ↔ D), the corresponding difference photocurrent signals are amplified and obtained in channels 1-4. With the usual assumptions for a balanced homodyne detection analysis, i.e., the intensity of the signal field being much smaller than the mean intensity of the LO field, and the photodetector pair being ideal (i.e. $\eta_L = \eta_R = 1$), a generic expression for the difference photocurrent $I(t) \propto I_L(t) - I_R(t)$ is Youn & Bae (2006)

$$I(t) = -\frac{ecK\sigma_{det}}{\sqrt{V}} \left[ E_S(\vec{r})e^{i\phi_{LO}}\hat{e}_L^* \cdot \hat{e}_S + E_S^*(\vec{r})e^{-i\phi_{LO}}\hat{e}_L \cdot \hat{e}_S^* \right], \tag{32}$$

where
$e$ is the electronic charge,
$c$ is the speed of light in vacuum,
$K$ is a constant dependent on the local oscillator's field intensity,
$\sigma_{det}$ is the area of the detector,
$V$ is the mode volume,
$E_S(\vec{r})$ is the complex amplitude of the signal electric field,
$\hat{e}_S$ and $\hat{e}_L$ are the Jones vectors associated with the polarization of the Signal and the LO field, respectively, and $\phi_{LO}$ is overall phase of the LO field (relative to signal).
Again, in the general case, both electric fields are arbitrarily polarized, i.e.,

$$\hat{e}_S = a_1 e^{i\delta_1}\hat{x} + a_2 e^{i\delta_2}\hat{y}, \tag{33}$$

$$\hat{e}_L = b_1 e^{i\phi_{Lx}}\hat{x} + b_2 e^{i\phi_{Ly}}\hat{y}, \tag{34}$$

where $a_i, b_i$ are real numbers and the coefficients satisfy $a_1^2 + a_2^2 = 1, b_1^2 + b_2^2 = 1$. In addition, if we also consider the vacuum field fluctuations that are quantum mechanically independent from the signal field, then its Jones polarization vector would be orthogonal to $\hat{e}_S$, and would thus be of the form

$$\hat{e}_V = -a_2 e^{i\delta_1}\hat{x} + a_1 e^{i\delta_2}\hat{y}. \tag{35}$$

Now, if a symmetric input polarization is employed in the local oscillator beam, i.e., $\hat{e}_L = 1/\sqrt{2}\,\hat{x} + 1/\sqrt{2}\,\hat{y}$, then evaluating Eq. (30) yields the components for the output polarization, $\hat{e}_L^{out}$, in quadrants A, B, C and D, with pre-chosen values of $\alpha$ (same distribution as portrayed in Fig. 5 and $\beta$. Table 1 lists these polarization states:

| Quadrant | $\alpha$ | $\beta$ | $(\hat{e}_{Lx}^{out}, \hat{e}_{Ly}^{out})$ |
|---|---|---|---|
| A | $\frac{\pi}{4}$ | $-\frac{\pi}{4}$ | $\left(\frac{1}{\sqrt{2}}, \frac{1}{\sqrt{2}}\right)$ |
| B | $\frac{\pi}{2}$ | $0$ | $\left(\frac{1}{\sqrt{2}}, \frac{i}{\sqrt{2}}\right)$ |
| C | $-\frac{\pi}{4}$ | $-\frac{\pi}{4}$ | $\left(-\frac{i}{\sqrt{2}}, -\frac{i}{\sqrt{2}}\right)$ |
| D | $\frac{\pi}{2}$ | $-\frac{\pi}{4}$ | $\left(-\frac{i}{\sqrt{2}}, \frac{1}{\sqrt{2}}\right)$ |

Table 1. Output polarization with HWP & QWP rotated by $\alpha$ & $\beta$, respectively.

Further, considering a suitable form of Eq. (32) that accounts for the effect of vacuum, as well as substituting $\hat{e}_L^{out}$ for $\hat{e}_L$, in the same, yields (on simplification)

$$I(t) = -\frac{2ecK\sigma_{det}}{\sqrt{V}}\Big(b_1(a_1 E_S - a_2 E_V)\cos(\delta_1 - \phi_{Lx} + \phi_{LO}) +$$

$$b_2(a_2 E_S + a_1 E_V)\cos(\delta_2 - \phi_{Ly} + \phi_{LO})\Big). \qquad (36)$$

Here, the $\vec{r}$ dependence in $E_S$ and $E_V$ has been dropped for convenience sake, and $\phi_{Lx}$ and $\phi_{Ly}$ refer to the phases of the polarization components of the LO *after* the modulation, as dictated by the wave-plate assembly. Also, for non-classical fields such as the squeezed vacuum state, the difference photocurrent that constitutes terms arising from LO quantum noise and the quadrature amplitude of the signal (enhanced by the power in LO) has a statistical average of zero, i.e., $\langle I(t)\rangle = 0$. For a weak coherent field however, $\langle I(t)\rangle$ is finite and this gives us a ground to make measurements merely on I(t), instead of the usual $\langle I^2(t)\rangle$. Finally, in Eq. (36), setting $\delta_1 + \phi_{LO} \to \delta_1$ and $\delta_2 + \phi_{LO} \to \delta_2$, or equivalently absorbing the effect of $\phi_{LO}$ in $\delta_1$ and $\delta_2$ by reducing it to zero and computing $\phi_{Lx}$ and $\phi_{Ly}$ from the last column of table 1, the expressions for the difference photocurrents are produced in channels 1-4 corresponding to quadrants A-D:

| Quadrant | $\phi_{Lx}^{out}$ | $\phi_{Ly}^{out}$ | $I_{out}^v$ | $I_{out}$ |
|---|---|---|---|---|
| A | $0$ | $0$ | $(a_1 E_s - a_2 E_v)\cos\delta_1 + (a_2 E_s + a_1 E_v)\cos\delta_2$ | $a_1\cos\delta_1 + a_2\cos\delta_2$ |
| B | $0$ | $\frac{\pi}{2}$ | $-(a_1 E_s - a_2 E_v)\sin\delta_1 - (a_2 E_s + a_1 E_v)\sin\delta_2$ | $-a_1\sin\delta_1 - a_2\sin\delta_2$ |
| C | $-\frac{\pi}{2}$ | $-\frac{\pi}{2}$ | $(a_1 E_s - a_2 E_v)\cos\delta_1 + (a_2 E_s + a_1 E_v)\sin\delta_2$ | $a_1\cos\delta_1 + a_2\sin\delta_2$ |
| D | $-\frac{\pi}{2}$ | $0$ | $-(a_1 E_s - a_2 E_v)\sin\delta_1 + (a_2 E_s + a_1 E_v)\cos\delta_2$ | $-a_1\sin\delta_1 + a_2\cos\delta_2$ |

Table 2. Photocurrent expressions for the four different quadrants. The last column corresponds to the case when vacuum fluctuations are totally neglected, i.e., when the signal field is dominant.

In general, an arbitrary (but fixed) value of the LO phase should be considered in Eq. (36). Then, replacing $\phi_{LO}$ by $\psi$ and an appropriate phase factor (so as to preserve the orthogonality between the *representation* of the signal polarization), the four channels' difference photocurrents for various (assumed) polarization states for the signal field can be found, as displayed in table 3.

| SNo. | $\hat{e}_s$ | A | B | C | D |
|------|-------------|---|---|---|---|
| 1 | $(1,0)$ | $\cos\psi$ | $-\sin\psi$ | $\cos\psi$ | $-\sin\psi$ |
| 2 | $(0,1)$ | $\cos(\psi-\frac{\pi}{4})$ | $-\sin(\psi-\frac{\pi}{4})$ | $\sin(\psi-\frac{\pi}{4})$ | $\cos(\psi-\frac{\pi}{4})$ |
| 3 | $\frac{1}{\sqrt{2}}(1,1)$ | $\sqrt{2}\cos\psi$ | $-\sqrt{2}\sin\psi$ | $\sin(\psi+\frac{\pi}{4})$ | $\cos(\psi+\frac{\pi}{4})$ |
| 4 | $\frac{1}{\sqrt{2}}(1,-1)$ | $0$ | $0$ | $\cos(\psi+\frac{\pi}{4})$ | $-\sin(\psi+\frac{\pi}{4})$ |
| 5 | $\frac{1}{\sqrt{2}}(1,i)$ | $\cos(\psi+\frac{\pi}{4})$ | $-\sin(\psi+\frac{\pi}{4})$ | $\sqrt{2}\cos\psi$ | $-\sqrt{2}\sin\psi$ |
| 6 | $\frac{1}{\sqrt{2}}(1,-i)$ | $\sin(\psi+\frac{\pi}{4})$ | $\cos(\psi+\frac{\pi}{4})$ | $0$ | $0$ |

Table 3. Measured photocurrent values for various $\hat{e}_s$ in the four quadrants. A phase factor of $\frac{\pi}{4}$ is present so that the representation of $\hat{e}_s = (1,0)$ is orthogonal to $\hat{e}_s = (0,1)$.

Thus, the comparative knowledge from this should make it possible to derive information about the six different signal polarization states, which also happen to be the most fundamental ($|H\rangle$, $|V\rangle$, $|+\rangle$, $|-\rangle$, $|R\rangle$, and $|L\rangle$) in quantum information science. For example, for an experimental run, if no modulation is observed in channels A & B simultaneously, then $\hat{e}_s = \frac{1}{\sqrt{2}}(1,-1)$. A note of caution: since a knowledge of the absolute phase is not possible, the above expressions have been derived considering a relative phase of $\delta = \delta_2 - \delta_1$ (refer to Eq. (33)); hence, the information is true up to a global phase. Further, using Table 3, we can now also compute error functions, $er_i(\psi)$, that indicate the deviation of the theoretical values from the *experimentally* measured values, i.e.

$$er_i(\psi) = \sum_{j=1}^{4}(n_{ij} - o_j)^2, \tag{37}$$

where $o_j$ is the actual value measured in the $j^{th}$ channel, and $n_{ij}$ is an element of the 6x4 matrix that is contained in Table 3, with rows indicated by the signal polarization and columns by the four quadrants: e.g., $n_{32} = -\sqrt{2}\sin\psi$. Thus, for a given observation, Eq. (37) yields six different plots as functions of $\psi$ and the unknown signal field polarization is indicated by the curve with the absolute minimum amongst the six.

## 4. Conclusion and discussion

For an ensemble of equally prepared signals, the well-established optical homodyne tomography method give us the quasi probability distribution of the signal field, and the quasi probability distribution defines the statistical characteristics of the signal field, such as the amplitude squeezed state and quadrature squeezed state Leonhardt (1997). This method, however, is not adequate for obtaining information on a signal field that is changing pulse by pulse. Our polarization-modulated homodyne detection scheme is able to obtain the polarization state of a signal field in a single-shot scheme. Although the quasi probability of the signal field for a given single pulse cannot be known, we can determine the varying polarization state of the signal field pulse by pulse. Our novel scheme can determine the varying polarization state of the signal and can be used in quantum information science.

We propose a novel homodyne detection method to measure a polarization state of a weak field. At first we introduced a novel scheme of polarization modulated balanced homodyne detection method. By inserting a set of wedged wave plate in the local oscillator port, we can modulated the relative phase of the two orthogonal components as well as the overall phase retardation of the local oscillator electric field. Using this spatially modulated local oscillator field in the homodyne detection scheme, we can obtain the amplitudes and the relative phase of the two orthogonal components of the signal electric field. It's a practical method to measure the polarization state of the weak signal in a single-shot scheme without any scanning time.

Note that, of course, other schemes may be able to measure the state of the polarization of a pulsed weak signal at once: for example, dividing the signal into many beams, measuring them with polarization-measurement setups, and finally deriving the polarization state of the input signal electric field. However, it is not practical, especially, for very weak beams because a divided beam is too weak for measurement, so the loss is not negligible. In our scheme, in some sense, we also divided the signal field spatially, but the strong local oscillator field plays a role in measuring the weak beam by a kind of amplification as in the usual homodyne detection scheme.

The second polarization-modulated homodyne detection scheme uses photodetectors that measure four simultaneous photocurrent signals corresponding to four different quadrant shaped combination of wave plates. We make a spatial phase modulation of the two orthogonal polarization components of the local oscillator electric field by inserting a system of wave plates in the path. Information about the Jones vector associated with the signal field can then be obtained, and characterization can be performed in real time, i.e., on a single shot basis.

The subtle aspect in our scheme is representing the polarization of photons, i.e., a qubit system, in a higher (four) dimensional space, thus allowing for a better discrimination between any two polarization states. To elaborate, if the quadrant-measured values are taken as components of a four-dim vector, then an orthogonality between two different signal polarization states is preserved in this new representation. Further, it includes the power of homodyne detection, which allows a measurement of a very weak or highly attenuated field, by amplifying it sufficiently. This new scheme might have application in bio-physics, where we have to measure the polarization change of the very weak beam from a single molecule.

## 5. Acknowledgements

I would like to thank Samyong Bae and Nitin Jain for their valuable work for this article.

## 6. References

H. P. Yuen and J. H. Shapiro, IEEE Trans. Inf. Theory 24, 657 (1978).
J. H. Shapiro, H. P. Yuen, and J. A. Machado-Matta, IEEE Trans. Inf. Theory 25, 179 (1979).
H. P. Yuen and J. H. Shapiro, IEEE Trans. Inf. Theory 26, 78 (1980).
B. Yurke, Phys. Rev. A 32, 311 (1985).
R. E. Slusher, L. W. Hollberg, B. Yurke, J. C. Mertz, and J. F. Valley, Phys. Rev. Lett. 55, 2409 (1985).
L-A. Wu, H. J. Kimble, J. H. Hall, and H. Wu, Phys. Rev. Lett. 57, 2520 (1986).
E. S. Polzik, J. Carri, and H. J. Kimble, Phys. Rev. Lett. 68, 3020 (1992).
D. T. Smithey, M. Beck, M. G. Raymer, and A. Faridani, Phys. Rev. Lett. 70, 1244 (1993).

K. Banaszek and K. W'odkiewicz, Phys. Rev. Lett. 76, 4344 (1996).

S. Wallentowitz and W. Vogel, Phys. Rev. A 53, 4528 (1996).

S. H. Youn, Y. T. Chough and K. An, J. Korean Phys. Soc. 39, 255, (2001).

T. Hirano, T. Konishi, and R. Namiki, quant-ph/0008037.

A. Yariv, *Quantum Electronics* (Wiley, New York, 1989).

U. Leonhardt, *Measuring the Quantum State of Light*, (Cambridge University Press, Cambridge, 1997).

S. Schiller, G. Breitenbach, S. F. Pereira, T. Muller, and J. Mlynek, Phys. Rev. Lett. 77, 2933 (1996).

D. S. Krahmer and U. Leonhardt, Phys. Rev. A 55, 3275 (1997).

H. R. Noh and S.H. Youn, J. Korean Phys. Soc. 43, 1029 (2003).

E. Hecht, *Optics* (Addison Wesley, New York, 2002).

S. H. Youn, J. Korean Phys. Soc. 47, 803 (2005)

W. H. Press *et al.*, *Numerical Recipes* second edition, (Cambridge University Press, Cambridge, England, 1993).

M. Munroe, D. Boggavarapu, M. E. Anderson and M. G. Raymer, Phys. Rev. A 52, R924 (1995).

C.H. Bennet and G. Brassard, Proceedings of IEEE Intl. Conf. on Computers Systems and Signal Processing, Bangalore India, December 175-179 (1984).

S.H. Youn and Nitin Jain, J. Korean Phys. Soc. 54, 29 (2009).

S.H. Youn and Samyong Bae, J. Korean Phys. Soc. 48, 397 (2006).

Daniel F. V. James, Paul G. Kwiat, William J. Munro, and A. G. White, Phys. Rev. A 64, 052312 (2001)

H-A. Bachor, J. Mod. Opt. 53, 5-6, 597 (2006).

# 8

# Spin Photodetector: Conversion of Light Polarization Information into Electric Voltage Using Inverse Spin Hall Effect

Kazuya Ando and Eiji Saitoh
*Institute for Materials Research, Tohoku University*
*Japan*

## 1. Introduction

Recent developments in optical and material science have led to remarkable industrial applications, such as optical data recording and optical communication. The scope of the conventional optical technology can be extended by exploring simple and effective methods for detecting light circular polarization; light circular polarization carries single-photon information, making it essential in future optical technology, including quantum cryptography and quantum communication.

Light circular polarization is coupled with electron spins in semiconductors (Meier, 1984). When circularly polarized light is absorbed in a semiconductor crystal, the angular momentum of the light is transferred to the semiconductor, inducing spin-polarized carriers though the optical selection rules for interband transitions [see Fig. 1(a)]. This process allows conversion of light circular polarization into electron-spin polarization, enabling the integration of light-polarization information into spintronic technologies.

If one can convert electron spin information into an electric signal, light circular polarization information can be measured through the above process. Recently, in the field of spintronics, a powerful technique for detecting electron spin information has been established, which utilizes the inverse spin Hall effect (ISHE) (Saitoh, 2006; Valenzuela, 2006; Kimura, 2007). The ISHE converts a spin current, a flow of electron spins in a solid, into an electric field through the spin-orbit interaction, enabling the transcription of electron-spin information into an electric voltage. This suggests that light-polarization information can be converted into an electric signal by combining the optical selection rules and the ISHE.

This chapter describes the conversion of light circular polarization information into an electric voltage in a Pt/GaAs structure though the optical generation of spin-polarized carriers and the ISHE: the photoinduced ISHE (Ando, 2010).

## 2. Optical excitation of spin-polarized carriers in semiconductors

When circularly polarized light is absorbed in a semiconductor, the angular momentum of the light is transferred to the material, which polarizes carrier spins in the semiconductor through the spin-orbit interaction (Meier, 1984). This optical generation of spin-polarized carriers has been a powerful technique for exploring spin physics in direct band gap semiconductors, such as GaAs. In GaAs, the valence band maximum and the conduction band minimum are at $\Gamma$ with an energy gap $E_g = 1.43$ eV at room temperature. The valence band ($p$ symmetry) splits into fourfold degenerate $P_{3/2}$ and twofold degenerate $P_{1/2}$ states, which lie $\Delta = 0.34$ eV below $P_{3/2}$ at $\Gamma$, whereas the conduction band ($s$ symmetry) is twofold degenerate $S_{1/2}$ as schematically shown in Fig. 1(a). In the fourfold degenerate $P_{3/2}$ state, holes can occupy states with values of angular momentum $m_j = \pm 1/2, \pm 3/2$, corresponding to light hole (LH) and heavy hole (HH) sates, respectively (see Fig. 1(a)). Let $|J, m_j\rangle$ be the Bloch states according to the total angular momentum $J$ and its projection onto the positive $z$ axis $m_j$. The band wave functions can be expressed as listed in Table 1, where $|S\rangle, |X\rangle, |Y\rangle$, and $|Z\rangle$ are the wave functions with the symmetry of $s$, $p_x$, $p_y$, and $p_x$ orbitals. The interband transitions satisfy the selection rule $\Delta m_j = \pm 1$, reflecting absorption of the photon's original angular momentum. The probability of a transition involving a LH or HH state is weighted by the square of the corresponding matrix element connecting it to the appropriate electron state, so that the relative intensity of the optical transition between the heavy and the light hole subbands and the conduction band induced by circularly polarized light illumination is 3. Thus absorption of photons with angular momentum +1 produces three spin-down ($m_j = -1/2$) electrons for every one spin-up ($m_j = +1/2$) electron, resulting in an electron population with a spin polarization of 50% in a bulk material, where the HH and LH states are degenerate. The relative transition rates are summarized in Fig. 1(b). Therefore, because of the difference in the relative intensity, a spin-polarized carriers can be generated by the illumination of circularly polarized light. Note that the resulting electron spin is oriented parallel or anti-parallel to the propagation direction of the incident photon.

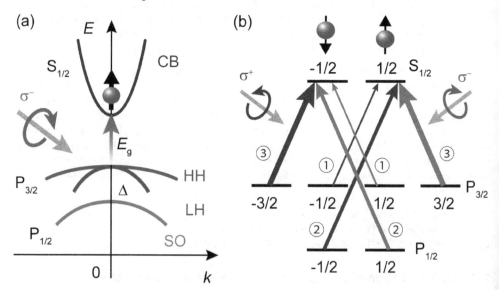

Fig. 1. (a) Optical generation of spin-polarized carriers in semiconductors. (b) Interband transitions for right and left circularly polarized light illumination.

| | $\lvert J, m_j \rangle$ | wave function |
|---|---|---|
| CB | $\lvert 1/2, 1/2 \rangle$ | $\lvert S \uparrow \rangle$ |
| | $\lvert 1/2, -1/2 \rangle$ | $\lvert S \downarrow \rangle$ |
| HH | $\lvert 3/2, 3/2 \rangle$ | $\lvert (1/2)^{(1/2)}(X + iY) \uparrow \rangle$ |
| | $\lvert 3/2, -3/2 \rangle$ | $\lvert (1/2)^{(1/2)}(X - iY) \downarrow \rangle$ |
| LH | $\lvert 3/2, 1/2 \rangle$ | $\lvert (1/6)^{(1/2)}[(X + iY) \downarrow + 2Z \uparrow] \rangle$ |
| | $\lvert 3/2, -1/2 \rangle$ | $\lvert -(1/6)^{(1/2)}[(X - iY) \uparrow - 2Z \downarrow] \rangle$ |
| SO | $\lvert 1/2, 1/2 \rangle$ | $\lvert -(1/3)^{(1/2)}[(X + iY) \downarrow - Z \uparrow] \rangle$ |
| | $\lvert 1/2, -1/2 \rangle$ | $\lvert (1/3)^{(1/2)}[(X - iY) \uparrow + Z \downarrow] \rangle$ |

Table 1. Wave functions for the conduction band (CB), heavy hole (HH), light hole (LH), and spin-orbit split-off band (SO).

## 3. Spin current and inverse spin Hall effect

A spin current is a flow of electron spins in a solid. One of the driving forces for a spin current is a gradient of the difference in the spin-dependent electrochemical potential $\nabla \mu_\sigma$ for spin up ($\sigma = \uparrow$) and spin down ($\sigma = \downarrow$). Here, $\mu_\sigma = \mu_\sigma^c - e\phi$, where $\mu_\sigma^c$ is the chemical potential. A current density for spin channel $\sigma$ is expressed as

$$\mathbf{j}_\sigma = \frac{\sigma_\sigma}{e} \nabla \mu_\sigma, \tag{1}$$

where $\sigma_\sigma$ is the electrical conductivity for spin up ($\sigma = \uparrow$) and spin down ($\sigma = \downarrow$) channel. Here, a charge current, a flow of electron charge, is the sum of the current for $\sigma = \uparrow$ and $\downarrow$ as $\mathbf{j}_c = \mathbf{j}_\uparrow + \mathbf{j}_\downarrow$:

$$\mathbf{j}_c = \frac{1}{e} \nabla \left( \sigma_\uparrow \mu_\uparrow + \sigma_\downarrow \mu_\downarrow \right). \tag{2}$$

This flow is schematically illustrated in Fig. 2(a). This flow carries electron charge while the flow of spins is cancelled. In contrast, the opposite flow of $\mathbf{j}_\uparrow$ and $\mathbf{j}_\downarrow$, $\mathbf{j}_s = \mathbf{j}_\uparrow - \mathbf{j}_\downarrow$, or

$$\mathbf{j}_s = \frac{1}{e} \nabla \left( \sigma_\uparrow \mu_\uparrow - \sigma_\downarrow \mu_\downarrow \right), \tag{3}$$

carries electron spins without a charge current. This is a spin current. In nonmagnetic materials, a spin current is expressed as $\mathbf{j}_s = (\sigma_N/2e)\nabla(\mu_\uparrow - \mu_\downarrow)$, since the electrical conductivity is spin-independent: $\sigma_\uparrow = \sigma_\downarrow = \sigma_N/2$.

Since charge $\rho$ is a conserved quantity, the continuity equation of charge is described as

$$\frac{d}{dt} \rho = -\nabla \cdot \mathbf{j}_c. \tag{4}$$

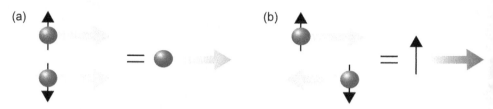

Fig. 2. (a) A schematic illustration of a charge current. (b) A schematic illustration of a spin current.

In contrast, spins are not conserved; a spin current decays typically in a length scale of nm to $\mu$m. Therefore, the continuity equation of spins are written as

$$\frac{d}{dt}M_z = -\nabla \cdot \mathbf{j}_s + T_z,\tag{5}$$

where $M_z$ is the $z$ component of magnetization. $z$ is defined as the quantization axis. Here, $T_z = e(n_\uparrow - \bar{n}_\uparrow)/\tau_{\uparrow\downarrow} - e(n_\downarrow - \bar{n}_\downarrow)/\tau_{\downarrow\uparrow}$ represents spin relaxation. $\bar{n}_\sigma$ is the equilibrium carrier density with spin $\sigma$ and $\tau_{\sigma\sigma'}$ is the scattering time of an electron from spin state from $\sigma$ to $\sigma'$. Note that the detailed balance principle imposes that $N_\uparrow/\tau_{\uparrow\downarrow} = N_\downarrow/\tau_{\downarrow\uparrow}$, so that in equilibrium no net spin scattering takes place, where $N_\sigma$ denotes the spin dependent density of states at the Fermi energy. This indicates that, in general, in a ferromagnet, $\tau_{\uparrow\downarrow}$ and $\tau_{\downarrow\uparrow}$ are not the same. In the equilibrium condition, $d\rho/dt = dM_z/dt = 0$, using the continuity equations, one finds the spin-diffusion equations:

$$\nabla^2(\sigma_\uparrow\mu_\uparrow + \sigma_\downarrow\mu_\downarrow) = 0,\tag{6}$$

$$\nabla^2(\mu_\uparrow - \mu_\downarrow) = \frac{1}{\lambda^2}(\mu_\uparrow - \mu_\downarrow),\tag{7}$$

where $\lambda = \sqrt{D\tau_{sf}}$ is the spin diffusion length. $D = D_\uparrow D_\downarrow(N_\uparrow + N_\downarrow)/(N_\uparrow D_\uparrow + N_\downarrow D_\downarrow)$ is the diffusion constant. The spin relaxation time $\tau_{sf}$ is given by $1/\tau_{sf} = 1/\tau_{\uparrow\downarrow} + 1/\tau_{\downarrow\uparrow}$. By solving the diffusion equations, one can obtain the spatial variation of spin currents generated by $\mu_\uparrow - \mu_\downarrow$. A spin current generated by $\mu_\uparrow - \mu_\downarrow$ decays as $e^{-\lambda/x}$. Thus a spin current play a key role only in a system with the scale of $\lambda$.

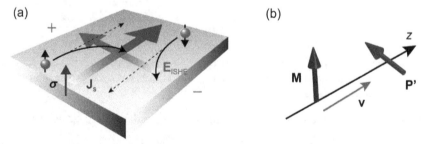

Fig. 3. (a) A schematic illustration of the inverse spin Hall effect. (b) Conversion of magnetic moment **M** into electric polarization **P'**.

Fig. 4. (a) A schematic illustration of the magnetic-field **H** generation from a charge current $j_c$ according to Ampere's law. (b) A schematic illustration of the electric-field **E** generation from a hypothetical magnetic-monopole current $j_m$ according to the electromagnetic duality and Ampere's law. (c) A schematic illustration of the electric-field **E** generation from a pair of hypothetical magnetic-monopole currents, $j_m$ and $-j_m$, or a spin current.

A spin current can be detected electrically using the inverse spin Hall effect (ISHE), conversion of a spin current into an electric field [see Fig. 3(a)]. The ISHE has the same symmetry as that of the relativistic transformation of magnetic moment into electric polarization, which is derived from the Lorentz transformation, as follows. Consider a magnet with the magnetic moment **M** moving at a constant velocity **v** along the $z$ axis with respect to an observer [see Fig. 3(b)]. This motion of the magnet is a flow of angular momentum, meaning an existence of a "spin current". In the observer's coordinate system, the Lorentz transformation converts a part of this magnetic moment **M** into an electric dipole moment **P'** as

$$\mathbf{P'} = -\frac{1}{\sqrt{1-(v/c)^2}}(\epsilon_0 \mathbf{v} \times \mathbf{M}), \qquad (8)$$

where $c$ and $\varepsilon_0$ are the light velocity and the electric constant, respectively. This indicates that electric polarization perpendicular to the direction of the magnetic-moment velocity is induced.

This electric-polarization generation can also be regarded as the spin-current version of Ampere's law as follows. As shown in Fig. 4(a), when a charge current $j_c$ flows, a circular magnetic field **H** is induced around the charge current, according to Ampere's law: rot**H** = $j_c$. If a hypothetical magnetic monopole flows, a circular electric field **E** is expected to be induced around the monopole current $j_m$ according to rot**E** = $j_m$ [see Fig. 4(b)], from the electromagnetic duality. Although this monopole has never been observed in reality, a spin current can be regarded as a pair of the hypothetical monopole currents flowing in the opposite directions along the spin current spatial direction. Therefore, a spin current may generate an electric field and this field is the superposition of the two electric fields induced by this pair of the monopole current, as shown in Fig. 4(c). This spin-current-induced electric field is identical to the field induced by the dipole moment described by Eq. (8).

In this way, electromagnetism and relativity predict that a spin current generates an electric field. According to Eq. (8), however, this electric field is too weak in a vacuum to be detected in reality. In a solid with strong spin-orbit interaction, in contrast, a similar but strong conversion between spin currents and electric fields appears, which is the ISHE.

In a solid, existence of a spin current can be modelled as that two electrons with opposite spins travel in opposite directions along the spin-current spatial direction $\mathbf{j_s}$, as shown in Fig. 3(a). Here, $\sigma$ denotes the spin polarization vector of the spin current. The spin-orbit interaction bends these two electrons in the same direction and induces an electromotive force $\mathbf{E_{ISHE}}$ transverse to $\mathbf{j_s}$ and $\sigma$, which is the ISHE. The relation among $\mathbf{j_s}$, $\mathbf{E_{ISHE}}$, and $\sigma$ is therefore given by (Saitoh, 2006)

$$\mathbf{E_{ISHE}} = D_{ISHE}\mathbf{J_s} \times \sigma, \tag{9}$$

where $D_{ISHE}$ is the ISHE efficiency. This equation is similar to Eq. (8) but this effect may be enhanced by the strong spin-orbit interaction in solids.

The ISHE was recently observed using a spin-pumping method operated by ferromagnetic resonance (FMR) and by a non-local method in metallic nanostructures (Saitoh, 2006; Valenzuela, 2006; Kimura, 2007). Since the ISHE enables the electric detection of a spin current, it will be useful for exploring spin currents in condensed matter.

## 4. Photoinduced inverse spin Hall effect: Experiment

The combination of the optical generation of spin-polarized carriers and the ISHE enables direct conversion of light-polarization information into electric voltage in a Pt/GaAs interface (Ando, 2010). Figure 5(a) shows a schematic illustration of the Pt/GaAs sample. Here, the thickness of the Pt layer is 5 nm. The Pt layer was sputtered on a Si-doped GaAs substrate with a doping concentration of $N_D = 4.7 \times 10^{18}$ cm$^{-3}$. The surface of the GaAs layer was cleaned by chemical etching immediately before the sputtering. Two electrodes are attached to the ends of the Pt layer as shown in Fig. 5(a). During the measurement, circularly polarized light with a wavelength of $\lambda = 670$ nm and a power of $I_i = 10$ mW was illuminated to the Pt/GaAs sample as shown in Fig. 5(a). In the GaAs layer, electrons with a spin polarization $\sigma$ along the light propagation direction are excited to the conduction band by the circularly polarized light due to the optical selection rule. Here, note that hole spin polarization plays a minor role in this setup, since it relaxes in ~ 100 fs, which is much faster than the relaxation time of ~ 35 ps for electron spin polarization (Hilton, 2002; Kimel, 2001). This spin polarization of electrons then travels into the Pt layer across the interface as a pure spin current. The injected spin current is converted into an electric voltage by the ISHE in the Pt layer due to the strong spin-orbit interaction in Pt (Ando, 2008). Here, note that the angle of the light illumination to the normal axis of the film plane is set at $\theta_0 = 65°$ to obtain the photoinduced ISHE signal, since the electric voltage due to the photoinduced ISHE is proportional to $j_s\sin\theta_0$ because of the relation $\mathbf{E_{ISHE}} \propto \mathbf{j_s} \times \sigma$, where the spin polarization $\sigma$ is directed along the light propagation direction. The difference in the generated voltage between illumination with right circularly polarized (RCP) and left circularly polarized (LCP) light, $V^R - V^L$, was measured by a polarization-lock-in technique using a photoelastic modulator operated at 50 kHz. The difference in the intensities between RCP and LCP light incident on the sample was confirmed to be vanishingly small. All the measurements were performed at room temperature at zero applied bias across the junction.

In-plane light illumination angle $\theta$ dependence of $V^R - V^L$ for the Pt/GaAs sample is shown in Fig. 5(b), where the in-plane angle $\theta$ is defined in Fig. 5(a). Figure 5(b) shows that $V^R - V^L$ varies systematically by changing the illumination angle $\theta$. Notable is that this variation is well reproduced using a function proportional to $\cos\theta$, as expected for the photoinduced ISHE. The relation of the ISHE, $E_{ISHE} \propto j_s \times \sigma$, indicates that the electric voltage due to the photoinduced ISHE is proportional to $|j_s \times \sigma|_x \propto \cos\theta$, since $\sigma$ and $j_s$ are directed along the light propagation direction and the $z$ axis, respectively. Here, $|j_s \times \sigma|_x$ denotes the $x$ component of $j_s \times \sigma$ [see Fig. 5(a)]. This electromotive force was found to be disappeared in a Cu/GaAs system, where the Pt layer is replaced by Cu with very weak ISHE, supporting that ISHE is responsible for the observed electric voltage.

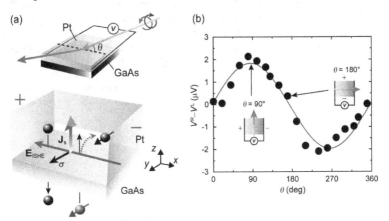

Fig. 5. (a) A schematic illustration of the Pt/GaAs hybrid structure and the photoinduced ISHE in the Pt/GaAs system. (b) In-plane illumination angle $\theta$ dependence of $V_R$-$V_L$ measured for the Pt/GaAs hybrid structure.

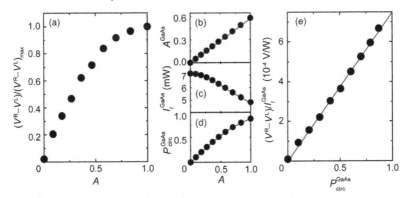

Fig. 6. (a) Ellipticity $A$ of the illuminated light dependence of $V_R - V_L$. (b) $A$ dependence of the ellipticity $A^{GaAs}$ of the light injected into the GaAs layer. (c) $A$ dependence of the intensity $I_t^{GaAs}$ of the light injected into the GaAs layer. (d) $A$ dependence of the degree of circular polarization $P_{circ}^{GaAs}$ of the light injected into the GaAs layer. (e) $P_{circ}^{GaAs}$ dependence of $(V^R - V^L)/I_t^{GaAs}$.

The observed electric voltage signal depends strongly on the ellipticity of the illuminated light polarization. Here, the ellipticity $A$ is defined as the ratio of the minor to major radiuses of the elliptically polarized light. Figure 6(a) shows the illuminated-light ellipticity $A$ dependence of $V^R - V^L$. As shown in Fig. 6(a), the $V^R - V^L$ signal increases with the ellipticity $A$ of the illuminated light. This supports that this signal is induced by the photoinduced ISHE, since the angular momentum component of a photon along the light propagation direction is zero (maximized) when $A = 0$ (1).

## 5. Photoinduced inverse spin Hall effect: Theory

The $A$ dependence of $V^R - V^L$ shown in Fig. 6(a) demonstrates that the electric voltage observed in the Pt/GaAs junction is induced by the circularly polarized light illumination. However, the variation of the electric voltage with respect to $A$ is not straightforward to understand; the $V^R - V^L$ signal is not linear to $A$. In the following, we discuss in detail on the experimental result by calculating the polarization of the light injected into the GaAs layer.

The propagation of light in a multilayer film is characterized by the optical admittance $Y^{s(p)}$ $= C^{s(p)}/B^{s(p)}$, where s(p) denotes s(p) polarized light. $B^{s(p)}$ and $C^{s(p)}$ are expressed as

$$\begin{pmatrix} B^{s(p)} \\ C^{s(p)} \end{pmatrix} = \begin{pmatrix} \cos\delta & (i\sin\delta)/\eta_1^{s(p)} \\ i\eta_1^{s(p)}\sin\delta & \cos\delta \end{pmatrix} \begin{pmatrix} 1 \\ \eta_2^{s(p)} \end{pmatrix}, \tag{10}$$

where $\delta = 2\pi n_1 d_1 \cos\theta_1 / \lambda$, $\eta_r^p = n_r(\varepsilon_0/\mu_0)^{1/2}/\cos\theta_r$, and $\eta_r^s = n_r(\varepsilon_0/\mu_0)^{1/2}\cos\theta_r$ ($r = 0$, 1, 2). Here, $n_0$, $n_1$, and $n_2$ are the complex refractive indices for air, Pt, and GaAs, respectively. $d_1$ is the thickness of the Pt layer and $\theta_r$ is the incident angle of the light defined as shown in Fig. 7. Using $B^{s(p)}$ and $C^{s(p)}$, the transmittance $T^{s(p)} \equiv I_t^{s(p)} / I_i^{s(p)}$ and the transmission coefficient $\tau^{s(p)} \equiv E_t^{s(p)} / E_i^{s(p)}$ are obtained as

$$T^{s(p)} = \frac{4\eta_0^{s(p)}\Re[\eta_2^{s(p)}]}{(\eta_0^{s(p)}B^{s(p)} + C^{s(p)})(\eta_0^{s(p)}B^{s(p)} + C^{s(p)})^*}, \tag{11}$$

$$\tau^s = \frac{2\eta_0^s}{\eta_0^s B^s + C^s}, \quad \tau^p = \frac{2\eta_0^p}{\eta_0^p B^p + C^p} \frac{\cos\theta_0}{\cos\theta_2}, \tag{12}$$

where $I_{i(t)}^{s(p)}$ and $E_{i(t)}^{s(p)}$ are the illuminated (transmitted) light intensity and the amplitude of the electric field of s(p) polarized light [see Fig. 7], respectively. Here, $\Re[\eta_2^{s(p)}]$ is the real part of $\eta_2^{s(p)}$. Using Eqs. (11) and (12) with the parameters shown in Table 2, the transmittance $T^{s(p)}$ and the transmission coefficient $\tau^{s(p)}$ for the Pt/GaAs system are obtained as shown in Table 3. The calculated transmission coefficients $\tau^{s(p)}$ show that the transmission of the s and p polarized light is different. This indicates that the ellipticity of the illuminated to the sample is changed during the propagation of the film. The relation between the ellipticity $A^{GaAs}$ of the light injected into the GaAs layer and the ellipticity $A$ of the illuminated light is shown in Fig. 6(b). Here, $A^{GaAs}$ is obtained using

$A^{\text{GaAs}} \simeq (\Re[\tau^s] / \Re[\tau^p])A$. From the value of the ellipticity $A$, the degree of circular polarization $P_{\text{circ}}$, the difference in the numbers between RCP and LCP photons, can be written as,

$$P_{\text{circ}} \equiv \frac{I^+ - I^-}{I^+ + I^-} = \frac{2A}{1 + A^2},\tag{13}$$

where $I^+$ and $I^-$ are the intensities of the RCP and LCP light, respectively. The degree of circular polarization $P_{\text{circ}}^{\text{GaAs}}$ of the light injected into the GaAs layer is shown in Fig. 6(d), which is obtained from the ellipticity shown in Fig. 6(b) using Eq. (13). Here, notable is that the degree of circular polarization $P_{\text{circ}}^{\text{GaAs}}$ of the light injected into the GaAs layer is proportional to the electron spin polarization generated by the circularly polarized light. The propagation of the circularly polarized light also changes the intensity of the light as $I_t^{\text{GaAs}} = T^s I_i^s + T^p I_i^p$. Figure 6(c) shows the light ellipticity $A$ dependence of the intensity $I_t^{\text{GaAs}}$ of the light injected into the GaAs layer obtained from

$$I_t^{\text{GaAs}} = \left( T^s \frac{A^2}{1 + A^2} + T^p \frac{1}{1 + A^2} \right) I_i.\tag{14}$$

Here, $I_i = I_i^s + I_i^p$ is the illuminated light intensity. Since the electric voltage due to the photoinduced ISHE is expected to be proportional to the intensity of the absorbed light, or the number of spin-polarized carriers generated by the circularly polarized light, one should calculate $(V^R - V^L) / I_t^{\text{GaAs}}$ to compare the electric voltage induced by the circularly polarized light for different $A$. The $P_{\text{circ}}^{\text{GaAs}}$ dependence of $(V^R - V^L) / I_t^{\text{GaAs}}$ is shown in Fig. 6(e). As shown in Fig. 6(e), $(V^R - V^L) / I_t^{\text{GaAs}}$ is proportional to $P_{\text{circ}}^{\text{GaAs}}$, or the electron spin polarization. This is consistent with the prediction of the photoinduced ISHE. Thus both the light illumination angle and light ellipticity dependence of the electric voltage support that the electric voltage is induced by the ISHE driven by photoinduced spin-polarized carriers.

| $n_0$ | $n_1$ | $n_2$ | $\theta_0$ (deg) | $d_1$ (nm) | $\lambda$ (nm) |
|---|---|---|---|---|---|
| 1.00 | $2.12 - 4.00i$ | $3.79 - 0.157i$ | 65 | 5 | 670 |

Table 2. The parameters used in the calculation. $n_0$, $n_1$, and $n_2$ are the complex refractive indices for air, Pt, and GaAs, respectively (Adachi, 1993; Ordal, 1983). $\theta_0$ is the incident angle of the illumination to the normal axis of the film plane. $d_1$ is the thickness of the Pt layer and $\lambda$ is the wavelength of the light.

| $T^s$ | $\tau^s$ | $T^p$ | $\tau^p$ |
|---|---|---|---|
| 0.249 | $0.168 + 0.0200i$ | 0.715 | $0.287 + 0.00589i$ |

Table 3. The transmittance $T^{s(p)}$ and the transmission coefficient $\tau^{s(p)}$ for the Pt/GaAs hybrid structure.

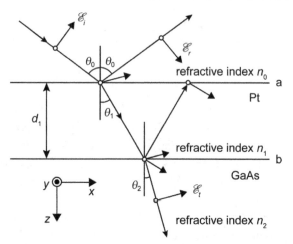

Fig. 7. The definition of $\theta_0$, $\theta_1$, and $\theta_2$.

The photoinduced ISHE allows direct conversion of the circular-polarization information $P_{circ}$ of the illuminated light into an electric voltage. The relation between $V^R - V^L$ and the circular-polarization information $P_{circ}$ of the illuminated light can be argued from the linear dependence of $(V^R - V^L) / I_t^{GaAs}$ on $P_{circ}^{GaAs}$ shown in Fig. 6(e). For simplicity, we assume that the imaginary parts of $n_2$ and $\tau^{s(p)}$ are negligibly small: $n_2 \equiv \Re[n_2]$ and $\tau^{s(p)} \equiv \Re[\tau^{s(p)}]$ [see Tables 2 and 3]. From Eqs. (11), (12), (13), and (14), one obtains

$$I_t^{GaAs} = \left( (\tau^s)^2 \frac{A^2}{1+A^2} + (\tau^p)^2 \frac{1}{1+A^2} \right) \frac{n_2 \cos\theta_2}{n_0 \cos\theta_0} I_i, \qquad (15)$$

$$P_{circ}^{GaAs} = \frac{2\tau^s \tau^p A}{(\tau^p)^2 + (\tau^s)^2 A^2}, \qquad (16)$$

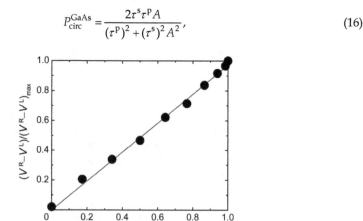

Fig. 8. The degree of circular polarization of the illuminated light ellipticity $A$ of the illuminated light $P_{circ}$ dependence of $V_R - V_L$.

and thus

$$V^R - V^L = \left( Q \frac{\tau^s \tau^P n_2 \cos\theta_2}{n_0 \cos\theta_0} I_i \right) P_{circ}. \tag{17}$$

Here, $Q \equiv (V^R - V^L) / (I_t^{GaAs} P_{circ}^{GaAs})$ is the proportionality constant as seen from Fig. 6(e) and we used $A^{GaAs} \simeq (\tau^s / \tau^P)A$. Since the proportionality constant $Q$ is proportional to $\sin\theta_2$ because of Eq. (9), we obtain

$$V^R - V^L = (Q'\tau^s \tau^P \cos\theta_2 \tan\theta_0 I_i) P_{circ}, \tag{18}$$

where $Q \equiv Q' \sin\theta_2 = Q'(n_0 / n_2)\sin\theta_0$. Equation (18) shows that, in spite of the inequalities of $I_t^{GaAs} \neq I_i$ and $P_{circ}^{GaAs} \neq P_{circ}$ due to the presence of the top Pt layer and oblique illumination, the output signal $V^R - V^L$ is proportional to the degree of circular polarization $P_{circ}$ of the illuminated light outside the sample. This indicates that the photoinduced ISHE can be used as a spin photodetector: the direct conversion of circular polarization information into electric voltage. This function is demonstrated experimentally in Fig. 8, in which $V^R - V^L$ is proportional to the degree of circular polarization of the illuminated light outside the sample.

## 6. Conclusion

The photoinduced inverse spin Hall effect provides a simple way for detecting light circular polarization through a spin current. This phenomenon enables the direct conversion of light-polarization information into electric voltage in a Pt/GaAs junction. This technique will be useful both in spintronics and photonics, promising significant advances in optical technology.

## 7. Acknowledgment

The authors thank to M. Morikawa, T. Trypiniotis, Y. Fujikawa, C. H. W. Barnes, and H. Kurebayashi for valuable discussions.

## 8. References

Adachi, S. (1993). *Properties of Aluminium Gallium Arsenide*, Inspec.
Ando, K., Takahashi, S., Harii, K., Sasage, K., Ieda, J., Maekawa, S. & Saitoh, E. (2008). Electric manipulation of spin relaxation using the spin Hall effect, *Physical Review Letters* Vol. 101: 036601.
Ando, K., Morikawa, M., Trypiniotis, T., Fujikawa, Y., Barnes, C. H. W. & Saitoh, E. (2010). Photoinduced inverse spin Hall effect Conversion of light-polarization information into electric voltage, *Applied Physics Letters* Vol. 96: 082502.
Hilton, D. J. & Tang, C. L. (2002). Optical orientation and femtosecond relaxation of spin-polarized holes in GaAs, *Physical Review Letters* Vol. 89: 146601.

Kimel, A. V., Bentivegna, F., Gridnev, V. N., Pavlov, V. V., Pisarev, R. V. & Rasing, T. (2001). Room-temperature ultrafast carrier and spin dynamics in GaAs probed by the photoinduced magneto-optical Kerr effect, *Physical Review* B Vol. 63:235201.

Kimura, T., Otani, Y., Sato, T., Takahashi, S. & Maekawa, S. (2007). Room-temperature reversible spin Hall effect, *Physical Review Letters* Vol. 98: 156601.

Meier F. & Zakharchenya, B. P. (1984). *Optical orientation*, North-Holland.

Ordal, M. A., Long, L. L., Bell, R. J., Bell, S. E., Bell, R. R., Alexander, J. R. W. & Ward, C. A. (1983). Optical properties of the metals Al, Co, Cu, Au, Fe, Pb, Ni, Pd, Pt, Ag, Ti, and W in the infrared and far infrared, *Applied Optics* Vol. 22: 1099.

Saitoh, E., Ueda, M., Miyajima, H. & Tatara, G. (2006). Conversion of spin current into charge current at room temperature: Inverse spin Hall effect, *Applied Physics Letters* Vol. 88: 182509.

Valenzuela, S. O. & Tinkham, M. (2006). Direct electronic measurement of the spin Hall effect, *Nature* Vol. 442: 176.

# 9

# Remote Optical Diagnostics of Nonstationary Aerosol Media in a Wide Range of Particle Sizes

Olga Kudryashova et al.*
*Institute for Problems of Chemical and Energetic Technologies SB RAS*
*Russia*

## 1. Introduction

Polydisperse gas flows with condensed particles suspended therein are widely spread in nature and play an important role in many branches of the modern engineering and technology. There is a necessity for estimating parameters of disperse media in the manufacturing processes, scientific research, and atmosphere sounding. In industry, the result of this estimation can be employed as the quality measure of a product. When designing and distributing multi-phase systems to study the processes of controlling their parameters, it is also needed to determine values of the quantities characterizing such systems.

Technological processes exploit submicron media alongside with those of large and medium sizes. In such a system the particle sizes vary from tenths of nanometers to tenths of micrometers, and the resultant product may possess significantly different properties, even at small size variations of the condensed phase of a substance used. In this case, it becomes necessary to restore the particle size distribution function taking into account the contribution of all the particles present in the medium. Such functionality may be required from a measuring device when selecting modes of operation or designing sprayer units, controlling the quality of various micro- and nanopowders, monitoring the ecological situation of an area, and studying the dustiness of shop floors. Furthermore, the experimental information on disperse aerosol parameters is needed for evaluating the adequacy degree of a mathematical model accepted to describe an actual process and serves as initial data to calculate working processes in certain devices (Zuyev et al., 1986). The overview of the existing devices to investigate aerosol media has however shown that it cannot be realized using the known instruments.

The available techniques of determining the condensed phase dispersiveness can conditionally be divided into two main groups based on: (i) estimating the particle size in samples selected from a medium under examination and (ii) direct noncontact determination of sizes of the particles present in the medium. The sampling methods have gained a wide distribution, but the essential drawbacks of such methods are both the

* Anatoly Pavlenko, Boris Vorozhtsov, Sergey Titov, Vladimir Arkhipov, Sergey Bondarchuk,
Eugeny Maksimenko, Igor Akhmadeev and Eugeny Muravlev
*Institute for Problems of Chemical and Energetic Technologies SB RAS, Russia*

introduction of disturbances into a medium under examination and the complexity of ensuring the representativeness of samples collected. To the noncontact methods of studying aerosols are optical techniques related (Table 1).

| Method | Problems to solve | Peculiarities |
|---|---|---|
| Spectral transparency | Estimation of the distribution function and concentration. The size range of the measured particles when using the probe radiation of the visible spectrum is 0.01–30 µm. | Requires determining the transmission coefficient within a wavelength interval of $\lambda_{min} \leq \lambda \leq \lambda_{max}$ where about 20–30 measurements need to be conducted at equidistant values of $\lambda$. The technical implementation involves broad-spectrum optical radiation sources and selective detectors. |
| Small-angle scattering | Estimation of the distribution function without preliminary information on physical properties of a substance. When using a laser with a wavelength of 0.63 µm, the size range of the measured particles is 2–100 µm. | The method is very demanding of the accuracy in determining the scattering indicatrix; in this connection, the problem of restoring the particle size distribution function may be solved incorrectly. The technical implementation is relatively simple. |
| Complete indicatrix | Estimation of the distribution function. When a laser with a wavelength of 0.63 µm is used, the size range of the measured particles is 0.2–100 µm. | Requires measurements in the entire range of scattering angles. |
| Lidar | Evaluation of the microstructure of an aerosol. | The back scattering is very small. Requires the application of an high-sensitive radiation detector. |

Table 1. The optical methods to study aerosols

The methods allow the high-speed characterization of a disperse medium directly during the process of its generation or evolution without introducing any changes into an object under study. Various values that are of interest to a researcher under the given process can be estimated from the change in the scattered, attenuated or reflected radiation passed through an aerosol (Arkhipov, 1987).

The main aerosol parameters are the condensed phase dispersiveness and its concentration. In order to measure simultaneously these characteristics in dynamics and consider the evolution process of an aerosol cloud as applied to media of a broad particle size range, it is necessary to employ several different methods combined into a unified measuring system. Following the analysis of the optical methods for measuring the aerosol dispersiveness, a decision was made to utilize the methods of small-angle scattering and spectral transparency. The chosen methods, mutually supplementing each other, permit (i) determining the dispersiveness of a nonsteady heterogeneous system with its high velocity of travel, (ii) conducting measurements at significant background light, (iii) recording particles of diameters between hundreds of nanometers (the spectral transparency method) and several tenths of micrometers (the small-angle method), and (iv) using less detectors with the possibility of their remote location. When classically implemented, these methods however rely on solving the inverse problem that is incorrect and therefore require a modification in a part of both mathematical result treatment and instrumentation, to increase the informativeness of data obtainable from experiments.

## 2. Modified method of spectral transparency

The study into the disperse parameters of aerosol media containing 1–100 µm particles has led to a modified method of spectral transparency to measure the mean size and concentration. The method is based on measuring spectral coefficients of particle cloud-induced attenuation of laser radiation with a limited set of probe radiation wavelengths and on calculating averaged attenuation efficiency factors of radiation $\overline{Q}$ (Vorozhtsov et al. 1997). This method does not make it possible to determine the particle size distribution function $f(D)$, but is suitable to measure mean particle sizes, particularly the mean volumetric-surface diameter $D_{32}$, because it has such merits as simple instrumentation and alignment and can diagnose high-temperature two-phase flows and other aerosol media of a high optical density.

The essence of this method consists in solving the inverse problem for the integral equation:

$$\tau_\lambda = \frac{\pi C_n l}{4} \int_0^\infty D^2 Q(D,\lambda,m) f(D) dD \, , \tag{1}$$

where $\tau_\lambda$ – the optical thickness; $C_n$ – the calculated particle concentration; $\lambda$ – the probe radiation wavelength; $Q$ – the attenuation efficiency factor for single particles; $l$ – the optical length of the probing; $D$ – the particle diameter; $m = n + i\wp$ – the complex refractive index of the particles material, $n$ – the refractive index; $\wp$ – the absorption coefficient; $f(D)$ – the particle size distribution function.

Most of the unimodal particle distributions occurring in the disperse media physics as well as those characteristic of two-phase flows of various substances are described by the gamma distribution that takes the form:

$$f(D) = aD^\alpha \exp(-bD) \, , \tag{2}$$

where $a$, $\alpha$ and $b$ – the distribution parameters.

By introducing the notion of the averaged attenuation efficiency factor $\overline{Q}$ in the form:

$$\overline{Q}(\lambda,m) = \frac{\displaystyle\int_0^\infty Q(\lambda,D,m)D^2 f(D)dD}{\displaystyle\int_0^\infty D^2 f(D)dD} \, , \tag{3}$$

and having replaced the calculated concentration $C_n$ by the mass concentration $C_m$ through

$$C_m = C_n \frac{\pi \rho_p}{6} \int\limits_0^\infty D^3 f(D) dD ,$$ (4)

we derive an expression for the optical thickness:

$$\tau_{\lambda i} = \frac{1.5 C_m l \overline{Q}(\lambda, m)}{\rho_p D_{32}} ,$$ (5)

where $\rho_p$ – the material density of the particles; $D_{32}$ – the mean volumetric-surface particle diameter that is calculated by the formula:

$$D_{32} = \frac{\int\limits_0^\infty D^3 f(D) dD}{\int\limits_0^\infty D^2 f(D) dD} ,$$ (6)

The physical model of the method relies on the interaction between laser radiation and a polydisperse medium through the Mie mechanism and on the conservation of the invariance of the averaged attenuation efficiency factor with respect to a form of the particle size distribution function. The correctness of this assumption is governed by the fact that $\overline{Q}$ is defined by integrals from $f(D)$ and, hence, $\overline{Q}$ is insensitive to the behavior of $f(D)$ in the particle size range under consideration (Prishovalko & Naumenko, 1972).

The averaged attenuation efficiency factor under certain conditions is independent on a form of the particle size distribution function $f(D)$ but is determined by the mean volumetric-surface particle diameter $D_{32}$ at a specified $\lambda$, $\overline{Q}_\lambda = f(D_{32})$, and is the most important feature determining the optical properties of polydisperse two-phase media.

The problem of estimating particle sizes by the present method reduces to measuring the optical density of a disperse medium at the two wavelengths $\lambda_1$ and $\lambda_2$ and to calculating averaged attenuation efficiency factors of laser radiation for the same wavelengths.

The relation of the experimentally measured optical thicknesses at the two wavelengths is equal to the relation of averaged attenuation efficiency factors, both representing the particle size function:

$$\frac{\tau_{\lambda i}}{\tau_{\lambda j}} = \frac{\overline{Q}_2 (D_{32}, m, \lambda_i)}{\overline{Q}_1 (D_{32}, m, \lambda_j)} = F_{ij} (D_{32}),$$ (7)

The measurement range of averaged particle sizes depends on selecting probe radiation wavelengths. Thus, at $\lambda_1 = 0.63$ and $\lambda_2 = 3.39$ µm the $D_{32}$ range is 0.5–4 µm.

The developed method employed the three wavelengths, $\lambda_1 = 0.63$, $\lambda_2 = 1.15$ and $\lambda_3 = 3.39$ $\mu$m, and the relations of the experimentally measured optical thicknesses at the three wavelengths, $\dfrac{\tau_{\lambda_2}}{\tau_{\lambda_1}}$, $\dfrac{\tau_{\lambda_3}}{\tau_{\lambda_1}}$, and $\dfrac{\tau_{\lambda_3}}{\tau_{\lambda_2}}$.

The averaged efficiency factors for each wavelength are calculated from precise formulae of the Mie theory which have the form of infinite weakly converging series obtained from the rigorous solution of the diffraction problem of electromagnetic fields on a sphere. The plots of the averaged efficiency factors $\overline{Q}_i(D_{32}, m, \lambda_i)$ versus the mean volumetric-surface diameter $D_{32}$ are shown in Figure 1.

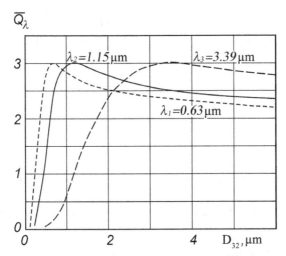

Fig. 1. The averaged attenuation efficiency factors of radiation $\overline{Q}$ plotted against the mean volumetric-surface particle diameter $D_{32}$

The efficiency factors for single particles were calculated according to the Mie theory using a logarithmic derivative of the Riccati-Bessel function (Deyrmendzhan, 1997). The optical constants for oxide particles were taken from Gyvnak and Burch (1965).

Thus, $D_{32}$ can be directly found from the experimentally measured $\tau_{\lambda i}$ and calculated values of $F(D_{32})$. The particle concentration in the measuring zone is determined by the formula:

$$C_m = \frac{\tau_{\lambda i} \rho_p D_{32}}{1.51 \overline{Q}(\lambda_i, D_{32})}, \tag{8}$$

where $i = 1; 2; 3$ on the condition that the optical length of the probing is known or experimentally established, with the values of $\overline{Q}_i$ being determined from the plots (Figure 1).

The practical realization of the multifrequency probing of two-phase media requires solving the problem of selecting radiation wavelengths at which the measurements should be conducted, so that they could be informative with respect to the particle size range (Pavlenko et al., 2005).

The main condition for selecting $\lambda_i$ during the method implementation is the explicit sensitivity of the dependence of $\overline{Q}$ on the mean volumetric-surface particle diameter $D_{32}$, with the maximum of the functions $\overline{Q}(D_{32})$ being well-described by the formula:

$$D_{32}^\circ = \frac{\lambda}{\pi}\left(1 + \frac{4}{n^2 - 1}\right),\tag{9}$$

where $D_{32}^\circ$ – the mean particle size at which the function $\overline{Q}(D_{32})$ has an extremum; $n$ – the aerosol particle refractive index.

Hence, when the variation range of the mean particles size of a medium under study is assigned *a priori*, the radiation wavelengths must be close to the values corresponding to the following expressions:

$$\lambda_1 = \pi D_{32}^{min}\left(1 - \frac{4}{n^2 + 3}\right),\tag{10}$$

$$\lambda_2 = \pi D_{32}^{max}\left(1 - \frac{4}{n^2 + 3}\right),\tag{11}$$

If $\lambda_2$ is lower than that determined by formula (11), the experiment will estimate the mean size of a small fraction of particles of a studied medium; and if $\lambda_1$ is greater that that determined by formula (10), the measured mean size will characterize a large fraction of the desired distribution.

For the selected wavelengths $\lambda_1$ and $\lambda_2$, the dependence $\overline{Q}(D_{32})$ has been calculated and the relation $\overline{Q}_{\lambda_2}/\overline{Q}_{\lambda_1} = F(D_{32})$ determined. Table 2 lists data on the relations of the averaged attenuation efficiency factors of radiation $\overline{Q}_{\lambda_2}/\overline{Q}_{\lambda_1} = F(D_{32})$ for various wavelength pairs, depending on $D_{32}$, from which the limits of measuring the mean particle sizes are established.

In the given case, the maximal measurable particle size $D_{32}$ will be determined by a value at which the function $F(D_{32})$ has a maximum. On the understanding that $\overline{Q}$ is unambiguously dependent on $D_{32}$ only in some interval of particle sizes, generally from zero up to the maximal value (Figure 1), and is an analog of the dependence of the efficiency factor $Q$ on the particle diameter $D$ for an arbitrary particle size distribution, $f(D)$, it follows that particles whose size exceeds $D_{32}^{max}$ out of the whole range of the disperse medium particle sizes are excluded from consideration. Here, we can introduce a notion of the active particle fraction in the distribution $f(D)$ for which there exists a strongly pronounced dependence of $Q$ on $D$ at different $\lambda$.

Thus, the laser measurement of the dispersiveness of two-phase media ascertains the mean size of the active particle fraction starting from zero up to some value that is dependent on the selection of $\lambda_2$ and at which the function $F(D_{32})$ reaches the maximum.

| $D_{32}$, μm | $F_{21}$ | $F_{31}$ | $F_{32}$ | $D_{32}$, μm | $F_{21}$ | $F_{31}$ | $F_{32}$ |
|---|---|---|---|---|---|---|---|
| 1.5 | 1.117 | 0.648 | 0.580 | 5.0 | 1.065 | 1.281 | 1.202 |
| 2.0 | 1.116 | 0.946 | 0.847 | 5.5 | 1.089 | 1.303 | 1.196 |
| 2.5 | 1.100 | 1.139 | 1.031 | 6.0 | 1.063 | 1.254 | 1.180 |
| 3.0 | 1.125 | 1.270 | 1.128 | 6.5 | 1.067 | 1.231 | 1.152 |
| 3.5 | 1.087 | 1.308 | 1.203 | 7.0 | 1.099 | 1.226 | 1.114 |
| 4.0 | 1.064 | 1.296 | 1.218 | 7.5 | 1.056 | 1.222 | 1.157 |
| 4.5 | 1.089 | 1.310 | 1.203 | 8.0 | 1.050 | 1.211 | 1.151 |

Table 2. The averaged attenuation efficiency factors for each wavelength: $\lambda_1 = 0.63$ μm, $\lambda_2 = 1.15$ μm, $\lambda_3 = 3.39$ μm

Based on the calculations of the dependences $\bar{Q}(D_{32})$ and $F(D_{32})$ for the corresponding wavelengths, the measurement ranges of $D_{32}$ ($D_{32}^{min}$, $D_{32}^{max}$) and the active particle fraction range ($D^{min}$, $D^{max}$) of the studied distributions $f(D)$ were determined and are presented in Table 3.

| $\lambda_1$, μm | $\lambda_2$, μm | $D_{32}^{min}$, μm | $D_{32}^{max}$, μm | $D^{min}$, μm | $D^{max}$, μm |
|---|---|---|---|---|---|
| 0.63 | 1.15 | 0 | 1.5 | 0 | 1.5 |
| 0.63 | 3.39 | 0 | 3.5 | 0 | 4.0 |
| 1.15 | 3.39 | 0.1 | 3.6 | 0.1 | 4.0 |
| 0.63 | 5.30 | 0.8 | 6.5 | 0 | 8.0 |

Table 3. Ranges of the determinable active fractions of particles

The theory of the modified method of spectral transparency has been well elaborated so far for disperse media that are characterized by the unimodal particle size distribution. However, as it follows from the literature sources, in some cases like heterogeneous combustion, plasma spraying, local man-made aerodisperse systems, there is a possibility for a bimodal particle size distribution.

In order to expand the modified method of spectral transparency for investigating disperse systems of a bimodal particle size distribution, a mathematical model of calculating characteristics of the laser radiation attenuation by such media has been devised (Potapov & Pavlenko, 2000).

From the physical point of view, bimodal disperse media can correctly be described in the form of the following analytical dependence:

$$f(D) = aD^{\alpha} \exp(-bD^{\beta}) + c \exp[-p(D-q)^2], \tag{12}$$

where $a$, $c$, $b$, $p$, $q$, $\alpha$, $\beta$ - the distribution parameters. Dependence (12) is a sum of the generalized gamma distribution and normal distribution. The position of the first maximum is at point $D_0^{(1)} = \alpha/b$ and that of the second − $D_0^{(2)} = q$.

In the calculation of $\overline{Q}(D_{32})$, a series of $D_{32}$ values and the corresponding distributions of $f(D)$ are preliminary assigned because the analytical dependence of $D_{32}$ on the parameters of distribution (12) cannot be established.

What is important is to assign a form of the bimodal distribution, its determinative characteristics, and the variation mechanism of $f(D)$ when scanning the argument of the function $\overline{Q}(D_{32})$.

The following model of assigning $f(D)$ seems appropriate:

- the form of the bimodal particle size distribution function is determined by such characteristics as:

$$V = D_0^{(2)} - D_0^{(1)}, \ W = f\left[D_0^{(1)}\right] \Big/ f\left[D_0^{(2)}\right]; \tag{13}$$

- the variation of $f(D)$ upon scanning the argument of the function $\overline{Q}(D_{32})$ occurs on the condition that $V$ and $W$ remain constant.

Let the parameters $\alpha$, $\beta$ and $q$ in equation (12) be represented as $\alpha = vk$, $b = 1$, $q = vk/b + V$, where $k = 1, 2, 3... z$. From the condition of normalization and from formula (13) was a system of equations relative to the parameters $a$ and $b$ in formula (12) derived

$$a(k)E^{11} + b(k)E^{12} = E^{13}, \tag{14}$$

$$a(k)E^{21} + b(k)E^{22} = E^{23}, \tag{15}$$

where $E^{23} = 1$; $E^{13} = 0$, a $E^{11}$, $E^{12}$, $E^{21}$ and $E^{22}$ are calculated by the formulae:

$$E^{11} = (vk/b)^{vk} \exp(-vk) - W\left(V + vk/b^{vk} \exp\left[-(bV + vk)\right]\right), \tag{16}$$

$$E^{12} = \exp\left(-pV^2\right) - W, \tag{17}$$

$$E^{21} = \int_0^\infty D^{vk} \exp(-bD)\,dD, \tag{18}$$

$$E^{22} = \int_0^\infty \exp\left[-p(D - (vk/b - V))^2\right] dD, \tag{19}$$

The parameters $a(k)$ and $b(k)$ obtained from solving equations (14) – (15) are used to calculate $f_k(D)$, and $D_{32}(k)$. The found values of $f_k(D)$ are employed in software to compute $\overline{Q}$.

For the bimodality characteristics $V = 1 \div 2$ and $W = 0.25 \div 4$, $\overline{Q}(D_{32})$ was calculated from the suggested model. The character of the dependence $\overline{Q}(D_{32})$ for bimodal and unimodal distributions was shown to be identical. In this case, the maximal value of $\overline{Q}$ is kept at the corresponding value of $D_{32}$ established from the theory:

$$D_{32} = \lambda\left[1 + 4\big/\left(n^2 - 1\right)\right]\big/\pi. \tag{20}$$

This indicates that the model of calculating $\overline{Q}(D_{32})$ for bimodal particle size distributions may be regarded as correct.

When the modified method of spectral transparency was applied, the error in determining the particle sizes was established to grow due to the uncertainty of the particle size distribution class. To enhance the accuracy of the particle size determination, the probing of disperse media requires *a priori* information on a class of particle size distribution, either bimodal or unimodal, and requires the established invariance of $\overline{Q}$ as a function of $f(D)$ within the limits of a certain distribution.

In the diagnostics of bimodal disperse media of a strongly pronounced functionality, when particle fractions are at a great distance from each other, it is possible to implement various mechanisms of the interaction between probing radiation and particles, on the basis of which the particle size of each separate fraction can be estimated. The bimodal distribution can be represented in the form:

$$f(D) = f_1(D) + f_2(D), \tag{21}$$

where $f_1(D)$ and $f_2(D)$ – the first and second functions that describe small and large particle fractions having the modal diameters $D_0^{(1)}$ and $D_0^{(2)}$ and the mean volumetric-surface diameters $D_{32}^{(1)}$ and $D_{32}^{(2)}$, respectively.

From the analysis of the range of the $\overline{Q}$ dependence on $D_{32}$ for various wavelengths, formulae for selecting the probe radiation wavelengths were derived:

$$\lambda_{i1} = D_{32}^{(i)}\big/\pi, \tag{22}$$

$$\lambda_{i2} = \lambda_{i1}\left(5 - D_{32}^{(i)}\big/4\right), \tag{23}$$

where $i = 1, 2$ – the number of particle size distribution functions.

With such a choice of wavelengths, there occurs a fractional interaction between the radiation and the polydisperse particles, and, in contrast to the common techniques, three scattering mechanisms are immediately brought about: Rayleigh scattering, Mie scattering, and scattering by large particles.

The interaction of the radiation of the wavelengths $\lambda = \lambda_{1j}$, where $j = 1, 2$ is the number of the radiation wavelengths needed for probing a separate fraction, with the first $f_1(D)$ and the second fractions $f_2(D)$ results in the Mie scattering and the scattering by large particles, respectively. When the radiation having the wavelengths $\lambda = \lambda_{2j}$ interacts with the first and second fractions, there occurs the Rayleigh scattering and Mie scattering, respectively.

Thus, when the radiation passes at the four wavelengths $\lambda_{ij}$ through a bimodal disperse medium, the optical thicknesses $\tau$ are expressed by the following dependences:

$$\tau_{1j} = \frac{1{,}5l}{\rho_p}\left[ M^{(1)} \frac{Q(\lambda_{1j})}{D_{32}^{(1)}} + M^{(2)} \frac{2}{D_{32}^{(2)}} \right], \tag{24}$$

$$\tau_{2j} = \frac{1{,}5l}{\rho_p}\left[ M^{(1)} \frac{J}{\lambda_{2j}} + M^{(2)} \frac{Q(\lambda_{2j})}{D_{32}^{(2)}} \right], \tag{25}$$

where

$$J = \frac{24\,\pi\,n\,\wp}{(n^2 - \wp^2 + 2)^2 + 4n^2\wp^2}. \tag{26}$$

The mean volumetric-surface diameters $D_{32}^{(i)}$ and particle mass concentrations $M^{(i)}$ for each separate fraction are determined from expressions (24) – (26).

To determine disperse characteristics of bimodal heterogeneous media of a strongly pronounced functionality, Ar, He–Ne and $CO_2$ lasers as well as collimated radiation-based semiconductor lasers can be employed.

To experimentally test the devised method of spectral transparency, the mean particle sizes of the combustion products produced from a model nozzle-free generator were estimated (Arkhipov et al. 2007).

The tests showed that the mean volumetric-surface diameter of the particles is 0.8–2.7 μm. Figure 2 illustrates typical experimental data on the dependences obtained.

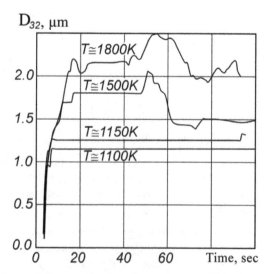

Fig. 2. Typical experimental data on the time dependences of $D_{32}$ for various average temperatures in a combustion chamber

The nonmonotonic behavior of the dependences for higher temperatures is governed by a strong nonstationarity of intraballistic parameters under these conditions.

Data on the particle sizes are qualitatively consistent with the calculation results for particle size distribution spectra at the exit from the nozzle-free fire channel.

## 3. High-selective optical integral method of estimating the particle size distribution function

A new optical technique has been developed for determining the particle size distribution function in aerosol media with a particle size range of 0.1–5 μm and is based on the spectral transparency method.

The classical method of spectral transparency to estimate the particle dispersiveness relates to inverse problems of aerosol optics (Arkhipov, 1987). The stable solution of the problem of determining aerosol parameters by the spectral transparency method is possible if the transparency is ascertained over the whole wavelength range (Shifrin, 1971). However, nowadays devices can measure the spectral transparency coefficient only at several wavelengths or in some region of wavelengths without selectivity, which undoubtedly reduces accuracy and capabilities of the method (Shaikhatarov et al., 1986).

The method of spectral transparency relies on measuring the spectral transmission factor of optical radiation of a two-phase flow in some range of wavelengths. The initial equation for the method of spectral transparency is as follows (Van de Hulst, 1961):

$$I(\lambda,t) = I_0(\lambda)\exp\left[-\frac{\pi C_n(t)l(t)}{4}\int_0^\infty Q\left(\frac{\pi D}{\lambda}, m(\lambda)\right)D^2 f(D)dD\right],\qquad(27)$$

where $I(\lambda,t)$ – the intensity of the radiation passed through the aerosol; $I_0(\lambda)$ – the probe radiation intensity; $C_n(t)$ – the calculated concentration of the aerosol condensed phase; $l(t)$ – the optical path length; $Q\left(\frac{\pi D}{\lambda}, m(\lambda)\right)$ – the attenuation efficiency factor of the probe radiation; $m(\lambda)$ – the complex refractory index of the aerosol condensed phase.

The problem of finding the aerosol particle diameter distribution $f(D)$ from expression (27) by the known radiation intensity values ($I(\lambda)$ and $I_0(\lambda)$) is incorrect. The said incorrectness is unavoidable for the spectral transparency method. It is suggested that the problem be solved using regularizing algorithms.

The regularizing algorithms are built in such a way as to attract additional a priory (with respect to the experiment on the basis of which the inverse problem is set) information on the desired function. With the aid of such information (details on smoothness of the desired solution, its monotoneness, convexity, pertain to the finitely parametric family, etc), such a solution is selected which is close to the true one in some specified sense. The regularizing algorithms allow a stable approximation to the true solution of an incorrect problem, meaning that the approximate solution tends to the true one as the measurement error goes to zero.

One of the regularization ways is the parametrization technique. For the given situation, it will consist in the fact that a form of distribution function (2) is believed to be known *a priori*, in which case the parameters $\alpha$ and $b$ are assigned by the coordinate-wise descent method, and the normalizing coefficient is to be found from the expression:

$$a = \left[ \int_{D_{min}}^{D_{max}} D^{\alpha} \exp(-bD)dD \right]^{-1}, \tag{28}$$

where $D_{min}$ and $D_{max}$ are the minimal and maximal diameters of the particles present in a medium under examination, respectively. Gamma distribution (2) for describing the particle size distribution was chosen in view of its universality as applied to aerosols, with one mechanism of producing a disperse phase (Arkhipov et al., 2006). The attenuation efficiency factor of the probe radiation ($Q\left(\dfrac{\pi D}{\lambda}, m(\lambda)\right)$) is calculated by the formula using specified parameters of the distribution function:

$$Q\left(\frac{\pi D}{\lambda}, m(\lambda)\right) = \frac{2}{\left(\dfrac{\pi D}{\lambda}\right)^{2}} \sum_{n=1}^{\infty} (2n+1)\operatorname{Re}(a_{n} + b_{n}), \tag{29}$$

where $a_{n}$ and $b_{n}$ are the Mie coefficients and are calculated by the formula:

$$a_{n}\left(\frac{\pi D}{\lambda}, m(\lambda)\right) = \frac{\left[ \dfrac{A_{n}\left(\dfrac{\pi D m(\lambda)}{\lambda}\right)}{m(\lambda)} + \dfrac{n}{\left(\dfrac{\pi D}{\lambda}\right)} \right] \operatorname{Re}\left[\xi_{n}\left(\dfrac{\pi D}{\lambda}\right)\right] - \operatorname{Re}\left[\xi_{n-1}\left(\dfrac{\pi D}{\lambda}\right)\right]}{\left[ \dfrac{A_{n}\left(\dfrac{\pi D m(\lambda)}{\lambda}\right)}{m(\lambda)} + \dfrac{n}{\left(\dfrac{\pi D}{\lambda}\right)} \right] \xi_{n}\left(\dfrac{\pi D}{\lambda}\right) - \xi_{n-1}\left(\dfrac{\pi D}{\lambda}\right)}, \tag{30}$$

$$b_{n}\left(\frac{\pi D}{\lambda}, m(\lambda)\right) = \frac{\left[ m(\lambda)A_{n}\left(\dfrac{\pi D m(\lambda)}{\lambda}\right) + \dfrac{n}{\left(\dfrac{\pi D}{\lambda}\right)} \right] \operatorname{Re}\left[\xi_{n}\left(\dfrac{\pi D}{\lambda}\right)\right] - \operatorname{Re}\left[\xi_{n-1}\left(\dfrac{\pi D}{\lambda}\right)\right]}{\left[ m(\lambda)A_{n}\left(\dfrac{\pi D m(\lambda)}{\lambda}\right) + \dfrac{n}{\left(\dfrac{\pi D}{\lambda}\right)} \right] \xi_{n}\left(\dfrac{\pi D}{\lambda}\right) - \xi_{n-1}\left(\dfrac{\pi D}{\lambda}\right)}, \tag{31}$$

where $A_n$ and $\xi_n$ are in recurrence relationships:

$$\xi_n\left(\frac{\pi D}{\lambda}\right) = \frac{2n-1}{\left(\frac{\pi D}{\lambda}\right)}\xi_{n-1}\left(\frac{\pi D}{\lambda}\right) - \xi_{n-2}\left(\frac{\pi D}{\lambda}\right),$$ (32)

$$\xi_0\left(\frac{\pi D}{\lambda}\right) = \sin\left(\frac{\pi D}{\lambda}\right) + i\cos\left(\frac{\pi D}{\lambda}\right),$$ (33)

$$\xi_{-1}\left(\frac{\pi D}{\lambda}\right) = \cos\left(\frac{\pi D}{\lambda}\right) - i\sin\left(\frac{\pi D}{\lambda}\right),$$ (34)

$$A_n\left(\frac{\pi D m(\lambda)}{\lambda}\right) = -\frac{n}{\left(\frac{\pi D m(\lambda)}{\lambda}\right)} + \left[\frac{n}{\left(\frac{\pi D m(\lambda)}{\lambda}\right)} - A_{n-1}\left(\frac{\pi D m(\lambda)}{\lambda}\right)\right]^{-1},$$ (35)

$$A_0\left(\frac{\pi D m(\lambda)}{\lambda}\right) = ctg\left(\frac{\pi D m(\lambda)}{\lambda}\right).$$ (36)

The limit of summation in expression (29) comes when the condition is fulfilled:

$$\left[\sum_{n=1}^{N}(2n+1)\mathrm{Re}(a_n+b_n)\right]\cdot 10^{-8} > (2(N+1)+1)\mathrm{Re}(a_{N+1}+b_{N+1}).$$ (37)

To solve the problem, the spectral transparency coefficient is used as experimental information and is calculated by the formula:

$$\tau_\lambda^e(t) = \ln\frac{I_0(\lambda)}{I(\lambda,t)}.$$ (38)

Afterwards, there is found a relation of the spectral transparency coefficients obtained from experiments for some wavelengths $\lambda_1$ and $\lambda_2$:

$$k_e(t) = \frac{\tau_{\lambda_1}^e(t)}{\tau_{\lambda_2}^e(t)}.$$ (39)

Then, a relation of the theoretically obtained coefficients of spectral transparency for various distribution functions (2) is calculated, in accordance with equation (27) for the wavelengths $\lambda_1$ and $\lambda_2$, by the formula:

$$k_d(t) = \frac{\tau_{\lambda_1}^d(t)}{\tau_{\lambda_2}^d(t)} = \frac{\displaystyle\int_{D_{min}}^{D_{max}} Q\left(\frac{\pi D}{\lambda_1}, m(\lambda)\right)D^2 f(D)dD}{\displaystyle\int_{D_{min}}^{D_{max}} Q\left(\frac{\pi D}{\lambda_2}, m(\lambda)\right)D^2 f(D)dD}.$$ (40)

The cumulative departure of the experimental data $k_e(t)$ from the theoretical $k_d(t)$ is further calculated for all the wavelengths used. In addition, mathematical studies showed that the selection method for pairs of the wavelengths $\lambda_1$ and $\lambda_2$ has no affect on the accuracy and performance stability of the algorithm suggested and, hence, any selection method can be employed. The comparison of the relations of the spectral transparency coefficients but not their absolute values was undertaken to get rid of the constant factor before the integral sign in expression (27) when calculating theoretical values of the transmission factors of a medium under study (40).

The final step is selecting such a form of the particle size distribution function wherein the departure of the experimental data from the theoretical is minimal.

The direct problem of estimating the spectral coefficient of the probe radiation attenuation for various parameters of the distribution function can thus be solved by the numerical methods.

The optical path length ($l$) of the probe radiation (say, with high-speed video shooting (Titov & Muravlev, 2008)) in a studied aerosol is then determined, and the mass concentration of the aerosol dispersed phase is calculated by the formula:

$$C_m(t) = \frac{\tau_\lambda^e(t)\rho_p D_{32}}{1.5l(t)\overline{Q}\left(\frac{\pi D}{\lambda},m(\lambda)\right)},$$ (41)

$$\overline{Q}\left(\frac{\pi D}{\lambda},m(\lambda)\right) = \frac{\displaystyle\int_{D_{min}}^{D_{max}} Q\left(\frac{\pi D}{\lambda},m(\lambda)\right)D^2 f(D)dD}{\displaystyle\int_{D_{min}}^{D_{max}} D^2 f(D)dD}.$$ (42)

To verify the operability of the method devised, an experiment to measure the attenuation of the optical radiation by a suspension of chemically pure submicron $Al_2O_3$ powder in distilled water was performed. Aluminum oxide was chosen because empirical dependences of refractive and absorption indices on probe radiation wavelength for this substance are known (Dombrovskiy, 1982). Neglecting the dependence and using constants as refractive and absorption indices for all the wavelengths was shown by preliminary experiments to result in errors.

The suspended aluminum oxide was exposed to ultrasound in order to grind the resultant agglomerates, following which the whole was placed into a glass cuvette for examination. Prior to measurement, the cuvette was left to stand undisturbed to settle down ungrindable agglomerates. The cuvette that is part of the measuring complex is displayed in Figure 3. The numbers in Figure 3 denote: 1 – glare shield, 2 – collimator outlet, 3 – cuvette, 4 – condenser, 5 – optical waveguide, 6 – direction of optical radiation passage. All the dimensions in Figure 3 are given in millimeters. The optical radiation path length in the medium studied (distance between the inner surfaces of the cuvette glasses) was 5.075 mm.

Before measuring the radiation transmission through the cuvette with suspended aluminum oxide, the radiation transmission through the cuvette filled with pure distilled water was estimated. The transmission spectrum of the cuvette with pure distilled water was used as the reference; it is necessary for all variations of the transmission spectrum of the suspension with respect to the reference spectrum to be caused only by the action of aluminum oxide nanopowder. The reference spectrum and the transmission spectrum of the studied suspension are shown in Figure 4.

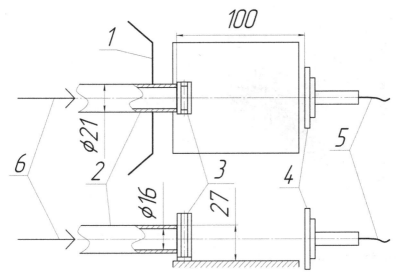

Fig. 3. The cuvette compartment of the measuring complex

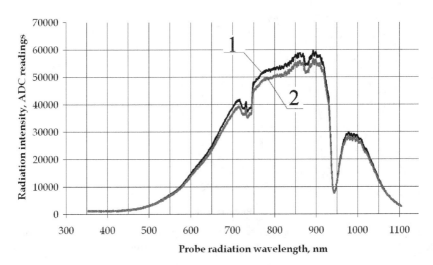

Fig. 4. The optical radiation transmission spectra: 1 – cuvette transmission spectrum with pure distilled water; 2 – cuvette transmission spectrum with Al$_2$O$_3$ suspension

From the spectral data presented was the spectral transparency coefficient calculated by formula (38); the resultant dependence of the spectral transparency coefficient on the wavelength is displayed in Figure 5. The moving average method was applied to the dependence shown in Figure 5 in order to eliminate high-frequency noises, resulting in dependence no.1 in Figure 6. The theoretical dependence of the spectral transparency coefficient on the probe radiation wavelength is no.2 in Figure 6, its difference from the experimental data appearing to be minimal.

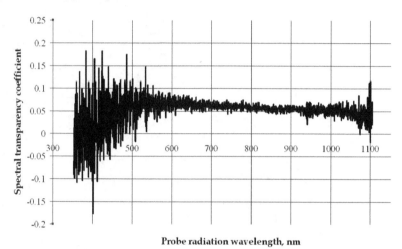

Fig. 5. Experimental spectral transparency coefficient

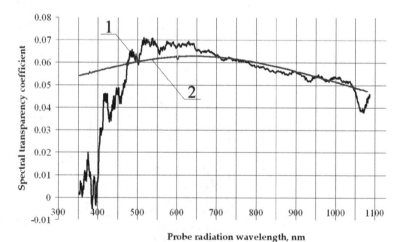

Fig. 6. Experimental and theoretical dependences of the spectral transparency coefficients on the probe radiation wavelength: 1 – experiment; 2 – calculation

The strong difference of the experimental data from the theoretical in the region of wavelengths up to 480 nm is due to the low-level wanted signal therein, as is well illustrated in Figure 4, which nearly coincides with the noise value. This can also be seen from the untreated dependence of the spectral transparency coefficient on the wavelength in Figure 5. This discrepancy can therefore be considered acceptable since instrumental implementation peculiarities are concerned regarding the measurement of optical radiation attenuation, and it does not characterize the algorithm applied.

The found theoretical dependence of the spectral transparency coefficient on the wavelength (curve 2, Figure 6) is consistent with the particle size distribution function shown in Figure 7.

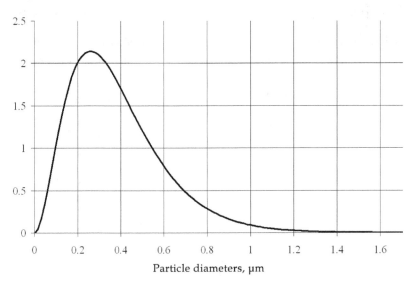

Particle diameters, μm

Fig. 7. The restored $Al_2O_3$ particle size distribution function

The particle size distribution function obtained (Figure 7) is in good agreement with the known data on the aluminum oxide nanopowder studied.

## 4. Modified method of small-angle scattering

To investigate the aerosol genesis from 2 to 100 μm, a method based on estimating disperse parameters of aerosols from the measured scattered radiation passed through a studied volume has been devised.

The method consists in finding the aerosol particle size range from the measured scattering indicatrix, by means of searching the corresponding parameters of the distribution function. Gamma distribution (2) was accepted as the reference function (Arkhipov, 1987).

Laser beam LS (Figure 8) propagates through a scattering layer with boundaries "1" and "2" to produce some illuminance on plane Y. The distance between the laser and the first boundary of the layer is $l_1$. The aerosol cloud particles present in the beam scatter radiation. As a consequence, the irradiance of plane Y is determined not only by the direct beam attenuated due to absorption and scattering, but also by the radiation scattered by the particles.

On the assumption of the uniform distributions of concentration and particles sizes in the aerosol cloud, the equation for scattered radiation flux coming on plane Y has the following form:

$$I(y) = \frac{\pi S}{4} \int_0^z \left[ I_0(x) B(x,y) F(x) \right] dx .$$ (43)

where $F(x) = \int_0^\infty Q_S(D,\theta(x)) D^2 f(D) dD$; $I_0(x) = I_0 \exp[-x C_n Q_{ext}]$ – the intensity of the radiation falling on point $x$; $I_0$ – the initial radiation intensity; $x$ – the distance between boundary "1" of the scattering layer and point $P$; $Q_{ext}$ – the attenuation coefficient.

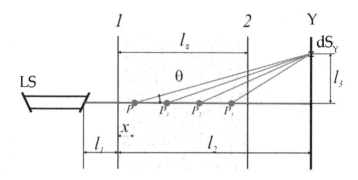

Fig. 8. Diagram of interaction between radiation and aerosol

The radiation scattered from a single particle in the small-angle region under the assumption that the particles are spherical is defined as an analytical dependence in the following form:

$$Q_S(\rho,\theta) = \frac{\rho^2}{4\pi} \cdot \left[ \frac{2 J_1(\theta \rho)}{\theta \rho} \right]^2 ,$$ (44)

where $\rho = \frac{\pi D}{\lambda}$ – the diffraction parameter (Mie parameter); $\theta$ – the radiation scattering angle; $J_1(\theta \rho)$ – the first-order Bessel function of the first kind.

The multiplier $B(x)$ that takes into account the scattered radiation attenuation pursuant to the Buger law is defined by the relation:

$$B(x,y) = \exp\left[ -Q_{ext} \frac{z-x}{\cos\theta(x,y)} \right] ,$$ (45)

where $\theta(x,y) = \operatorname{arctg}(y / (l_2 - x))$; $y = l_3$.

The estimation of $f(D)$ from the measured scattering indicatrix $I_e(y)$ reduces to searching the parameters $\{a,b\}$ of distribution (2) and calculating the functional:

$$\Omega = \min_{a,b}\left\{\sum_{i=1}^{n}\left|I_e(y_i) - I(y_i)\right|\right\}, \tag{46}$$

where $I_e(y_i)$ ($i = 1,2,...,n$) – the measured values of the scattering indicatrix for discrete values of the detector positions; $I(y_i)$ – the values calculated from (43).

It is possible to restore the droplet size distribution function with a sufficient accuracy if the condition is fulfilled (Gritsenko & Petrov, 1979; Belov et al., 1984):

$$\ln\frac{I_0}{I} \leq 1.5, \tag{47}$$

where $I_0$, $I$ – the illuminance in the central beam before and after its passing through the scattering volume.

To implement the title method, a laser setup has been assembled which comprises (Figure 9): a 1 m³ measuring chamber; a radiator – HeNe laser with 0.632 μm wavelength and 5 mW power; a recording unit consisted of 8 photodiodes placed on the same array; a measuring 8-channel amplifier; ADC and PC; software to record and process measurement information in order to determine calculated and mass functions of particle size distribution, mean volumetric-surface diameter, and aerosol particles concentration.

The laser radiation is 90° angle oriented toward one of the volume faces, 80 Hz frequency modulated, and directed through a scattering medium.

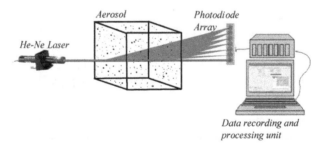

Fig. 9. The laser setup diagram

The optical radiation flux scattered at different angles is recorded by the photodiode array that is located in the plane perpendecular to the laser beam. The photodiode array enables recording the scattered radiation at angles of 0.3−20° relative to the laser beam.

The optical path length is estimated with the aid of a video camera or is set to a fixed value equal to a measuring volume for a steady-state generation process of a two-phase flow.

The aluminum powder measurement results obtained using the setup are collected in Table 4; Figure 10 shows mass functions of the particle size distribution.

| Measurement | Distribution coefficient $\alpha$ | Distribution coefficient $b$ | $D_{32}$, µm | $D_{43}$, µm |
|---|---|---|---|---|
| 1 | 1.14 | 0.7 | 5.91 | 7.3 |
| 2 | 0.3533 | 0.55 | 6.09 | 7.91 |
| 3 | 0.53 | 0.53 | 6.66 | 8.54 |
| 4 | 1.24 | 0.64 | 6.6 | 8.2 |
| 5 | 1.1 | 0.7 | 5.8 | 7.2 |

Table 4. The laser method study results for the fine aluminum powder

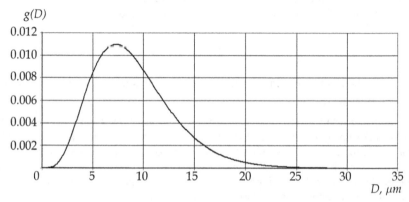

Fig. 10. The mass particle size distribution for the fine aluminum powder

The calculated distribution function can readily be converted into the mass distribution function by the formula (Arkhipov et al., 2006):

$$g(D) = \frac{m}{m_0} f(D), \tag{48}$$

where $m$ – the weight of the particles of size $D$; $m_{10} = \int_0^\infty mf(D)dD$.

The spread of the aluminum powder dispersiveness estimation results obtained using different methods is not greater than 15%, as demonstrated in Table 5.

An aerosol water cloud produced by pulsed generation was also studied. The results for water and an aqueous glycerol solution are given in Figures 11–13.

| | Fine aluminum powder | |
|---|---|---|
| | $D_{32}$, µm | $D_{43}$, µm |
| Sieve analysis | 7.2 | 8.3 |
| Microscope | 5.7 | 6.44 |
| Modified method of small-angle scattering | 6.2 | 7.45 |

Table 5. Comparison of the results of different methods

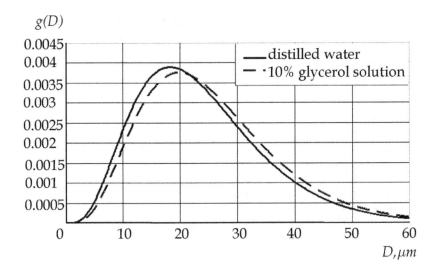

Fig. 11. Mass droplet distribution after spraying (time, 1 s)

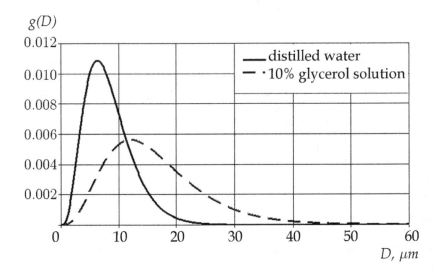

Fig. 12. Mass droplet distribution after spraying (time, 6 s)

The modified small-angle scattering method developed makes it possible to restore the particle size distribution function even from three measured points of the scattering indicatrix.

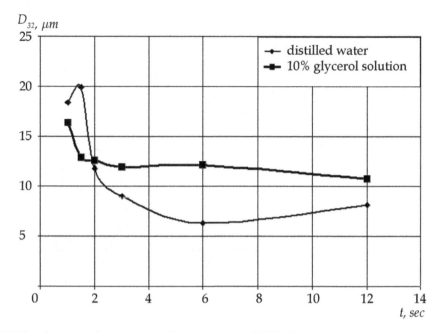

Fig. 13. The change in the mean size $D_{32}$ of water and 10% glycerol solution droplets as a function of the aerosol cloud generation time

## 5. Conclusion

The optical diagnostics methods for heterogeneous flows have been reviewed and analyzed as the most promising with respect to measuring the dispersiveness and concentration of aerosol particles in dynamics, including the cases of high-velocity flows, significant background illuminance, and similar phenomena accompanying technological processes and affecting measuring instrumentation.

The authors have suggested and implemented a modified method of spectral transparency for estimating the aerosol dispersiveness, which is distinct both in simplicity of instrumental implementation and in possibility of diagnosing two-phase flows of a high optical density.

A new method of determining the particle size distribution function has been developed and is based on the classical method of spectral transparency using information on the probe radiation attenuation in a wide range of wavelengths; the operability of the method has been verified.

The inverse problem of aerosol optics has mathematically been solved for the small-angle scattering method in a part of results processing in order to eliminate incorrect solutions when restoring the particles size distribution function; the possibility of studying the dynamics of generation and propagation of aerosols using the algorithm suggested has been demonstrated.

The developed measuring complex employs data recording and processing algorithms obtained for the first time and combines the modified method of spectral transparency, the high-selective optical integral method of estimating the particle size distribution function, and the modified method of small-angle scattering. The combined application of the methods devised makes it possible to estimate disperse parameters of aerosol media of any nature and of a wide particle size range (0.1–100 μm) under conditions of their high-velocity genesis with a high time resolution. The measuring complex permits studying the genesis of two-phase flows and evaluating the effects on the fractional composition by such processes as particle coagulation, sedimentation, and evaporation.

# 6. References

Arkhipov, V. (1987). Laser diagnostics methods of heterogeneous flows (in Russian), Tomsk University Press, Tomsk

Arkhipov, V.; Akhmadeev, I.; Bondarchuk, S.; Vorozhtsov, B.; Pavlenko, A. & Potapov, M. (2007). A modified method of spectral transparency to measure aerosol dispersiveness (in Russian), *Atmospheric and Oceanic Optics*, Vol. 20, No. 1, pp. 48–52, ISSN 0869-5695

Arkhipov, V.; Bondarchuk, S.; Korotkikh, A. & Lerner, M. (2006). Production technology and disperse characteristics of aluminum nanopowders (in Russian). *Gornyi Zhurnal. Special Issue. "Nonferrous Metals"*, No.4, pp. 58–64, ISSN 0372-2929

Belov, V.; Borovoi, A. & Volkov, S. (1984). On the small-angle method under single and multiple scattering (in Russian). Izvestia AN SSSR Fizika Atmosfery i Okeana, Vol.20, No.3, pp.323–327

Deyrmendzhan, D. (1997). Electromagnetic radiation scattering by spherical polydisperse particles (in Russian), Mir, Moscow

Dombrovskiy, L. (1982). On a possibility of determining the dispersed composition of a two-phase flow by small-angle light scattering (in Russian), *Teplophysika Vysokikh Temperatur*, No.3, pp.549–557, ISSN 0040-3644

Gritsenko, A. & Petrov, G. (1979). On the role of the multiple scattering in the inverse optics problems of coarse-dispersed aerosols (in Russian). *Optics and Spectroscopy*, Vol.46, No.2, pp.346–349, ISSN 0030-4034

Gyvnak, D. & Burch, D. (1965). Optical and infrared properties of Al2O3 at elevated temperatures. *Journal of Optical Society of America*, Vol. 55, No.6, pp. 625-629

Pavlenko, A.; Arkhipov, V.; Vorozhtsov, B.; Ahmadeev, I. & Potapov, M. (2005). Informative radiation wavelengths under laser diagnostics of aerosol media (in Russian), *XII Joint International Symposium "Atmospheric and Oceanic Optics. Atmospheric physics"*, Tomsk: IAO SB RAS, pp. 148

Potapov, M. & Pavlenko, A. (2000). A mathematical model of calculating characteristics of the laser radiation attenuation for bimodal media (in Russian), *Proceedings of the 1st All-Russian conference "Measurements, automation and simulation in industry and scientific research"*, Biysk, pp. 135–139, ISSN 2223-2656

Prishivalko, A. & Naumenko, E. (1972). The light scattering by spherical particles and polydisperse media (in Russian), IF AN BSSR, Minsk

Shaikhatarov, K.; Lapshin, A.; Stolyarov, A. & lapshina, T. (1986). Disperse media photometer, Pat. SU 1435955 A1 G 01J 1/44

Shifrin, K. (1971). The study of substance properties from single scattering (in Russian), In: Theoretical and Applied Problems of Light Scattering, B.I. Stepanov, A.P. Ivanov, (Ed.), pp. 228−244, Nauka I Tekhnika, Minsk

Titov, S. & Muravlev, E. (2008). The use of a Videosprint/C/G4 digital camera in the studies of dynamic processes (in Russian), *All-Russian Conference "Perspectives of Designing and Applying High-energy Condensed Materials"*, pp. 173−179, ISBN 978-5-9257-0134-8, Biysk, Russia, 25-26 Sep 2008

Van de Hulst, G. (1961). Light scattering by fine particles, Moscow

Vorozhtsov, B.; Potapov, M.; Pavlenko, A.; Lushev, V.; Galenko, Yu. & Khrustalev, Yu. (1997). A multifrequency laser measuring complex of monitoring atmospheric and industrial aerosols (in Russian), *Atmospheric and Oceanic Optics*, Vol. 10, No. 7, pp. 928-832, ISSN 0869−5695

Zuyev, E.; Kaul, B.; Samokhvalov, I.; Kirkov, K. & Tsanev, V. (1986). *The laser probing of industrial aerosols* (in Russian), Nauka, Novosibirsk

# 10

# Shape of the Coherent Population Trapping Resonances Registered in Fluorescence

Sanka Gateva and Georgi Todorov

*Institute of Electronics, Bulgarian Academy of Sciences*

*Bulgaria*

## 1. Introduction

There has been permanent interest in investigations of new magnetic sensors and their various applications (Edelstein, 2007). Last years there is a rapid progress in development of magneto-optical sensors because of their sensitivity and potential for miniaturization. Magnetometers, based on magneto-optical sensors have high sensitivity - comparable to, or even surpassing this of the SQUIDs (Superconducting Quantum Interference Devices) (Kominis et al., 2003; Dang et al., 2010; Savukov, 2010; Knappe, 2010). Microfabrication of components using the techniques of Micro-Electro-Mechanical Systems (MEMS) developed for atomic clocks (Knappe, 2004) gives the opportunity for building small, low consuming, low cost and non-cryogenic (as SQUIDs) sensors (Griffith et al., 2010). Coherent optical effects can be applied for magnetic field detection and offer perspectives for development of high-precision optical magnetometers (Cox et al., 2011; Kitching et al., 2011). These magnetometers are appropriate for geomagnetic, space, nuclear and biological magnetic field measurements (cardio and brain magnetic field imaging), environmental monitoring, magnetic microscopy, investigations of fundamental physics, etc. Coherent magneto-optical resonances have many applications not only in magnetometry, but in high-resolution spectroscopy, lasing without inversion, laser cooling, ultraslow group velocity propagation of light, etc. (Gao, 2009).

Magneto-optical resonances can be prepared and registered in different ways (Budker&Romalis, 2007 and references therein). Most frequently Coherent–Population-Trapping (CPT) is observed when two hyperfine levels of the ground state of alkali atoms are coupled by two laser fields to a common excited level. When the frequency difference between the laser fields equals the frequency difference between the two ground states, the atoms are prepared in a non-absorbing state, which can be registered as a fluorescence quenching and transparency enhancement in spectral interval narrower than the natural width of the observed optical transition (Arimondo, 1996).

In degenerate two-level systems coherent states can be created by means of Hanle effect configuration (Alzetta et al., 1976). In this case the coherent non-absorbing state is prepared on two Zeeman sublevels of one hyperfine level by monochromatic laser field (the so called single frequency CPT). Hanle configuration is important for performing significantly simplified experiments and to build practical devices as well.

For a lot of applications in medicine, quantum information experiments, materials characterization, read out of stored memory domains etc., magnetic field imaging is essential (Bison et al., 2009; Johnson et al., 2010; Kimble, 2008; Lukin, 2003; Mikhailov et al., 2009; Romalis, 2011; Xia et al., 2006). The most common techniques for monitoring the flow phenomena involve fluorescence detection (Xu et al., 2008) because it is sensitive, non-invasive, and allows in-line analysis of multiple step processes.

For all applications, where narrow signals and high signal-to-noise ratios are important, ensuring reliable operation requires good knowledge of the resonance shape and the internal and external factors influencing it.

In this chapter, the influence of different factors on the shape of the CPT resonances obtained by means of the Hanle effect configuration and registered in fluorescence is analysed in uncoated, room temperature vacuum cells from point of view of cell diagnostics and building high-sensitive magneto-optical sensors.

## 2. Shape of the single frequency CPT resonances

### 2.1 Experimental shapes

### 2.1.1 Experimental set-up

The experimental set-up geometry is shown in Fig. 1. The resonances were measured in uncoated vacuum cells containing a natural mixture of Rb isotopes at room temperature (22°C). A single-frequency linearly polarized in z direction diode laser beam (2 mm in diameter, 22 mW in power) was propagating along the cell's axis x. Its frequency and emission spectrum were controlled by observing fluorescence from a second Rb vapor cell and a Fabry-Perot spectrum analyzer. A magnetic field $B_{scan}$, created by a solenoid, was applied collinearly to the laser beam. As the shape of the CPT resonance is very sensitive to stray magnetic fields (Huss et al., 2006), the gas cell and the solenoid were placed in a 3 layer μ-metal magnetic shield. The fluorescence was detected perpendicular to the laser beam direction and to the light polarization vector (in y direction) by a photodiode. In all experiments it was placed close to the front window of the cell because the CPT resonances registered in fluorescence are sensitive to the position of the photodiode along the cell (Godone et al., 2002). The signals from the photodiode were amplified and stored in a PC, which also controlled the magnetic field scan. The main sources of errors were the accuracy of the laser power measurements and the poor signal-to-noise ratio at low laser powers.

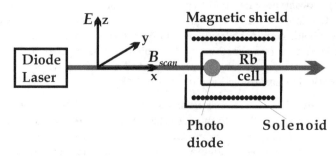

Fig. 1. Experimental setup geometry.

The resonances were examined in 4 Rb vacuum cells with internal dimensions: cell **A** (length $l_A$=4.5 cm, diameter $d_A$=2.9 cm); cell **B** (length $l_B$=2.0 cm, diameter $d_B$=2.0 cm); cell **C** (length $l_C$=4.4 cm, diameter $d_C$=1.0 cm) and cell D (length $l_D$=4.3 cm, diameter $d_D$=2.1 cm).

All investigations were performed on the degenerate two-level system of the ($F_g$=2 – $F_e$=1) transition of the $^{87}$Rb $D_1$ line because the $D_1$ line consists of hyperfine transitions at which only dark resonances can be observed, the fluorescence of the $F_g$=2 – $F_e$=1 transition is practically not overlapping with another and it has the highest contrast (55%) (Dancheva et al., 2000).

As the spectra in the Hanle effect configuration were measured by tuning the magnetic field, all experimental spectra are plotted as function of the magnetic field. The conversion of magnetic field to frequency is straightforward on the basis of the Zeeman splitting between adjacent magnetic sublevels of the ground level ~0.7 MHz/G (Steck, 2009).

### 2.1.2 Experimental shapes

The measurement of the CPT resonances in four uncoated Rb vacuum cells with different dimensions (the dimensions of the cells **A**, **B**, **C** and **D** are given in Section 2.1.1) has shown that they are different not only in width but also in shape. In Fig. 2 the normalized CPT signals measured in cell **A** and cell **D** are compared in 1.25 G and 40 mG tuning ranges. In cell **A** the resonance width (FWHM) is about 50 mG, while in cell **D** it is 250 mG (Fig. 2a). In cell **D** the resonance has triangular shape (Fig. 2b scattered circles curve), while in cell **A** the resonance has a narrow Lorentzian structure of the order of few mG centred at zero magnetic field superimposed on a broad pedestal of the order of a few hundred mG (Fig. 2 dashed line curve).

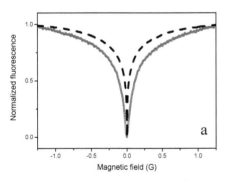 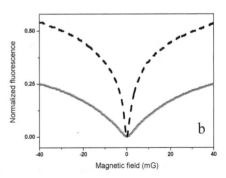

Fig. 2. Normalized CPT resonance shapes in cell **A** (dash line curve) and cell **D** (scattered circles curve) in 1.25 G (a) and 40 mG (b) tuning range. Power density 75 mW/cm².

In order to obtain information about the processes leading to the narrow structure formation, measurement of the resonance shapes in all cells in dependence on the laser beam diameter, laser power density, additional constant magnetic fields, and the laser frequency position relative to the centre of the Doppler broadened line profile were performed (Alipieva et al., 2003). The analysis of the shapes, measured at the same geometry of excitation and registration, has shown that the origin of the narrow structure is connected with processes in the vacuum cell. The width of the resonances is different in different cells and it is about 2 orders smaller than the transit-time broadening, which is of the order of 100 kHz (Thomas & Quivers, 1980). The width of the narrow Lorentzian structure of the CPT resonance is in agreement with the assumption that the broadening is mainly affected by the relaxation processes due to atomic collisions with the walls of the cell.

In Fig. 3 the measured narrow resonance widths of the four cells are plotted as function of the mean distance between two collisions with the cell walls L. The red line is the calculated resonance width (FWHM) $\Delta_L = 1/2\pi\tau$ determined by the mean time between two collisions with the cell walls $\tau=L/v= 4V/Sv$ (Corney, 1977) where v is the mean thermal velocity, V is the volume, S is the surface area of the cell and L is the mean distance between 2 collisions with the cell walls. The time of atom-wall interaction during the collision is small enough (Bouchiat & Brossel, 1966) and is neglected. The measured values of the resonance width are in good agreement with the evaluated for three of the cells (cell **A**, **B** and **C**). Only in cell **D** there is no narrow structure and in Fig. 3 this is designated by a dashed line at the corresponding to cell **D** mean distance between collisions with the cell walls. The different dimensions of the cell can not be the reason for the existence of the narrow resonance. The estimated mean path between two collisions with the cell walls L is 2.1 cm in cell **A** and 1.6 cm in cell **D**. As the measurements are made at the same temperature, the mean velocity of the atoms is the same. Then, the evaluated difference between the widths of the eventual narrow resonances is of the order of 25-30% and is detectable.

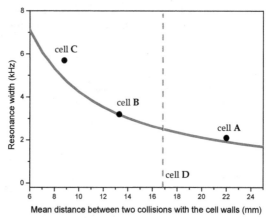

Fig. 3. FWHM of the narrow resonances (black circles) as function of the dimensions of the cell (mean distance between two collisions with the cell walls) and the calculated dependence of the width of the resonance, result of relaxation on the cell walls (red line). Laser beam diameter 2 mm, $P_{las}= 75$ mW/cm².

In Fig. 4 the power dependences of the amplitudes and widths of the narrow and the wide structures of the CPT resonances measured in cell **A** are shown. The amplitudes of both signals increase with power. The amplitude of the wide structure increases fast with power and at intensities larger than 150 mW/cm² saturates. The amplitude of the narrow structure is lower and does not saturate up to a power density of 1 W/cm².

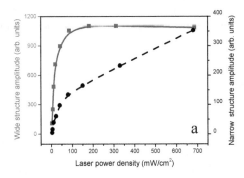 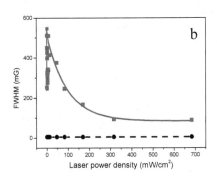

Fig. 4. Amplitudes (a) and widths (b) of the narrow (black circles) and wide (blue squares) structure of the CPT resonance in dependence on the laser power density. (The lines are only to guide the eyes.)

The width of the narrow structure does not depend on the laser power, while the wide structure width does - at low powers it increases, and at laser power densities higher than 1 mW/cm² , there is no power broadening, but a resonance-line narrowing with the power - an exponential decrease to 80 mG. A possible explanation of this narrowing in our experiment is the high pumping rate. The CPT signal width is proportional to the pumping rate $\Omega^2/\gamma^*$ ($\gamma^*$ is the total decay rate of the excited state defined not only by the spontaneous decay but by all relaxation processes) (Levi et al., 2000). When the laser power density is increased, the saturation of the absorption cannot be neglected, the decay rate of the excited state $\gamma^*$ increases due to the stimulated emission and a resonance width narrowing is observed.

The comparison of the narrow structure registered in fluorescence and transmission (Gateva et al., 2005) shows that the narrow structure width is the same in fluorescence and transmission. In fluorescence, the narrow resonances can be observed in the whole range of laser powers, while in transmission, in a very small range of powers because of the absorption saturation.

The measured dual structure resonance shapes and the power dependences of the two components make reasonable the assumption that the signal in these cells is formed by different subensembles of atoms interacting with different light fields – the laser beam and a weak light field, for example, of scattered light.

This assumption is confirmed by the series of experiments with cell **A** with an expanded beam filling the whole volume of cell **A**. The dependence of the shape of the CPT resonances on the laser beam diameter is given in Fig. 5a. (The diameter of the laser beam with laser power density 0.3 mW/cm² was changed by different diaphragms in front of the cell.) The resonances broaden with decreasing the diameter, but there is a narrow range (about 3 mG in width) around the zero magnetic field, where their shapes are Lorentzian and equal in width. In the limits of the accuracy of the measurement this width is equal to the narrow component width and to the non power broadened CPT resonance of the expanded to 30 mm laser beam (Fig. 5b). In Fig. 5b the power dependence of the resonance width with

30 mm laser beam is shown. The initial intensity of the laser beam was reduced by a neutral filter from maximum to 1/128. In this case, the resonance width is determined by the relaxation processes due to atomic collisions with the walls of the cell and the power broadening.

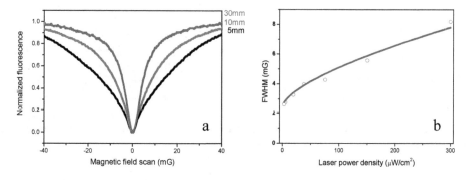

Fig. 5. CPT resonances with expanded laser beam:
(a) shape of the resonances at different diameters of the laser beam and reduced laser power density 0.3 mW/cm²;
(b) power dependence of the resonance width with 30 mm laser beam diameter.

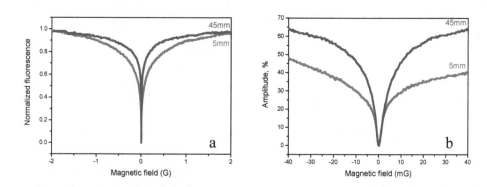

Fig. 6. CPT resonances registered at 5 mm and 45 mm along the cell in two different tuning ranges: a) 2 G and b) 40 mG. Beam diameter 2 mm.

In Fig. 6 the resonance shapes registered with laser beam diameter 2 mm at 5 mm and 45 mm distance from the input window are given. The comparison of the shapes of the resonances in a 2 G tuning range (Fig.6a) shows, that the resonance at the end of the cell is narrower. This result coincides with the theoretical and experimental results of Godone et al. (Godone et al., 2002) and is due to the lower laser power at the end of the cell and the smaller power broadening.

The comparison of the signals in a 10 mG range shows that there is a narrow range (about 3 mG in width) around the zero magnetic field, where their shapes coincide.

## 2.2 Theoretical description

The shape and width of the CPT resonances and their dependence on the input power has been studied experimentally and theoretically in many papers (Auzinsh, 2009a,b and references therein). Most of the experimental investigations were performed at low laser power, the shape of the resonances was Lorentzian and the dependence of the resonance width $\Gamma$ on the laser power density was linear up to a few $mW/cm^2$. The slope of this dependence was smaller than the predicted by the theory (Arimondo, 1996):

$$\Gamma = \gamma_\varphi + \Omega^2/\gamma_f \tag{1}$$

where $\gamma_\varphi$ is the ground state coherence relaxation rate; $\Omega$, the Rabi frequency; $\gamma_f$, the population decay rate from the excited state into the ground states.

There are different decoherence processes influencing the resonance shape – transit time broadening, population exchange, atom-atom and atom-wall collisions, transverse laser intensity distribution etc. The investigations of the resonance shapes at different conditions have shown that not always the power dependence of the resonance width is linear (Figueroa et al., 2006; Javan et al., 2002; Ye&Zibrov, 2002).

An analytical expression for the FWHM of the EIT resonances $\Gamma_{EIT}$ in the case of a Doppler broadened medium in the linear with respect to the probe field approximation, when the population exchange between the ground states is the main source of decoherence, was obtained by Javan et al. (Javan et al., 2002):

$$\Gamma_{EIT}^2 = \frac{\gamma_\varphi}{\gamma_f}\Omega^2(1+x)\left\{1+\left[1+\frac{4x}{(1+x)^2}\right]^{1/2}\right\} \tag{2}$$

under the condition $\left(\gamma_f / w_D\right)^2 <<x<< \gamma_f / \gamma_\varphi$. The parameter $x = \Omega^2 / \Omega_{in\,hom}^2 = \Omega^2\gamma_f / 2\gamma_\varphi w_D^2$ represents the degree of optical pumping within the inhomogeneous linewidth $w_D$. The dependence of $\Gamma_{EIT}$ on the laser power density is not linear at low intensities -when $x<<1$, $\Gamma_{EIT} \approx \sqrt{\frac{2\gamma_\varphi}{\gamma_f}}.\Omega$, and linear at high intensities - when $x>>1$, $\Gamma_{EIT} \approx \frac{\Omega^2}{w_D}$.

The investigation of the influence of the light beam transverse intensity distribution on the line shape has shown that due to the Gaussian shape of the beam, at high pumping rates the shape of the resonances is no more Lorentzian (Pfleghaar et al., 1993). The obtained by Levi et al. (Levi et al., 2000) analytical lineshape is:

$$\Pi(\delta) = \frac{1}{2}\ln\frac{\Omega^2 / \gamma_f\gamma_\varphi}{1+(\delta / \gamma_\varphi)^2} \tag{3}$$

where $\delta$ is the two-photon detuning, while by Taichenachev et al. (Taichenachev et al., 2004) it is:

$$R_G = \pi r_0^2 (S_0 - \delta \arctan[S_0\delta / (1+S_0 +\delta^2)]+\frac{1}{2}\ln\{(1+\delta^2) / [(1+S_0)^2 +\delta^2]\}) \tag{4}$$

where $S_0 = \dfrac{\Omega^2}{\gamma_f \cdot \gamma_\varphi}$ .

At high power densities ($S_0 \gg 1$) it is

$$R_G \propto \{1 - (\delta/S_0)\,\arctan(S_0/\delta)\} \tag{5}$$

These equations describe well the standard (bichromatical) CPT resonance registered in absorption.

To understand the reason for the observed peculiarities in the shapes and widths of the Hanle - CPT resonances in different cells we have performed a numerical modelling of the influence of the experimental conditions. The theoretical description is based on the standard semiclassical approach: the atomic system is described by the statistical operator $\hat{\rho}$ in density matrix representation (Landau & Lifshitz, 1965) and laser and magnetic fields are considered classically.

The Hamiltonian of the system $\hat{H}$ is sum of the operator of the free atom $\hat{H}_0$, the operator of magneto-dipole interaction $\hat{H}'$ and the operator of interaction with the laser radiation $\hat{V}$. Terms describing the atomic relaxation and excitation transfer are added to the Liuville/Neuman equation:

$$\dot{\hat{\rho}} = -\frac{i}{h}\left[\hat{H},\hat{\rho}\right] + \left(\dot{\hat{\rho}}\right)_{relax} + \left(\dot{\hat{\rho}}\right)_{tr} + \hat{N} \tag{6}$$

where $\hat{H} = \hat{H}_0 + \hat{H}' + \hat{V}$ .

At weak magnetic fields the magneto-dipole interaction is smaller than the hyperfine splitting and the hyperfine splitting can be included in the main Hamiltonian $\hat{H}_0$. In this case the magneto-dipole operator can be written as: $\hat{H}' = \mu_B g_{F_s}\left[\vec{F}_s, \vec{H}\right]$, where $\mu_B$ is the Bohr magneton, $g_{F_s}$ is the Lande factor, $F_s$ is the total moment of the atom in the $s$ state, ($s = f, \varphi$ ; $f$ - upper, $\varphi$ - lower state), and $\vec{H}$ is the vector of the magnetic field. The complete magnetic field includes the scanning magnetic field and a residual magnetic field $H_{str}$ .

The operator of interaction with the laser radiation $\hat{V}$ is described with the scalar product of the vector operator of the dipole moment $\vec{d}$ and the electric vector $\vec{E}$ of the resonant to the atomic transition $\varphi \to f$ light:

$$\hat{V} = -\left(\vec{d}.\vec{E}\right), \qquad \vec{E} = E\vec{e}_Q \exp\{-i(\omega_{las}t - kz)\} + c.c. \tag{7}$$

where: $\vec{e}_Q$ ($Q = 0, \pm1$) are the circular components of the laser field, $\omega_{las}$ is the frequency of the laser light and $\vec{k}$ is the wave vector.

Applying the irreducible tensor operator (ITO) formalism, the density matrix $\rho_{nn'}$ in (n,n') representation can be decomposed into polarization moment (PM) representation with tensor components $\rho_q^k$ (Dyakonov & Perel, 1966)

$$\rho_q^k = (2F_{\rho'} + 1)^{1/2} \sum_{n,n'} (-1)^{F_\rho - n} \begin{pmatrix} F_\rho & k & F_{\rho'} \\ -n & q & n' \end{pmatrix} \rho_{nn'} \qquad (8)$$

Here $\rho = f, \varphi, \xi$ denote the components $(f_q^k)$ and $(\varphi_q^k)$ of the upper $(f, F_f)$ and lower $(\varphi, F_\varphi)$ levels respectively and the optical components $(\xi_q^k)$.

The parentheses denote the 3j-symbol. Since $q = n - n'$, the $\rho_q^k$ components with $q = 0$ are linear combination of the sublevels population. For example, for k=0, $\rho_0^0 = \sum_n \rho_{nn}$ is the total population and for k=2, q=0, $\rho_0^2 = \sum_{n=-F_\rho}^{F_\rho} f_{nn} \dfrac{3n^2 - F_\rho(F_\rho + 1)}{[(2F_\rho + 3)(F_\rho + 1)(2F_\rho - 1)]^{1/2}}$ is the longitudinal alignment, etc.

The relaxation of the atomic system is described phenomenologically by sum of two terms and in the common case includes radiation and collisional processes. Besides the clear physical meaning of the tensor components in this $(k,q)$ representation, the relaxation matrix is diagonal in the main cases and all relaxation parameters for different $\rho$, $\gamma_\rho(k,q) = \gamma_\rho(k)$ $(\rho = f, \varphi, \xi)$ for the k-th tensor component of the upper $(f)$ and lower $(\varphi)$ levels and the decay of the optical coherence $(\xi)$ depend on the rank $k$ of the components only. The decay of the lower level $\gamma_\varphi(0)$ includes the time-of-flight influence. The second term $(\dot{\rho})_{tr}$ describes the excitation transfer by spontaneous emission $\Gamma_{F_f F_\varphi}(\kappa)$ from the upper level to the lower one and includes the transfer of Zeeman coherence.

According to Ducloy & Dumont (Ducloy & Dumont, 1970) and taking into account Dyakonov & Perels's normalization of the irreducible tensor operators (Dyakonov & Perel, 1966)

$$\Gamma_{F_f F_\varphi}(k) = (-1)^{F_f + F_\varphi + k + 1} [\gamma_f(0)(2F_f + 1)(2F_\varphi + 1)(2J_f + 1)] \cdot$$
$$\cdot \begin{Bmatrix} F_f & F_\varphi & 1 \\ J_\varphi & J_f & I \end{Bmatrix}^2 \sqrt{(2F_f + 1)(2F_\varphi + 1)} \begin{Bmatrix} F_f & F_f & k \\ F_\varphi & F_\varphi & 1 \end{Bmatrix} \qquad (9)$$

Here $\gamma_f(0)$ is the total probability of decay of the upper state to the lower states.

It is worth to note that the relaxation constant $\Gamma_{F_f F_\varphi}(k)$ describes the "losses" in the channel $F_f \to F_\varphi$. If the branching ratio $\Gamma_{F_f F_\varphi}(0) / \gamma_f(0)$ is close to 1, the atomic system is closed. In the particular case of 87Rb D$_1$ line transition $F_f = 1 \to F_\varphi = 2$, the ratio $\Gamma_{F_f F_\varphi}(0) / \gamma_f(0) = 5/6$.

For the matrix element of the dipole transitions between hf states in ITO representation one can write (Alexandrov et al., 1991):

$$d_{F_f F_\varphi} = \|d_{f\varphi}\| (-1)^{2F_\varphi + J_f + I + 1} [(2F_f + 1)(2F_\varphi + 1)]^{1/2} \begin{Bmatrix} I & J_f & F_f \\ 1 & F_\varphi & J_\varphi \end{Bmatrix} (-1)^{F_\varphi + m} \begin{Bmatrix} F_\varphi & 1 & F_f \\ -m & Q & \mu \end{Bmatrix} e_Q \qquad (10)$$

The brackets denote 6j-Wigner symbols, $\|d_{f\varphi}\|$ is the reduced matrix element of the dipole $F_f \to F_\varphi$ transition.

The initial conditions are defined by the operator $\hat{N} = N_s(2F_s + 1)W(v)$, which describes the population of the resonant to the laser light levels. It is supposed that the atomic ensemble is in equilibrium and the velocity distribution of the atoms is described by a Maxwell function $W(v)$.

$$W(v) = (u\sqrt{\pi})^{-1} e^{(-v^2/u^2)} \tag{11}$$

Taking into account the stated above, the system of equations describing the ground state $(\varphi)$, the exited state $(f)$ and the optical coherency $(\xi)$ for arbitrary angular moments is:

$$\dot{f}_q^k + \gamma_f(k)f_q^k = i\mu_B g_f h^{-1}\left\{qH_0 f_q^k + \left[\frac{1}{2}(k+q)(k-q+1)\right]^{1/2}H_1 f_{q-1}^k - \right.$$
$$\left. - \left[\frac{1}{2}(k-q)(k+q+1)\right]^{1/2}H_{-1}f_{q+1}^k\right\} + L_q^k + (2F_f+1)N_f W(v)\delta_{k0}\delta_{q0} \tag{12a}$$

$$\dot{\varphi}_q^k + \gamma_\varphi(k)\varphi_q^k = i\mu_B g_\varphi h^{-1}\left\{qH_0\varphi_q^k + \left[\frac{1}{2}(k+q)(k-q+1)\right]^{1/2}H_1\varphi_{q-1}^k - \right.$$
$$\left. - \left[\frac{1}{2}(k-q)(k+q+1)\right]^{1/2}H_{-1}\varphi_{q+1}^k\right\} + M_q^k + (2F_\varphi+1)N_\varphi W(v)\delta_{k0}\delta_{q0} + \Gamma_{F_f F_\varphi}(k)f_q^k \tag{12b}$$

$$\dot{\xi}_q^k + \left(\gamma_\xi(k) + i\omega_0\right)\xi_q^k = ih^{-1}\sum_{k'Q}(-1)^Q H_{-Q}(-1)^{F_f + F_\varphi + q}(2k'+1)\begin{pmatrix} k' & 1 & k \\ -q' & Q & q \end{pmatrix}$$
$$\left[\mu_B g_\varphi(\varphi\|j\|\varphi)\begin{Bmatrix} F_f & F_\varphi & k' \\ 1 & k & F_\varphi \end{Bmatrix} + (-1)^{k+k'}\mu_B g_f(f\|j\|f)\begin{Bmatrix} F_\varphi & F_f & k' \\ 1 & k & F_f \end{Bmatrix}\right]\xi_{q'}^{k'} + G_q^k \tag{12c}$$

Where:

$$L_q^\kappa = ih^{-1}(2F_\varphi+1)^{-1/2}\sum_{\kappa'q'Q} E_{-Q}C_{qq'Q}^{\kappa\kappa'}\left[d\xi_{q'}^{\kappa'} + d^*(\xi_{q'}^{\kappa'})^*(-1)^{\kappa+\kappa'+q'}\right]$$

with $C_{qq'Q}^{\kappa\kappa'} = (-1)^{2F_\varphi+q'}(2F_f+1)^{1/2}(2\kappa'+1)\begin{Bmatrix} \kappa' & 1 & \kappa \\ F_f & F_f & F_\varphi \end{Bmatrix}\begin{pmatrix} k' & 1 & \kappa \\ -q' & Q & q \end{pmatrix};$

$$M_q^\kappa = (-1)^\kappa ih^{-1}(2F_\varphi+1)^{-1/2}\sum_{\kappa'q'Q} E_{-Q}(-1)^{\kappa'}B_{qq'Q}^{\kappa\kappa'}\left[d\xi_{q'}^{\kappa'} + d^*(\xi_{q'}^{\kappa'})^*(-1)^{\kappa+\kappa'+q'}\right]$$

with $B_{qq'Q}^{\kappa\kappa'} = (-1)^{2F_f+q'}(2F_\varphi+1)^{1/2}(2\kappa'+1)\begin{Bmatrix} \kappa' & 1 & \kappa \\ F_\varphi & F_\varphi & F_f \end{Bmatrix}\begin{pmatrix} \kappa' & 1 & \kappa \\ -q' & Q & q \end{pmatrix};$

$$G_q^\kappa = ih^{-1}(2F_f+1)^{-1/2}d^*\sum_{\kappa'q'Q} E_{-Q}\left[S_{qq'Q}^{\kappa\kappa'}f_{q'}^{\kappa'} + (-1)^{\kappa+\kappa'}R_{qq'Q}^{\kappa\kappa'}\varphi_{q'}^{\kappa'}\right]$$

with $\quad R_{qq'Q}^{\kappa\kappa'} = (-1)^{2F_{\varphi}+q'}(2F_f+1)^{1/2}(2\kappa'+1)\begin{Bmatrix} \kappa' & 1 & \kappa \\ F_f & F_{\varphi} & F_{\varphi} \end{Bmatrix}\begin{pmatrix} \kappa' & 1 & \kappa \\ -q' & Q & q \end{pmatrix}$

and $\quad S_{qq'Q}^{\kappa\kappa'} = (-1)^{2F_f+q'}(2F_{\varphi}+1)^{1/2}(2\kappa'+1)\begin{Bmatrix} \kappa' & 1 & \kappa \\ F_{\varphi} & F_f & F_f \end{Bmatrix}\begin{pmatrix} \kappa' & 1 & \kappa \\ -q' & Q & q \end{pmatrix}$

Here $\omega_0$ is the resonance frequency for the given transition and $d \equiv \|d_{f\varphi}\|$, $(\rho\|j\|\rho) = [(2F_{\rho}+1)(F_{\rho}+1)F_{\rho}]^{1/2}$.

This basic system of equations (Eqs. 12) is practically analogous to the well known equations (Decomps et al., 1976; Dyakonov&Perel, 1966).

The system of Eqs. 12 is specified for the atomic transition $F_f=1\to F_{\varphi}=2$ and the experimental geometry. The quantization axis is chosen parallel to the electric vector $E(\equiv E_z=E_0)$ of the laser field (see the experimental setup in Fig. 1). The scanned magnetic field $H_{scan}(\equiv H_x)$ is perpendicular to this axis. The influence of the laboratory magnetic field is taken into account including two orthogonal components $H_z$ and $H_y$ into the equations.

Using a rotating wave approximation (RWA) and assuming a single-frequency laser field (Eq. 7), the system of equations is reduced to an algebraic one. The obtained system of equations is solved numerically at different parameters: Rabi frequency ($\Omega = d\cdot E/\hbar$), relaxation constant of the levels $f$ and $\varphi$ ($(\gamma_f(k),\gamma_{\varphi}(k))$, spontaneous emission transfer coefficients $\Gamma_{f\varphi}(k)$ and stray field components $(H_y,H_z)$. All these parameters and the magnetic fields are expressed in units $\gamma_f(0)$. The initial (unbroadened by the laser field) low level population relaxation rate constant $\gamma_{\varphi}(0)$ is determined by the time of flight of the atoms. To describe better the experimental conditions the program is additionally modified to take into account the Gaussian intensity distribution of the irradiating beam and the Maxwell velocity distribution of the atoms. The solution for a given variable ($\rho_q^k$) is taken after summarizing the partial solutions for the sub-ensemble of atoms with velocities in a given interval. The integration step is varied and the integration region is chosen to be > 40 $\gamma_f(0)$. The obtained solutions for a given power are summarized on the laser beam intensity distribution.

The intensity of the registered in our experiments fluorescence $I_{F_eF_g}(\vec{n}(\theta,\varphi))$ from the upper to the lower level in direction $\vec{n}(\theta,\varphi)$ is described with:

$$I_{F_FF_{\varphi}}(\vec{n}) = C_0(-1)^{F_e+F_g}(2F_f+1)^{-1/2}\left|d_{f\varphi}^2\right|\sum(2\kappa+1)\begin{Bmatrix} 1 & 1 & \kappa \\ F_f & F_f & F_{\varphi} \end{Bmatrix}\sum_q(-1)^q f_q^{\kappa}\Phi_{-q}^{\kappa}(\vec{n}(\theta,\varphi)) \quad (13)$$

where $f_q^k$ are the tensor components describing the upper level, $\Phi_q^{\kappa}(\vec{n}(\theta,\varphi))$ is the observation tensor (Alexandrov et al., 1991), $\theta$ is the angle between the laser light polarization and the direction of registration (the inclination angle) and $\varphi$ is the azimuth angle (Alexandrov et al., 1991; Dyakonov&Perel, 1966).

In a dipole approximation the fluorescence is defined only by the tensor components $f_q^k$ with rank $k \le 2$. The unpolarized fluorescence intensity $I_{f\varphi}^{unpol}$ for $\theta=0$ can be written as:

$$I_{f\varphi}^{unpol} = C_0\left[\frac{f_0^0}{\sqrt{2F_f+1}} + (-1)^{F_f+F_\varphi+1}\sqrt{30}\begin{Bmatrix}1 & 1 & 2 \\ F_f & F_f & F_\varphi\end{Bmatrix}f_0^2\right] \qquad (14)$$

Only the two tensor components for the upper level $f_0^0$ and $f_0^2$ describe the signal observed in this case. In linear approximation, both components do not depend on the magnetic field. The magnetic field dependence is a result of the transfer of coherence created on the low level and it is a typical nonlinear effect.

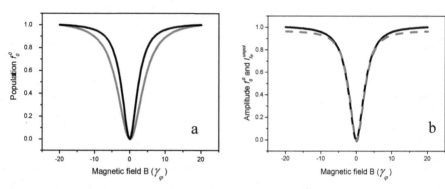

Fig. 7. Comparison of the theoretical upper level population $f_0^0$ : (a) calculated with (black line) and without (red line) integration over the Doppler velocity distribution and (b) with the unpolarized intensity $I_{f\varphi}^{unpol}$ (dashed red line).

It is well known that the Maxwell velocity distribution leads to narrowing of the CPT resonance (Firstenberg et al., 2007) and that the resonance shape remains quasi Lorentzian. The change in the upper level population shape and width after velocity integration over the Doppler distribution is illustrated in Fig. 7a (Petrov et al., 2007).

The main part of the unpolarized fluorescence intensity is determined by the upper level population $f_0^0$. The influence of the longitudinal alignment $f_0^2$ on the resonance shape is mainly on the wings of the resonance (Fig. 7b). It should be noted that the comparison of the numerical calculated shape of the population and intensity with a Lorentzian shape shows that there is a small specific structure in the vicinity of the zero magnetic field. This small structure is result of the High Rank Polarization Moments (HRPM) influence on the observables tensor components (see bellow).

The influence of the Gaussian distribution of the laser beam intensity on the shape of the CPT resonances is more essential. This has been discussed in a large number of papers (Knappe et al., 2001; Levi et al., 2000; Pfleghaar et al., 1993; Taichenaichev et al., 2004) and we will not illustrate it here.

The photon reabsorption is not included in this model, because in our experiments the medium is optically thin on the length scale of the laser light-atom interaction zone (Matsko et al., 2001).

The numerically calculated resonance shape which takes into account the Gaussian distribution of the laser beam intensity, the experimental geometry of the excitation and the velocity distribution of the atoms coincides with the shape of the fluorescence resonance in cell **D** (Fig. 8a) and the pedestal of the resonance in cell **A** (Fig. 8b). This shape is in very good coincidence with the shape calculated with Eq. 5 (Taichenachev et al., 2004). The difference between the experimental and theoretical shapes in cell **A** (Fig. 8b) is practically a Lorentzian which width is of the order of $\Delta_L=1/(2\pi\tau)$, defined by the mean time $\tau$ between two atom collisions with the cell's walls. This is the narrow structure from Fig. 2.

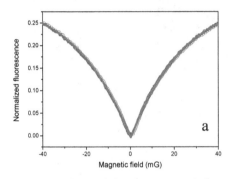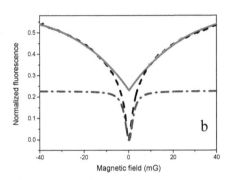

Fig. 8. Comparison of the experimental shapes with theoretical shapes:
a) experimental in cell **D** (scattered circles curve) and theoretical (solid curve);
b) experimental in cell **A** (dash line curve), theoretical (solid curve) and the difference between them (dash-dot line curve).

The comparison of the theoretical and experimental shapes in Fig. 8a shows that there is a small difference around zero magnetic field. Similar difference between the theoretical and experimental shapes was reported earlier (for example Pfleghaar et al., 1993; Taichenachev et al., 2004) and in all these cases the influence of the HRPM was not taken into account.

When the laser power density (the resonance excitation) is increased, together with the transfer of the quadrupole coherence from the ground to excited state, the influence of the multiphoton interactions will increase and high order (high rank) coherences will be created.

Our theoretical evaluations of the influence of the HRPM on the shape of the CPT resonances (Gateva et al., 2007, 2008a) proved that HRPM conversion (Okunevich, 2001) cause the CPT resonance shape peculiarities at the center of the resonance.

For a plane wave the difference from the Lorentzian shape is small and it is only around the center (at 0 magnetic field) of the resonance, with the shape and amplitude of this difference corresponding to multiphoton resonances. Although the moments of rank 4 (hexadecapole moments) (Decomps et al., 1976) do not influence directly the spontaneous emission, they are converted into the $f_q^2$ components of the upper level of rank 2 thus influencing the spontaneous emission.

In Fig. 9 the scheme of the connection of different rang polarization moments and the "conversion path" in population and longitudinal alignment for our scheme of excitation is illustrated (Polischuk et. al., 2011).

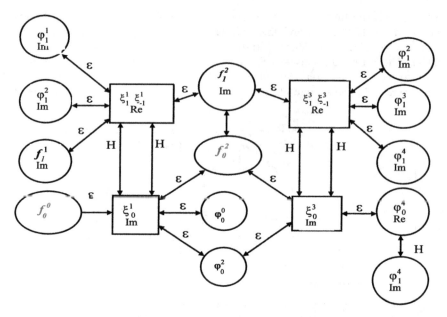

Fig. 9. Partial diagram of the mutual conversion of the tensor components forming the population ($f_0^0$) and the longitudinal alignment ($f_0^2$) ($\varepsilon$ describes the interaction with the laser field, H – with the scanned magnetic field).

In Fig. 10 the calculated shape of the CPT resonances at different Rabi frequencies is given. At relatively low powers (Fig. 10, curve 1) the shape is close to Lorentzian; as the power is increased, the resonance flattens at the centre and an inverse narrow structure appears (Fig. 10, curves 2-4).

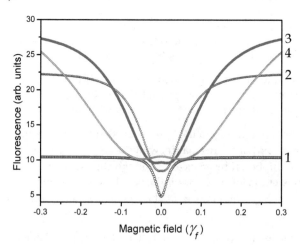

Fig. 10. Calculated shape of the CPT resonances at different Rabi frequencies $\Omega=\gamma_f$(1); $\Omega=3\gamma_f$ (2); $\Omega=5\gamma_f$(3); $\Omega=10\gamma_f$(4).

At low power density of 1.7 mW/cm² (for a laser line width 50 MHz, corresponding to reduced Rabi frequency $\Omega$=1 MHz), the measured CPT resonance shape can not be visually distinguished from a Lorentzian, but in the difference between the experimental shape and its Lorentzian fit there is a specific structure at the center of the shape (Fig. 11a). This structure is similar to the difference between the calculated resonance and its Lorentzian fit at $\Omega$=1 MHz (Fig. 11b).

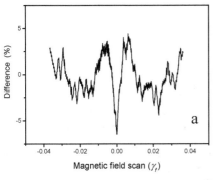
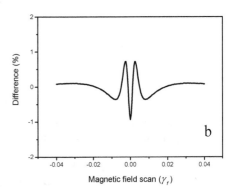

Fig. 11. Difference between:
(a) the experimental resonance shape and its Lorentzian fit
(b) the calculated resonance shape and its Lorentzian profile fit (b) at $\Omega$=1 MHz. (The magnetic field scan is presented in $\gamma_f$ units ($\gamma_f$ =6 MHz is the radiation width of the upper level).

In the case of a strong laser field, at power density 1 W/cm² (corresponding to a reduced Rabi frequency $\Omega$ =25 MHz), when the measured resonance has a complex shape, the comparison of the experimental and theoretical shapes is more complicated because the amplitude of the narrow resonance increases with power (Section 2.1.2).

To observe the influence of the HRPM, the narrow structure must be eliminated. As the narrow structure is very sensitive to magnetic fields perpendicular to the polarization vector $E(\equiv E_z)$ and $B_{scan}(\equiv B_x)$ (Alipieva et al., 2003), magnetic fields $B_y$ of the order of 10 mG destroy it, while the theoretical evaluations and experimental results show, that at this magnetic fields the changes in the broader structure due to the HRPM influence are practically not affected. The comparison of the shape of the resonances observed at $B_y$=0 (Fig. 12 curve 1) and the shape of the resonance at 32 mG (Fig. 12 curve 2) shows that at this power density the amplitude of the CPT resonance in a magnetic field is reduced twice and an additional inverted structure (about 5% in amplitude) is observed. At higher magnetic fields this inverted structure disappears (Fig. 12, curve 3).

The developed numerical model which takes into account the geometry of the excitation, the velocity distribution of the atoms and the Gaussian distribution of the laser beam intensity describes only the broad pedestal, not the narrow component from Section 2.1.2. The observed in fluorescence narrow structure of the resonance is with Lorentzian shape, it is not radiation broadened and its amplitude increases with the laser power. Its width (FWHM) $\Delta_L$ does not change with the laser beam diameter and corresponds to a relaxation time, equal to that due to atomic collisions with the cell walls.

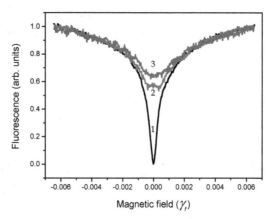

Fig. 12. CPT signals measured at different transverse to the polarization vector and propagation direction magnetic field $B_y$: 0 mG (curve 1); 32 mG (curve 2); 72 mG (curve 3).

Narrow structures in the CPT resonances are reported as result of diffusion-induced Ramsey narrowing (Xiao et al., 2006, 2008). In this case a sharp central peak on a broad pedestal was observed in the Electromagnetically Induced Transparency (EIT) resonance shapes. The broad pedestal is associated with the single pass interaction time and is power broadened. The sharp central peak is the central Ramsey fringe, which adds coherently for all Ramsey sequences. Its width changes with the laser beam diameter. At low laser power, small beam diameter and low buffer gas pressure the sharp central peak is not Lorentzian in shape and is insensitive to power broadening. At high laser intensity the central peak loses its contrast and is Lorentzian in shape and power broadened.

The described characteristics of the narrow structure in Fig. 2 and 8 show that it is not result of diffusion-induced Ramsey narrowing. The resonance shapes, measured at different geometries of excitation and registration (Gateva et al., 2008b, 2011), show that the narrow structure at the centre of the resonance can be considered as a result of a weak field – atom interaction, probably scattered light in the whole cell volume.

One of the possible scattering processes influencing the CPT resonance shape is Rayleigh scattering because in this case the scattered light maintains coherence with the incident beam and in this way a diffusion of the coherent light is created. For example, Rayleigh scattering has been used for studying the properties of cold atoms (Datsyuk et al., 2006) and optical lattices (Carminati et al., 2003 and references therein).

### 2.3 CPT resonances and Rayleigh scattering

For linear polarization of the incident light, the power of the Rayleigh-scattering light $P_R$ registered at 90° to the laser beam direction is (Boyd, 1992; Measures, 1984)

$$P_R(\lambda,\theta) = P_0 \; \alpha \; N \; \sigma(\lambda) \sin^2 \theta \qquad (15)$$

where $P_0$ is the incident light power, $\alpha$ is the registration efficiency coefficient, N is the scattering particles density, $\sigma(\lambda,\theta) = \sigma(\lambda) \sin^2 \theta$ is the differential cross-section of Rayleigh scattering of polarized light with wavelength $\lambda$ in the plane perpendicular to the beam direction and $\theta$ is the angle between the laser light polarization and the direction of registration. According to Equation (9) the Rayleigh scattering light has a maximum at $\theta = 90°$ and is zero at $\theta = 0°$.

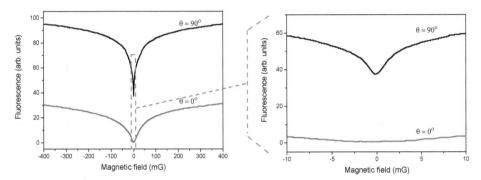

Fig. 13. Shape of the measured CPT resonances given in two different scales at angles $\theta = 0°$ and 90°.

In Fig. 13 are given the shapes of the CPT resonances measured at angles $\theta = 0°$ and 90° when the laser is tuned on the $F_g = 2 - F_e = 1$ transition of the $^{87}Rb$ $D_1$ line. The results are presented in two different scales. The comparison of the resonances on the broader scale shows that at the wings they are with the same shape and that this shape is in good coincidence with the calculated with the model from Section 2.2.

The comparison of the shapes in Fig. 13 on the narrower scale shows, that at $\theta = 90°$ there is a narrow structure. This narrow structure is Lorentzian in shape with width of the order of the defined by the mean time between 2 collisions with the cell walls $\Delta_L$. The amplitude of this narrow structure (Fig. 14a) has the typical $\sin^2 \theta$ dependence on $\theta$ and it is maximum at $\theta = 90°$ and minimum at $\theta = 0°$.

In Rb vacuum cell at room temperature the Rb pressure is of the order of $4.10^{-5}$ Pa and at this pressure the Rayleigh scattered light from Rb atoms can not be registered in our experiment (Boyd, 1992; Measures, 1984) – this is the case of cell **D**. However, if the vacuum cell is not pumped very well, there will be some residual air, water and oil vapor, as well as rare but relatively strongly scattering submicron particles, which will scatter the light – this is the case of cells **A**, **B** and **C**. These cells are very old (more than 20 years), manufactured with an oil diffusion pump connected by a liquid nitrogen cold trap to the cell. Cell **D** is a new one, which was sealed off some months before the experiments. It was manufactured using an oil-free vacuum installation – a turbomolecular pump and ion pump. Checking the vacuum with a Tesla coil (high voltage transformer) shows that there is no light emission (glowing) in cell **D**, while the cell glass is glowing in cell **A**, cell **B** and cell **C**.

A demonstration of the existence of Rayleigh scattered light is the measured angular dependence of the scattered light power when the laser is tuned out of line. It has the typical $\sin^2 \theta$ dependence from Equation 9 with maximum at $\theta = 90°$ and minimum at $\theta = 0°$ (Fig. 14b).

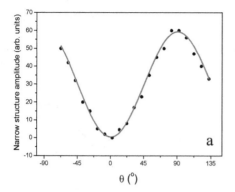
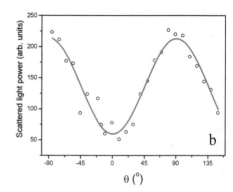

Fig. 14. Angular dependence on $\theta$, the angle between the direction of observation and laser light polarization at 90° to the laser beam of:
(a) the narrow Lorentzian structure amplitude;
(b) the measured scattered light power when the laser is out of line.

Qualitatively, the observed narrow resonance can be described by including into the numerical model from Section 3.2 a weak, resonant, polarized light field in the cell volume. Such an approach corresponds to the supposition that the signal in the cell is formed by two subensembles of atoms interacting with different light fields. The general solution for the fluorescence intensity is sum of the numerical integrations of the density matrix equations for each of the subensembles.

For the subensemble interacting with the laser beam the evaluations following the model described in Section 2.2 have shown that this part of the signal practically does not depend on the observation direction. Although the components of the observation tensor $\Phi_q^2(\vec{n})$ depend on the direction of observation, the performed evaluations for the investigated transition have shown that the main contribution to the intensity of the unpolarized fluorescence has the population $f_0^0$ and $I_{F_f F_\varphi}$ practically does not depend on the observation direction.

For the subensemble interacting with the weak Rayleigh scattered light, the angular distribution of the Rayleigh scattered light has to be included in the excitation tensor. This angular dependence in the excitation reflects on all tensor components $f_q^k$ and in this way on the fluorescence signal $I_{F_f F_\varphi}(\vec{n})$.

As the scattered light is very low in power, the resonances are with Lorentzian shape and not power broadened. When the Rabi frequency of the light field $\Omega$ is small enough, so that $\Omega^2 / \gamma_f < \gamma_\varphi$ (Eq. 1), there is no power broadening. So long as the mean free path of the atom is longer than the cell dimensions, $\gamma_\varphi$ is defined by the mean time between two successive collisions with the cell walls. Since the weak field is due to Rayleigh scattering, the amplitude of the Lorentzian will depend on the density of the scattering particles and can be used as indicator of the level of the vacuum cleanness of the cell.

Another confirmation that the shape of the resonances is sum of the resonance shapes of different subensembles of atoms is done in experiment where the shape of the resonances is measured at two different geometries of observation (Gateva et al., 2008b, 2011). In the first case, the photodiode is on the cell wall and the observation field of view is considerably larger than the laser beam diameter (Fig. 15, solid curve), but the angle of view is sufficiently small to ensure satisfactory angular resolution of observation. In the second case a lens is used in front of the photodiode in order to restrict the observation field of view just to the laser beam volume (Fig. 15, scattered circles curve). Because of the limited dimensions of the µ-metal shield, these measurements were performed in a smaller cell, cell **B**. The comparison of the shape of the resonances in these two cases shows that the broad pedestal doesn't change, while the shape and the width of the narrow structure change (Fig. 15). In the first case (Fig. 15, solid curve), when the fluorescence mostly of atoms out of the laser beam is registered, the narrow structure is Lorentzian in shape and its width $\Delta_L$ is defined by the relaxation on the cell walls. In the second case (Fig. 15, scattered circles curve), when fluorescence mostly from the laser beam volume is registered, the measured narrow structure is narrower than $\Delta_L$. For explanation of such narrowing the influence of the diffusion-induced Ramsey narrowing (Xiao et all., 2006), which is result of the impurities in the cell and/or on the cell walls, has to be taken into account. Diffusion induced Ramsey narrowing resonances have been measured in Na vacuum cell, too (Gozzini et al., 2011).

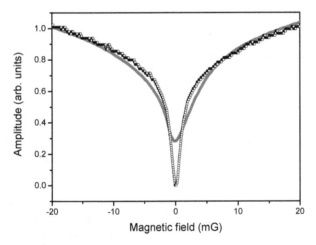

Fig. 15. Narrow structure of the CPT resonance at different geometries of registration: when the photodiode is on the cell wall (solid curve) and when the laser beam is projected by a lens on the photodiode (scattered circles curve).

This experiment can explain the difference in the narrow resonance shapes in the works of Alipieva et al. and Xiao et al. (Alipieva et al., 2003; Xiao et al., 2006). When the resonance is registered in transmission (EIT) only atoms from the laser beam volume contribute to the signal. In this case the diffusion-induced Ramsey narrowing is responsible for the narrow structure. If the signal is registered in fluorescence (Alipieva et al., 2003) the signal is mainly from atoms out of the laser beam volume, which interact with the Rayleigh scattered light. In this case the narrow component is Lorentzian and its width is defined by the relaxation time due to atomic collisions with the cell walls.

## 3. Conclusion

The investigation of the influence of different factors on the shape of the CPT resonances excited in Hanle effect configuration in uncoated, room temperature vacuum cell and registered in fluorescence has shown that the resonance has a complex shape. The performed numerical calculations, which take into account the Gaussian distribution of the laser beam intensity, the experimental geometry of the excitation, the velocity distribution of the atoms, the high rank polarization moment, the stray magnetic fields influence, and the Rayleigh scattering, describe very well the experimental shapes. The results show that the developed theoretical modelling can be applied for analysis of the influence of the different factors and analysis of the cell quality. The Rayleigh scattering can be used for analysis of the vacuum cleanness of the cell and the quality of the cell walls.

On the other hand these narrow resonances can be applied for building magneto-optical sensors. From the point of view of applications in magnetometry, where narrow signals and high signal-to-noise ratios are important, the narrow Rayleigh structure of the CPT resonance measured in fluorescence offer good possibilities: is not power broadened, its width do not depend on the position of the detector and its amplitude increases with the power. The complex structure of the resonance expands the range of the measured magnetic fields.

## 4. Acknowledgements

The authors are pleased to acknowledge the financial support of part of the investigations by the Bulgarian NCSR (Grant No. DO02-108/22.05.2009).

## 5. References

Alexandrov, E. B.; Chaika, M. P. & Hvostenko, G. I. (1991). *Interference of atomic states,* Springer-Verlag, ISBN-10: 354053752X, ISBN-13: 978-3540537526

Alipieva, E.; Gateva, S.; Taskova, E. & Cartaleva, S. (2003). Narrow structure in the coherent population trapping resonance in Rb, *Optics Letters,* Vol.28, No.19, pp. 1817-1819, ISSN 0146-9592

Alzetta, G.; Gozzini, A.; Moi, L. & Orriols, G. (1976). An experimental method for the observation of r.f. transitions and laser beat resonances in oriented Na vapour, *Il Nuovo Cimento B Series 11,* Vol.36, No.1, pp. 5-20, ISSN 0369-3546

Arimondo, E. (1996). Coherent population tapping in laser spectroscopy, *Progress in Optics,* Vol.35, pp. 257-354, ISSN 0079-6638

Auzinsh, M.; Budker, D. & Rochester, S.M. (2009a). Light-induced polarization effects in atoms with partially resolved hyperfine structure and applications to absorption, fluorescence, and nonlinear magneto-optical rotation, *Physical Review A - Atomic, Molecular, and Optical Physics,* Vol. 80, No. 5, art. no. 053406, pp. 1-22, ISSN 1050-2947

Auzinsh, M.; Ferber, R.; Gahbauer, F.; Jarmola, A. & Kalvans, L. (2009b). Nonlinear magneto-optical resonances at D1 excitation of Rb 85 and Rb87 for partially resolved hyperfine F levels, *Physical Review A - Atomic, Molecular, and Optical Physics,* Vol. 79 No. 5, art. no. 053404, pp. 1-9, ISSN 1050-2947.

Bison, G; Castanga, N.; Hofer, A.; Knowles, P.; Schenker, J.L.; P., Schenker, J.-L.; Kasprzak, M.; Saudan, H. & Weis, A. (2009). A room temperature 19-channel magnetic field mapping device for cardiac signals. *Applied Physics Letters*, Vol.95, No.17, art. no. 173701, pp. 1-3, ISSN 0003-6951

Bouchiat, M.A. & Brossel, J. (1966). Relaxation of optically pumped rb atoms on paraffin-coated walls. *Physical Review*, Vol.147, No., pp. 41-54, ISSN 0031-899X

Boyd, R. W. (1992). *Nonlinear Optics*, Academic Press, ISBN 0123694701

Budker, D. & Romalis, M. (2007). Optical magnetometry, *Nature Physics*, Vol.3, No.4, pp. 227-234, ISSN 17452473

Carminati, F.-R.; Sanchez-Palencia, L.; Schiavoni, M.; Renzoni, F. & Grynberg, G. (2003). Rayleigh Scattering and Atomic Dynamics in Dissipative Optical Lattices, *Physical Review Letters*, Vol.90, art.no. 043901, ISSN 0031-9007

Corney, A. (1977). *Atomic and Laser Spectroscopy*, Clarendon, Oxford, ISBN 9780198511380

Cox, K.; Yudin, V.I.; Taichnachev, A.V.; Novikova, I. & Mikhailov, E.E. (2011). Measurements of the magnetic field vector using multiple electromagnetically induced transparency resonances in Rb vapor, *Physical Review A - Atomic, Molecular, and Optical Physics*, Vol.83, No.1, art. no. 015801, pp. 1-4, ISSN 1050-2947.

Dancheva, Y.; Alzetta, G.; Cartaleva, S.; Taslakov, M. & Andreeva, Ch. (2000). Coherent effects on the Zeeman sublevels of hyperfine states in optical pumping of Rb by monomode diode laser, *Optics Communications*, Vol.178, No.1, pp. 103-110, ISSN 0030-4018

Dang, H.B.; Maloof, A.C. & Romalis, M.V. (2010). Ultrahigh sensitivity magnetic field and magnetization measurements with an atomic magnetometer, *Applied Physics Letters*, Vol.97, No.15, art. no. 151110, ISSN 0003-6951

Datsyuk, V. M.; Sokolov, I. M.; Kupriyanov, D. V. & Havey, M. D. (2006). Diffuse light scattering dynamics under conditions of electromagnetically induced transparency *Physical Review A - Atomic, Molecular, and Optical Physics*, Vol.74, art.no. 043812, ISSN 0031-9007

Decomps B., Dumont M. & Ducloy M. (1976). *Laser Spectroscopy of Atoms and Molecules*, Springer-Verlag, Berlin, ISBN 3-540-07324-8

D'yakonov, M. I. & Perel, V. I. (1966). On the theory of the gas laser in a magnetic field, *Optics and Spectroscopy*, Vol.20, No.3, pp. 472-480, ISSN 0030-400X

Ducloy, M. & Dumont, M. (1970). Etude du transfert d'excitation par emission spontanee, *Le journal de Physique*, Vol.31, pp. 419-427, ISSN 0302-0738

Edelstein, A. (2007). Advances in magnetometry. *Journal of Physics Condensed Matter*, Vol.19, No.16, art. no. 165217, ISSN 0953-8984.

Firstenberg, O.; Shuker, M.; Ben-Kish, A.; Fredkin, D.R.; Davidson, N. & Ron, A. (2007). Theory of Dicke narrowing in coherent population trapping. *Physical Review A - Atomic, Molecular, and Optical Physics*, Vol. 76, No. 1, art. no. 013818, pp.1-6, ISSN 1050-2947.

Figueroa, E.; Vewinger, F.; Appel, J. & Lvovsky, A.I. (2006) . Decoherence of electromagnetically induced transparency in atomic vapor. *Optics Letters*, Vol.31, No.17, pp. 2625-2627, ISSN 0146-9592

Gao, J.-Y.; Xiao, M. & Zhu, Y. (eds.) (2009). *Atomic coherence and Its Potential Application*, Bentham Science Publishers Ltd., ISBN: 978-1-60805-085-7

Gateva, S.; Alipieva, E. & Taskova, E. (2005). Power dependence of the coherent-population-trapping resonances registered in fluorescence and transmission: Resonance-width narrowing effects. *Physical Review A - Atomic, Molecular, and Optical Physics*, Vol. 72, No.1, art.no. 025805, pp.1-4, ISSN 1050-2947

Gateva, S.; Petrov, L.; Alipieva, E.; Todorov, G.; Domelunksen, V. & Polischuk, V. (2007). Shape of the coherent population-trapping resonances and high-rank polarization moments. *Physical Review A - Atomic, Molecular, and Optical Physics*, Vol. 76, No. 2, art. no. 025401, pp. 1-4, ISSN 1050-2947

Gateva, S.; Alipieva, E.; Petrov, L.; Taskova, E. & Todorov, G. (2008a). Single frequency coherent-population-trapping resonances for magnetic field measurement. *J. Optoelectronics and Advanced Materials*, Vol. 10, No.1, pp. 98-103, ISSN 1454-4164

Gateva, S.; Alipieva, E.; Domelunksen, V.; Polischuk, V.; Taskova, E.; Slavov, D.; Todorov, G. (2008b). Shape of the coherent-population-trapping resonances registered in fluorescence at different experimental geometries. *Proceedings of SPIE - The International Society for Optical Engineering*, Vol. 7027, art. no. 70270I, pp. 1-12, ISBN 9780819472410.

Gateva, S.; Gurdev, L.; Alipieva, E.; Taskova, E. & Todorov, G. (2011). Narrow structure in the coherent population trapping resonances in rubidium and Rayleigh scattering. *Journal of Physics B: Atomic, Molecular and Optical Physics*, Vol. 44, No. 3, art. no. 035401, pp. 1-6 , ISSN 0953-4075

Godone, A.; Levi, F.; Micalizio, S. & Vanier, J. (2002). Dark-line in optically-thick vapors: Inversion phenomena and line width narrowing, *The European Physical Journal D*, Vol.18, No.1, pp. 5-13, ISSN 1434-6060

Gozzini, S.; Marmugi, L.; Lucchesini, A.; Gateva, S.; Cartaleva, S. & Nasyrov, K. (2011). Narrow structure in the coherent population trapping resonance in sodium, *Physical Review A - Atomic, Molecular, and Optical Physics*, Vol. 84, No. 1, art. no. 013812, pp. 1-9, ISSN 1050-2947

Griffith, W.C.; Knappe, S. & Kitching, J. (2010). Femtotesla atomic magnetometry in a microfabricated vapor cell, *Optics Express*, Vol.18, No.26, pp. 27167-27172, ISSN 1094-4087.

Huss, A.; Lammegger, R.; Windholz, L.; Alipieva, E.; Gateva, S.; Petrov, L.; Taskova, E. & Todorov, G. (2006). Polarization-dependent sensitivity of level-crossing, coherent-population-trapping resonances to stray magnetic fields. *Journal of the Optical Society of America B: Optical Physics*, Vol.23, no.9, pp. 1729-1736, ISSN 0740-3224

Javan, A.; Kocharovskaya, O.; Lee, H. & Scully, M.O. (2002). Narrowing of electromagnetically induced transparency resonance in a Doppler-broadened medium. *Physical Review A - Atomic, Molecular, and Optical Physics*, Vol.66, No. 1, pp. 138051-138054, ISSN 1050-2947

Johnson, C.; Schwindt, P.D.D. & Weisend, M. (2010). Magnetoencephalography with a two-color pump-probe, fiber-coupled atomic magnetometer. *Applied Physics Letters*, Vol.97, No.24, art. no. 243703, ISSN 0003-6951

Kimble, H.J. (2008). The quantum internet. *Nature*, Vol.453, pp. 1023-1030, ISSN 0028-0836

Kitching, J.; Knappe, S. & Donley, E.A. (2011). Atomic sensors - A review. *IEEE Sensors Journal*, Vol. 11, No.9, art. no. 5778937, pp. 1749-1758. ISSN 1530-437X.

Knappe, S.; Wynands, R.; Kitching, J.; Robinson, H. G. & Hollberg, L. (2001). Characterization of coherent population-trapping resonances as atomic frequency references, *Journal of the Optical Society of America B: Optical Physics*, Vol.18, No.11, pp. 1545-1553, ISSN 0740-3224

Knappe, S.; Shah, V.; Schwindt, P.D.D.; Hollberg, L.; Kitching, J.; Liew, L. & Moreland, J. (2004). Microfabricated atomic clock, *Applied Physics Letters*, Vol. 85, No. 9, pp. 1460-1462, ISSN 0003-6951

Knappe, S.; Sander, T.M.; Kosch, O.; Wiekhorst, F.; Kitching, J. & Trahms, L. (2010). Cross-validation of microfabricated atomic magnetometers with superconducting quantum interference devices for biomagnetic applications, *Applied Physics Letters*, Vol. 97, art. no. 133703, pp. 1-3, ISSN 0003-6951

Kominis, I. K.; Kornack, T. W.; Allred, J. C. & Romalis, M. V. (2003). A subfemtotesla multichannel atomic magnetometer. *Nature*, Vol. 422, No.6932, pp. 596-599, ISSN 0028-0836

Landau, L.D. & Lifshitz, E.M. (1965). *Quantum Mechanics: Non-Relativistic Theory*, Pergamon Press, ISBN 978-0-080-20940-1

Levi, F.; Godone, A.; Vanier, J.; Micalizio, S. & Modugno, G. (2000). Line-shape of dark line and maser emission profile in CPT. *The European Physical Journal D*, Vol.12, No.1, pp. 53-59, ISSN 1434-6060

Lukin, M.D. (2003). Colloquium: Trapping and manipulating photon states in atomic ensembles. *Reviews of Modern Physics*, Vol.75, pp. 457-472, ISSN 0034-6861

Matsko, A. B.; Novikova, I.; Scully, M. O. & Welch, G. R. (2001). Radiation trapping in coherent media. *Physical Review Letters*, Vol.87, No.13, art. no. 133601, ISSN 1079-7114

Measures, R. M. (1984). *Laser Remote Sensing: Fundamentals and Applications*, Wiley ISBN 0471081930 (ISBN13: 9780471081937), New York

Mikhailov, E.; Novikova, I,; Havey, M.D. & Narducci, F.A. (2009). Magnetic field imaging with atomic Rb vapour. *Optics Letters*, Vol.34, No.22, pp. 3529-3531, ISSN 0146-9592

Okunevich, A. I., (2001). On the possibility of registration of hexadecapole transversal components of the atoms in fluorescence. *Optics and Spectroscopy*, Vol.91, No.2, pp. 193-200, ISSN 0030-400X

Petrov, L.; Slavov, D.; Arsov, V.; Domelunksen, V.; Polischuk, V. & Todorov, G. (2007) High rank polarization moments in a Doppler broadened 87Rb transition, *Proceedings of SPIE - The International Society for Optical Engineering*, Vol. 6604, pp. 66040H1-66040H5, ISBN 9780819467423

Pfleghaar, E.; Wurster, J.; Kanorsky, S.I. & Weis, A. (1993). Time of flight effects in nonlinear magneto-optical spectroscopy, *Optics Communications*, Vol. 99, No. 5-6, pp. 303-308. ISSN 0030-4018.

Polischuk, V.; Domelunksen, V.; Alipieva, E. & Todorov, G. (2011). MatLab based modelling of nonlinear interaction of Rb87 atoms with polarized radiation. *unpublished*

Romalis, M.V. & Dang, H.B. (2011). Atomic magnetometers for materials characterization. *Materials Today*, Vol.14, no.6, pp. 258-262, ISSN 1369-7021

Savukov, I. (2010). Ultra-Sensitive Optical Atomic Magnetometers and Their Applications, In: *Advances in Optical and Photonic Devices*, Ki Young Kim, (Ed.), 329-352, InTech, ISBN 978-953-7619-76-3, Croatia

Steck, D.A. (2009) *Rubidium 87 D Line Data* (available online http://steck.us/alkalidata)

Taichenachev, A.V.; Tumaikin, A. M.; Yudin, V. I.; Stähler, M.; Wynands, R.; Kitching, J. & Hollberg L. (2004). Nonlinear-resonance line shapes: Dependence on the transverse intensity distribution of a light beam. *Physical Review A - Atomic, Molecular, and Optical Physics*, Vol.69, art. no. 024501, ISSN 1050-294711

Thomas, J. E. & Quivers, W. W. (1980). Transit-time effects in optically pumped coupled three-level systems, *Physical Review A - Atomic, Molecular, and Optical Physics* 22(5), 2115-2121, ISSN 1050-294711

Xia, H.; Baranga, A.B.-A.; Hoffman, D. & Romalis, M.V. (2006). Magnetoencephalography with an atomic magnetometer. *Applied Physics Letters*, Vol.89, art. no. 211104, ISSN 0003-6951

Xiao, Y.; Novikova, I.; Phillips, D. F. & Walsworth, R. L. (2006) Diffusion-Induced Ramsey Narrowing. *Physical Review Letters*, Vol.96, art. no. 043601, ISSN 1079-7114

Xiao, Y.; Novikova, I.; Phillips, D. F. & Walsworth, R. L. (2008). Repeated interaction model for diffusion-induced Ramsey narrowing, *Optics Express*, Vol.16, pp. 14128-14141, ISSN 1094-4087

Xu, S.; Crawford, C.W.; Rochester, S.; Yashchuk, V.; Budker, D. & Pines, A. (2008). Submillimeter-resolution magnetic resonance imaging at the Earth's magnetic field with an atomic magnetometer. *Physical Review A - Atomic, Molecular, and Optical Physics*, Vol. 78 No.1, art. no. 013404, pp. 1-4, ISSN 1050-294711.

Ye, C.Y. & Zibrov, A. S. (2002). Width of the electromagnetically induced transparency resonance in atomic vapor. *Physical Review A - Atomic, Molecular, and Optical Physics*, Vol. 65, No. 2, art.no. 023806, pp. 1-5, ISSN 1050-2947

# 11

# Photodetectors in Calorimeters for the Linear Collider

Jaroslav Cvach and CALICE Collaboration
*Institute of Physics of the ASCR, v.v.i., Praha*
*Czech Republic*

## 1. Introduction

The next high energy accelerator that will be built after the Large Hadron Collider currently operating at CERN is expected to be a Linear Collider (LC). It is envisaged to be an electron-positron colliding beams facility with the centre-of-mass energy of 500–3000 GeV (ILC, 2007, Vol. 3; CLIC, 2011). The detector at the LC will have to deal with a large dynamic range in particle energy, complexity of final states and small signal-to-background ratio. The track density in collimated jets can be as high as one per $mm^2$ at a radius of 1.5 cm, and the accelerator induced backgrounds produce typical hit densities of the order of $0.03/mm^2$ per bunch crossing at a radius of 1.5 cm, and $0.003/cm^2$ at a radius of 30 cm. The physics requirements demand detector performance parameters which must be substantially better than at previous experiments on accelerators at the large electron-positron collider in CERN and at the Stanford linear collider in SLAC. The goal is to achieve the energy resolution for jets $\sigma/E < 3\%$ in wide range of energies starting from the mass of the intermediate bosons W and Z, $E \sim m_W, m_Z \sim 80$ GeV to several hundred GeV. This condition is often translated into the formula used for the energy resolution of the sandwich calorimeter $\sigma/E \sim 30\%/\sqrt{E}$ (jet energy $E$ is in GeV).

The experimental technique proposed to reach these goals was formulated ten years ago and is now known as the particle flow algorithm (PFA) (Brient & Videau, 2001; Cvach, 2002; Morgunov, 2002). The PFA combines the information from tracking and calorimetry to obtain an optimal estimate of the flow of particles from the interaction vertex and of the original parton four-momenta. The subdetectors must have excellent spatial granularity to enable a PFA algorithm which resolves energy depositions of almost overlapping particles, combines redundant measurements properly (e.g. of electrons in tracking and the electromagnetic calorimeter or of charged pions in tracking and calorimetry) and provides other corrections (e.g. calorimeter software compensation). The calorimeter for experiments at the LC must be realized as a dense and hermetic sampling calorimeter with a very high granularity, where one can efficiently separate the contributions of different particles in a jet and use the best suited detector to measure their four-momenta.

In this article we review the performance of modern photodetectors used in already built calorimeter prototypes with plastic scintillator as the active medium. Calorimeters are constructed as a sandwich with an absorber between the scintillator layers to decrease the calorimeter size.

## 2. Calorimeter prototypes optimized for the PFA

The CALICE collaboration (Calice, n.d.) has undertaken an intensive R&D program starting in 2001 to prove feasibility of the particle flow approach for the improvement of detector energy resolution by a factor of approximately two for jets produced in e+e- collisions at centre of mass energies 50-1000 GeV. The corresponding effort was done both on the subdetector side as well as on the software side – reconstruction and simulation programs. We shall concentrate in this article on calorimeters optimized for the PFA. For effective functioning of a calorimeter, the vertex tracker and tracker standing in front of calorimeters are important. We refer e.g. to (ILC, 2007, Vol. 4) where relevant solutions for trackers are discussed in detail.

The first completed prototype of the electromagnetic calorimeter ECAL was SiW ECAL (Repond et. al., 2008). The large silicon diode sensors of 1x1 cm$^2$ area made on high resistive Si wafers 0.5 mm thick define the size of the active calorimeter cell. A sensitive area of 18x18 cm$^2$ of silicon was inserted between absorber tungsten plates 1.2-4.8 mm thick. In total 30 layers of Si + W represent 24 radiation lengths $X_0$. The total number of cells was 9720. As this calorimeter does not use photodetectors, we refer for further details to the latest publication (Adloff et al., 2010b).

As an alternative to the SiW ECAL, a scintillator ECAL was built which used scintillator strips read by Multi-Pixel Photon Counter (MPPC) photodetectors. The absorber is also tungsten and the sensitive area was 18x18 cm$^2$. The scintillator strips are 1 cm wide and oriented alternatively in $x$ and $y$ directions to make effectively 1 cm$^2$ cells comparable to SiW ECAL. Thus good comparison between two ECAL realisations can be done. We refer in detail about the scintillator ECAL in section 4.

The first completed prototype of the hadron calorimeter was a 1-m$^3$ steel-scintillator calorimeter with SiPMs as photodetectors that provide analog amplitude. The optimisation studies for the PFA defined (Thomson, 2009) as the best tile size 3x3 cm$^2$. For the economical and practical reasons, the prototype was built from tiles of three different sizes made of 5 mm thick scintillator placed between steel absorber plates 16 mm thick. 38 sandwich layers have a depth of 1 m and represent 5.3 interaction lengths. The details are given in section 5.

Recently two versions of calorimeter with the scintillator replaced by the gas detector – Resistive Plate Chamber (RPC) – were completed and are currently tested in beam. As the RPC provides digital amplitude from a 1 cm$^2$ cell, we call this type of calorimeter digital. Further details are given in references (Bilki et al., 2009; Laktineh, 2011).

## 3. Silicon photomultiplier

The CALICE collaboration contributed to the first massive use of a novel multi-pixel silicon photodetector operated in the Geiger mode. There are several producers able to deliver this photodetector, e.g. Hamamatsu Photonics Japan (first commercial producer), MEPHI/PULSAR and Dubna/Micron in Russia, Photonique in Switzerland, SENSL in Ireland, ITC IRST Trento in Italy, Zecotek in Singapur, MPI Semiconductor Laboratory Munich in Germany, SensL Cork in Ireland, STmicroelectronics in Italy, Novel Device Laboratory Beijing in China, RMD Boston in U.S.A., and possibly others.

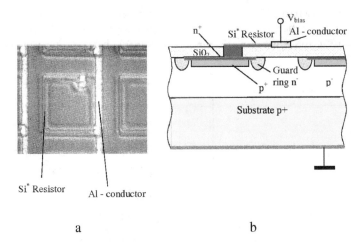

Fig. 1. (a) View of a pixel in an array of the silicon photomultiplier. (b) A cross section of a pixel of SiPM showing the layers. The Si* resistor serves as a quenching resistor, the SiO2 resistor reduces the inter pixel cross talk.

We shall here report about experimentation with the Silicon PhotoMultiplier (SiPM) from MEPHI/PULSAR which delivered more than 10000 pieces for the Analog Hadron CALorimeter (AHCAL) and the Tail Catcher and Muon Tracker (TCMT) and the Pixelated Photon Counter from Hamamatsu whose 2160 pieces of the "1600-pixel MPPC" were used in the Scintillator Electromagnetic CALorimeter (ScECAL).

The silicon photomultiplier is an array of small pixels connected on a common substrate (see Fig. 1). Each cell has its quenching resistor of the order of a MΩ. A common bias voltage is applied to all cells 10-20% above the breaking voltage. The cells fire independently when a carrier is liberated in the depletion layer or by a photon arriving on the surface or thermally. The output signal equals to the sum of fired cells. At small signals the detector works as an analog photodetector, at large signals the detector saturates due to the limited number of pixels (Bondarenko et al., 2000; Buzhan et al., 2001) The photon detection efficiency is $\varepsilon = QE$ $\varepsilon_{geom}$, where quantum efficiency $QE$ is 0.5-0.8 and $\varepsilon_{geom}$ is geometrical factor giving the effective sensitive area of the pixel. The geometrical factor increased from originally several percents up to 80% nowadays (Musienko, 2011). Each pixel works as a digital device – several photons hitting the same pixel produce the same signal as a single photon. Optical cross-talk between pixels causes adjacent pixels to be fired. This increases gain fluctuation, increases noise and excess noise factors. Due to the Geiger discharge, the pulse is short, typically several ns long. The photodetector for the AHCAL prototype was developed in collaboration of Moscow Institutes MEPhI and ITEP with PULSAR factory near Moscow and with support from DESY Hamburg. They called it silicon photomultiplier because of its high gain. The photosensitive area of 1.1 mm x 1.1 mm holds 1156 pixels each having an area 32 µm x 32 µm (sensitive area 24 µm x 24 µm). The single pixel signal is determined by the total charge collected during the Geiger discharge $Q = C_{pixel} \Delta V$ where $\Delta V$ is the voltage above the breakdown voltage $V_{break}$. For the pixel capacitance $C_{pixel} = 50$ fF and $\Delta V = 3$V the signal reaches $Q = 150$ fC $\sim 10^6$ electrons. This defines the gain of $\sim 10^6$. The working voltage equals to $V_{bias} = V_{break} + \Delta V \sim 50$-60 V. The cross talk increases with $\Delta V$, the typical value is

20%. The Geiger discharge time is ~ 500 ps. The pixel recovery time is defined by the product $C_{pixel}\, R_{pixel}$ , where $R_{pixel}$ is the pixel quenching resistor (400 kΩ - 10 MΩ). The quenching resistor interrupts the discharge in a pixel. For smaller resistor values the quenching time is ~ 20 ns.

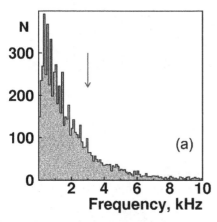

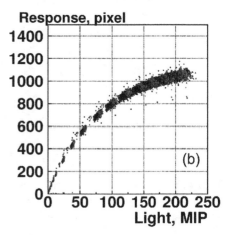

Fig. 2. (a) Distribution of noise above ½ MIP threshold for 10000 SiPMs produced in MEPHi/PULSAR. The arrow at 3 kHz shows selection cut. SiPMs with higher noise were discarded. (b) The response curve for the selected SiPMs. The curve shows the SiPM signal in number of fired pixels as a function of the LED light expressed in MIPs.

The aim was to keep it below 100 ns with the signal width of ~ 10 ns. For short recovery times comparable to the signal width, the SiPM can fire repeatedly causing undesirable inter pixel crosstalk. The pixels are connected in parallel to one readout channel therefore the output pulse is proportional to the number of detected photons (Danilov, 2007b).

More than 10000 SiPM have been produced by the MEPhI/PULSAR group and were tested at ITEP. The SiPMs were illuminated with calibrated light from a UV LED (Ultra-Violet Light Emitting Diode). The light was brought by a Kuraray Y11 WLS fibre. For each SiPM the working voltage was chosen individually to fire 15 pixels for LED light pulse corresponding to the signal which produces a minimum ionizing particle (MIP) - in practice a muon - in a scintillator tile. The gain, the dark rate, the inter-pixel crosstalk, the noise above a threshold of the ½ MIP and the non-linear response function were measured for all SiPMs. The SiPMs had to fulfil selection criteria on gain > $4.10^5$, noise at ½ MIP threshold < 3 kHz, cross talk probability < 0.35 and dark current < $2.10^{-6}$ A. As an example in Fig. 2a the noise distribution is shown, the arrow gives the position for the selection cut.

SiPMs were tested for radiation hardness on the proton accelerator at ITEP Moscow at a beam energy of 200 MeV. The dark current of SiPMs increased with the proton flux $\Phi$ as expected. At doses $\Phi$ ~ $10^{10}$ protons/$cm^2$ the detector does not see individual photoelectron peaks and still resolves MIPs at $\Phi$ ~ $10^{11}$ protons/$cm^2$. This radiation hardness is sufficient for doses the AHCAL will obtain at the LC (Danilov, 2007a).

Photodetectors for the ScECAL were developed by cooperation between the Shinshu University and the Hamamatsu Photonics Company. The MPPC has a photon detection area of 1 mm x 1 mm with 1600 pixels. The delivery was realised in two batches. The smaller batch in 2006 comprised 532 pieces for test purposes of a small calorimeter in the DESY test beam. The main delivery of ~ 2000 pieces was carried out in 2008. Some of the MPPCs from the DESY test set up (448 pieces) were used also in the ScECAL. The characteristics of MPPCs were measured in the pulse mode in a test stand using LED flashes controlled by a pulser. The MPPC temperature was kept at 25°C in a thermostat. The distribution of gain and crosstalk for three MPPC groups is shown in Fig. 3 (Sakuma, 2010). The MPPC properties depend on the production batch but within the same batch they exhibit low spread. The different MPPC gain between two production batches was equalized in the calorimeter by adjusting the individual bias voltage setting for each MPPC. This is done by the front-end ASIC chip.

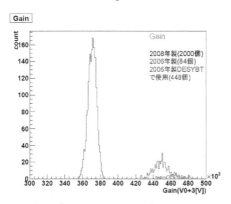
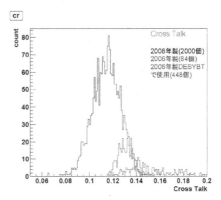

Fig. 3. Distribution of gain (left) and cross-talk (right) for MPPCs produced in Hamamatsu and used in the ScECAL. The production was done in years 2006 and 2008. 448 MPPCs from 2006 and all from 2008 were used in the ScECAL.

The SiPM becomes saturated at large photoelectron intensities $N_{ph}$ due to the limited number of pixels $N_{avail}$. From the probability considerations the number of fired pixels $N_{pix}$ increases exponentially with $N_{ph}$ as $N_{pix} = N_{avail} (1 - exp (-N_{ph} / N_{avail} ))$. The measured dependence of the response curve $N_{pix} (N_{ph})$ for 10000 SiPMs is given in Fig. 2b. The inverse of this dependence is used as a correction function of the measured signal $N_{pix}$ to get the effective number of photoelectrons $N_{ph}$. For more sophisticated treatment of the saturation correction see section 8.

## 4. Scintillator electromagnetic calorimeter

The most important role of the Electromagnetic Calorimeter (ECAL) is to identify photons in jets. The majority of photons are produced by two-photon decay of the neutral pions. The distance between two photons decreases with increasing pion energy and it is several centimetres at the distance 210 cm from the decay point for pion energy several tens of GeV. Therefore, the ECAL granularity around 1 cm will allow resolution of photons from pions up to energies of about 50 GeV.

The ScECAL was built with 1 cm x 4.5 cm scintillator strips 0.3 cm thick assembled from 18 pieces in 4 rows into a plane. Each strip is wrapped in reflecting foil to isolate it from the neighbour and to improve the light collection. Scintillation light is read by a wavelength shifting (WLS) fibre placed in the middle of the strip along its longer side to improve the homogeneity of the light collection. The calorimeter is made of 30 layers with strips alternately rotated by 90 degrees. In between two scintillator planes, tungsten absorber plane 0.35 cm thick, is inserted (see Fig. 4).

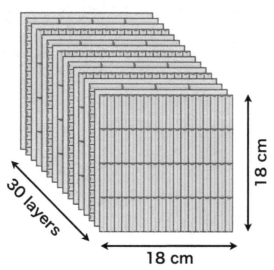

Fig. 4. Scintillator ECAL structure. The scintillator strips (green) are arranged in 4 rows by 18 pieces to form a plane. WLS fibers in yellow lead light to MPPC (red spot) at the strip end. The W absorber planes are shown in grey.

The WLS double clad fibre contains Y11 wavelength shifter and emits light at 550 nm. The fibre is read at one end by a MPPC. The MPPC has a photon detection area of 1 mm x 1 mm with 1600 pixels. The MPPCs were soldered on a flat cable and mounted at the end of each fibre. Test were also made where the MPPCs were attached directly to the tile as the loss in the light collection from the missing WLS can be partly regained by a better match in peak sensitivity of MPPC (at 400 nm) to the scintillation light. The number of photoelectrons dropped to one half but a significant simplification in construction can be achieved.

The readout of the calorimeter uses the same electronics as the AHCAL and it is described in the next section. The basic eighteen channel front-end ASIC chip reads out in this case one row of scintillators. The prototype of the ScECAL had 2160 MPPCs and it was built in collaboration of Shinshu, Kobe, Tsukuba, Tokyo, Niagata Universities in Japan and Kyungpook National University in Korea (Kotera, 2010).

The calorimeter has been tested in muon and electron beams at energies 1–32 GeV, pion beam 2-60 GeV energy (including neutral pions) in Fermilab in 2008-9. The detector was calibrated using muons to measure the energy response of a MIP. The light yield of 23 photoelectrons was measured for a minimum ionizing particle. The data were further corrected for the saturation (see section 8). The response of the whole calorimeter to several electron beam energies is shown in Fig. 5. The data show clear Gaussian behaviour for individual energy.

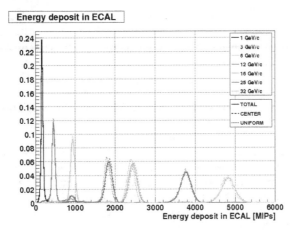

Fig. 5. The corrected energy spectra measured with ScECAL at energies 1–32 GeV.

The data were then purified for small contamination from other particles in the beam namely muons and electrons. The energy distribution was fitted for each energy with a Gaussian function including 90% of data around the Gaussian maximum to further suppress residual contamination (lowest energy point at 1 GeV required a more refined procedure). The deposited energy in the calorimeter is given as the mean of Gaussian function from the fit and the resolution by the standard deviation $\sigma$.

The mean of the Gaussian for each beam momentum is displayed in Fig. 6 (left). The points were fitted by a linear function forced to go through the origin. The points differ from the fit at most by 6% at the highest energies. The sources of the nonlinearity are temperature dependence of sensors, spread in saturation correction of sensors and spread in calibration

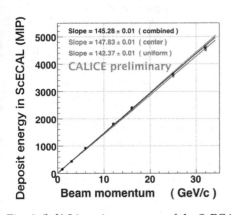

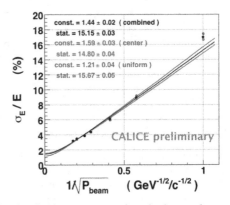

Fig. 6. (left) Linearity response of the ScECAL. Blue (red) data correspond to the beam shot uniformly on (at the center of) the front face of the calorimeter, black are data combined together. (right) The energy resolution of ScECAL as a function of the inverse of the square root of the beam momentum. The formula fitted to data points is quadratic and it is explained in the text.

factors for strips. The sources of the nonlinearity are being further investigated. The calorimeter energy resolution $\sigma_E/E$ is plotted in Fig. 6 (right) as a function of $\sqrt{(1/P_{beam})}$. Data at different beam momenta were fitted to the quadratic formula $\sigma_E/E = \sigma_{const} \oplus \sigma_{stat}/\sqrt{(P_{beam})}$, where $\sigma_{const}$ ($\sigma_{stat}$) represent constant and stochastic term in the calorimeter energy resolution and $A \oplus B = \sqrt{(A^2+B^2)}$. Non-zero $\sigma_{const}$ describes the effect of non-uniformities in the calorimeter (mechanical or due to physics, i.e. different response to electrons and hadrons). The value of $\sigma_{stat}$ depends on the ratio of scintillator and absorber thicknesses and it is called the 'stochastic coefficient'. It is due to the sampling fluctuations of particle showers in the active medium. The values obtained from tests are $\sigma_{const}$ = (1.44±0.02)% and $\sigma_{stat}$ = (15.15±0.03)%. The value $\sigma_{const}$ is slightly worse than the value 1.07% obtained for a similar calorimeter using 1 cm x 1 cm Si diodes (Repond et al., 2008). The value $\sigma_{stat}$ is better than the value 16.53% from the SiW calorimeter. The fact that the ScECAL uses 1 times fewer channels which are cheaper to produce and shows a similar performance makes this type of calorimeter a very attractive candidate for the electromagnetic calorimeter in the detector at the LC.

## 5. Analog hadron calorimeter

The most important task of the hadron calorimeter is to separate the contributions from deposits of charged and neutral hadrons in the hadron shower and to measure the energy of neutral particles. The choice of parts for the physics prototype was a compromise between performance, price and availability. The scintillator chosen comes from UNIPLAST Russia (polystyrene + 1.5% PTO + 0.01% POPOP), the WLS fibre of 1 mm diameter is Kuraray Y11(200) in a groove with ¼ circle for the smallest tiles and with 1 loop for the larger tiles. Before bending the fibre was heated to ~ 80° C.

One layer of the sampling structure of the tile AHCAL physics prototype calorimeter consists of a 16 mm thick absorber plate made from stainless steel and a steel cassette of 96 cm x 96 cm area with two 2 mm thick steel walls to house a plane made from 5 mm thick scintillator tiles. The calorimeter is built as a cube of approximately 1 m³ volume. The total number of 38 layers corresponds to 5.3 interaction lengths $\lambda_{int}$. The view of one plane with scintillator tiles is shown in Fig. 7 left. In the centre where the test beam passes through, one hundred small 3 cm x 3 cm tiles is surrounded by three rows of tiles 6 cm x 6 cm and finally at the perimeter by one row of tiles 12 cm x 12 cm. The tiles at 4 corners were removed as the occupancy is expected to be low. Each tile is coupled via a WLS fibre inserted in a groove to a SiPM via an air gap (Fig. 7 right). The tiles are covered from top and bottom with a super-radiant foil VM2000 from 3M. The tile side edges were matted in order to provide better homogeneity of the light on the photodetector.

One detector plane comprises 216 scintillator tiles closely packed inside a frame made from aluminium bar 10 mm x 8 mm. The SiPMs are connected to the front-end electronics via 50 $\Omega$ micro-coax cables that carry both signal and bias voltage. Each tile is also connected to a clear fibre which brings calibration light from LED. Both coax cables and fibres lay on a support FR4 plate covered with a mylar foil for insulation. It means that SiPMs are not soldered on a PCB as this construction is too fragile to carry the mass of the scintillator. The coax cables are connected on one side of the plane to the front-end electronics, eighteen optical fibres are attached on the other side to one LED mounted on the calibration and monitoring board (Adloff et al., 2010a).

Fig. 7. (left) Scintillator tile layer of the AHCAL calorimeter prototype. The tile of three different sizes can be recognized as well as the WLS fibers of the circular shape inside the tiles. The photodetector SiPM is placed in the tile corner and powered and read-out via connections to micro-coax cable below the scintillators. (right) Scintillator tile 3 cm x 3 cm with a WLS fiber in a groove attached via an air gap to the photodetector SiPM placed in the upper left corner of the tile. Two SiPM contacts are used to bring the bias voltage and the readout of the signal. The other end of the WLS fiber is covered by a reflector to improve the overall light yield.

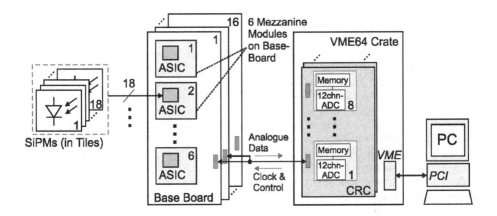

Fig. 8. Schematic view of the readout and data acquisition system of the AHCAL

A schematic view of the readout system is shown in Fig. 8. Eighteen SiPMs are connected to the same ASIC (de La Taille et al., 2005) which has for each SiPM preamplifier, shaper and sample-and-hold-circuit. It also allows individual bias voltage settings for each SiPM. It is made in the 0.8 μm Complementary Metal-Oxide-Semiconductor (CMOS) technology. The signals from twelve Application-Specific Integrated Circuits (ASICs) are fed into the CALICE Readout Card (CRC) and digitized by 16-bit Amplitude Digital Converters (ADCs). The data are stored via a PC, which controls also the data taking. The construction of the physics prototype and the readout and data acquisition system is a common effort of DESY, Hamburg University, Imperial College London, ITEP, Lebedev Physics Institute and MEPHI Moscow, LAL Orsay, and Institute of Physics Prague.

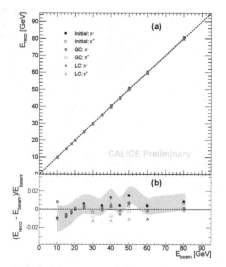
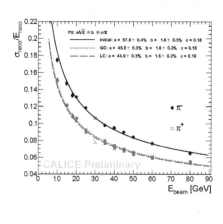

Fig. 9. (left a) Linearity of the AHCAL to pions. (left b) Relative residuals to beam energy versus beam energy. (right) Relative energy resolution of the AHCAL for pions versus beam energy. Black points and lines correspond to analysis without compensation, blue (red) points and curves show results of analysis after local (global) compensation.

The AHCAL was extensively exposed to beams of electrons and hadrons in years 2007-9. All detectors showed high reliability. The calibration and monitoring procedures were validated. At the beginning the calorimeter was calibrated to get the overall energy scale. With help of the wide muon beam each calorimeter cell was exposed to have at least 500 reconstructed muon tracks in the cell. These data were used to obtain the MIP conversion factor expressing the charge measured (in units of ADC counts) by the SiPM in units of MIPs. With low intensity LED light the individual photons (or pixels, see Fig. 10 in section 7) were recorded by the same data acquisition chain and also measured in ADC counts. The operation voltage of SiPMs was set at the value of ~ 13 pixels/MIP to equalize the SiPM response. The factor converting MIP to energy in MeV was obtained from the Monte Carlo simulation and allowed to express the charge measured by SiPM in energy scale in MeV. This energy was finally corrected for the SiPM saturation. The SiPMs proved their reliability and gain stability over the whole period and the whole HCAL showed its robustness during transport between laboratories at DESY, CERN and Fermilab. During the test beam periods at CERN the data were collected in the pion beam in the energy range 8-80 GeV and $e^{\pm}$ beams 6-45 GeV. The pion data were purified, reconstructed and calibrated using the standard CALICE chain (Adloff et al., 2011). To improve the linearity and the energy resolution of the AHCAL, two software correction methods were developed which compensate the AHCAL higher response to electrons e/π = 1.19. The algorithms use information on the shower substructure and reweight higher local energy deposits by suitable factors on event-by-event basis. The linearity and energy resolution of the complete CALICE setup (ECAL + AHCAL + TCMT) is shown in Fig. 9 for data (black) and data after software compensation correction by the local compensation (blue) and global compensation (red) methods. The response is linear within ± 1.5% with slight improvement due to compensation. The resolution formula fitted to data is $\sigma_E/E = \sigma_{const} \oplus \sigma_{stat}/\sqrt{(E_{beam})} \oplus 0.18/E_{beam}$, $E_{beam}$ in GeV. The last term describes the noise width estimation of the test beam setup. In this case the effect of compensation is significant. The values of $\sigma_{stat}$ = 57.6% (44.9%, 45.8%) for the data without (with local, global) the software compensation show that on average the compensation methods improve the energy resolution by 21%. The constant term $\sigma_{const}$ = 1.6%. (Chadeeva, 2011)

## 6. Tail catcher and muon tracker

The TCMT is a sandwich calorimeter placed behind the AHCAL. It has also a volume of 1 m³, but coarser structure than the AHCAL. Steel absorber plates have thickness 2 and 10 cm, the active layer consists of 100 cm long, 5 cm wide extruded scintillator strips 0.5 cm thick. The first 8 sections behind the AHCAL have a similar longitudinal segmentation as the AHCAL and will supplement the AHCAL measurement for the tail-end of the hadron showers. The last 8 coarser sections serve as a prototype of a possible muon detector for any design of the ILC detector. It represents 5.8 interaction lengths.

The scintillation light is read out by parallel WLS fibres of 1.2 mm diameter inserted in the co-extruded holes. The fibre is read at one end by a SiPM of the same type as in the AHCAL. The strips and SiPMs are inserted into cassettes made from 1 mm thick steel, the cassettes are inserted alternately in the $x$ and $y$ directions between absorber plates. The TCMT uses 320 SiPMs in total. Each cassette has inside a LED driver board with 20 UV LEDs, one per strip. The TCMT has been constructed by Fermilab, Northern Illinois University, and DESY (Chakraborty, 2005).

## 7. Monitoring and calibration

During the data taking with the calorimeter it is important to monitor the stability of the entire read-out system starting from the quality of light transfer between scintillator and photodetector to the read-out electronics. The photodetector gain G significantly changes with temperature as $dG/dT \sim -1.7\%/K$ and with the bias voltage as $dG/dU \sim 2.5\%/0.1V$ (Eigen, 2006), therefore the temperature of the photodetector must be periodically recorded.

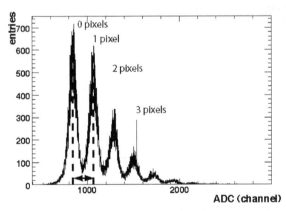

Fig. 10. SiPM response to low intensity light. The first peak corresponds to the pedestal, other peaks are signals from one, two and three photoelectrons. The peaks are equidistant, the distance indicated by ↔ is proportional to SiPM gain. The scale is in ADC counts corresponding to the charge of the SiPM signal (1 ADC count is 0.25 pC).

Also the stability of the SiPM response curve in the whole dynamic range must be known for offline corrections. These tasks are achieved with the calibration and monitoring system.

The photodetector gain and its saturation are monitored by UV LED flashes ($\lambda$ = 400 nm) which are delivered to each tile via clear fibres. For the response curve measurement, the LED pulses must be tuneable in intensity from a few photoelectrons to thousands photoelectrons on the SiPM. The upper range value is given by the maximal expected rate of particles in a calorimeter cell which was obtained from shower simulations and is about 100 particles (MIPs). For the gain setting of about 13 photoelectrons/MIP, we arrive at the highest signals of ~ 1300 photoelectrons.

The LED light pulse is ~ 10 ns wide and of trapezoidal shape. Particle ionization in the scintillator tile produces typically one order of magnitude shorter light pulses. The LED light pulse intensity is proportional to the pulse length. To get intensity on the level of a thousand of photoelectrons it is easier to use longer pulses. The pulse length is limited by the after-pulsing property of the photodetector and by choosing the value of the quenching resistor, the afterpulsing can be almost avoided. The low intensity light is used for gain calibration and uses the unique sensitivity of SiPM to single photoelectrons (see Fig. 10).

The scheme of the electronic circuit for LED light pulses – LED driver – is displayed in Fig. 11. The design is particularly optimized for the generation of nearly rectangular fast pulses (rise and fall time ~1 ns) with variable amplitude in a large dynamical range. For these reasons we abandoned the classical approach of discharging a capacitor to the LED (Apuhn et. al., 1997). We adopted a scheme using the IXLD02 integrated circuit developed for semiconductor lasers. Based on the Tcalib signal from the calorimeter control, three signals are derived. The ENable, DRVpulse, and invDRVpulse control the output stage of LED driver. The ENable pulse wakes up the circuit. The LED is reversely polarized until the arrival of the DRVpulse which switches on the transistor Q1 for about 10 ns and the LED makes a flash. The role of the invDRVpulse is to improve the rise and fall time of the LED flash by switching complementary the transistor Q2 and since that, LED is quickly reversed till the end of ENable. The intensity of the LED light amplitude is controlled by the Vcalib amplitude in the control current sink CCS. The LED light emission characteristics do not change with the increase of the LED current in the range 10 µA to 10 mA (Polak, 2006).

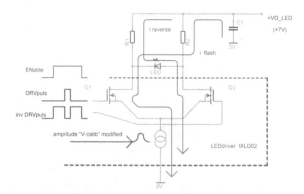

Fig. 11. The principle scheme of the LED driver. IXLD02 is the IC which produces rectangular pulses based on signals Ebable and DRV. The invDRVpulse improves the pulse shape. The pulse amplitude is controlled by the Vcalib value.

In order to reduce light fluctuations among LEDs, LEDs were sorted into groups with a similar light-intensity profile. One LED supplied 18 tiles and a monitoring PIN diode. Since PIN photodiodes have a gain of unity, an additional charge-sensitive preamplifier for the PIN photodiode readout was envisaged. The presence of a high-gain preamplifier directly on the board in the vicinity of the power signals for LEDs, however, has turned out to be a source of cross talk. One calorimeter plane of 216 tiles required 12 LEDs and 12 PIN diodes. The electronics providing LED pulses is laid on the Calibration and Monitoring Board (CMB) and located along one side of the calorimeter plane opposite to the read-out electronics and bias voltage supply. Another task of the CMB consists of reading out the temperature sensors via a 12 bit ADC. The temperature values are sent to the slow control system via a CANbus interface. One CMB operates seven temperature sensors, two sensors directly located on the readout board and five sensors distributed across the centre of the cassette. The sensors consisting of integrated circuits of type LM35DM produced by National Semiconductor are placed in a 1.5 mm high SMD socket. Their absolute accuracy is < 0.6°C. A microprocessor PIC 18F448 in association with a CAN controller interface (PCA82C250) provides the communication of the CMB with the slow control system. We have not observed any noise pickup in the CMB.

## 8. Saturation correction

As already mentioned in section 3, this photodetector is highly non-linear device due to finite number of pixels, the finite recovery time given by the value of internal resistor $R_{pixel}$ and also due to the crosstalk. The number of fired pixels $N_{pix}$ in the simplest case can be approximated by the formula $N_{pix} = N_{avail} ( 1 - \exp (-\varepsilon N_{ph} / N_{avail} ))$ for number of photons $N_{ph}$ impinging uniformly and simultaneously on the total number of available pixels $N_{avail}$ of the photocathode, $\varepsilon$ is the photon detection efficiency (defined in section 3). This simplest form has been used for the MPPCs of the ScECAL (Kotera, 2010). The number of pixels obtained from the fit to the measured response curve was $N_{avail} = 2424$ for photodetector with 1600 pixels. The large value of $N_{avail}$ from the fit was attributed to the short recovery time of some pixels. The pixel then can fire for a second time on late photons radiated from the WLS fibre. In ref. (Adloff et al., 2011) where properties of SiPMs in the AHCAL are described, the sum of two exponentials was used which takes into account possible non homogeneity in the light transmission between the WLS fibre and the photodetector photocathode. The formula can in principle be extended to the sum of more exponentials to take into account other possible non-homogeneities. The response function for each SiPM was inverted and the correction function for one SiPM is shown in Fig. 12. The response of all SiPMs of the AHCAL clearly showed that on average only 80% of the photodetector pixels are used. This was explained by the different geometry cross-section of the WLS fibre with 1 mm diameter that covers effectively 78.5% of the square surface of the photocathode 1x1 mm$^2$.

The detailed study of the parameter $N_{avail}$ was thereafter done for all SiPMs using only the response at the highest LED light intensities using the single exponential $N_{pix} = N_{avail} [1- \exp(-(X-C)*B)]$, X is the voltage on the LED. The constant C takes into account the LED starts to emit at voltage $X = C$. Fits to data allowed a value of $N_{avail}$ to be obtained for 83% of SiPMs with a spread of 0.7% around the value 1156 pixels * 0.8 = 928 pixels (Zalesak, 2011). For the remaining SiPMs the nominal value of $N_{avail} = 928$ pixels can be used.

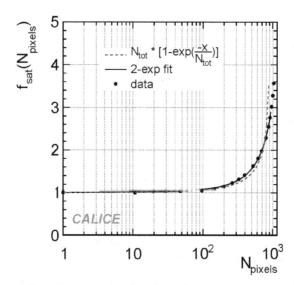

Fig. 12. The SiPM non-linearity correction function. The points are values for one SiPM. The solid line is double exponential fit, the dashed line is the single exponential fit, $N_{tot} = N_{avail}$ used in the text.

## 9. New developments

The experience with the physics prototypes will be used in the design of calorimeters for the full size detector at the LC. E.g. the hadron calorimeter will occupy volume of ~ 80 m³ and with scintillator tiles 30 mm x 30 mm it will have ~ 3.2 10⁶ channels (Abe et al., 2010). The readout electronics will be integrated in the calorimeter volume to minimize the dead space. The electronics has to have low power consumption to minimize the demands on cooling. The detector parts must be designed to allow the industrial production. As the next step technological prototypes are being designed and built and will meet most of these requirements.

### 9.1 Hadron calorimeter

The design of a half-octant of AHCAL is shown in Fig. 13. A single detection layer, one of 48 making the full depth of the calorimeter, consists typically of 16 HCAL Base Units (HBU). The HBU (Fig. 14) is the smallest component of the calorimeter with 144 scintillator tiles. To minimize the depth of the calorimeter, the tile thickness was decreased from 5 to 3 mm. Each tile will be equipped with a new type of SiPM (Sefkow, 2006).

The WLS fibre will be placed in a straight groove in the tile to avoid the time-consuming process of fibre bending (see Fig. 15). The thinner scintillator tile affects the SiPM design. We plan to use CPTA MRS APDs with the number of pixels reduced to ≤ 800 (Buanes et al., 2010). These SiPMs have smaller cross-talk and noise and lower sensitivity to temperature and voltage variations than SiPMs from PULSAR used in the AHCAL physics prototype. The latest tests in the positron beam showed a very good functionality with a signal of ~ 15 pixels/MIP.

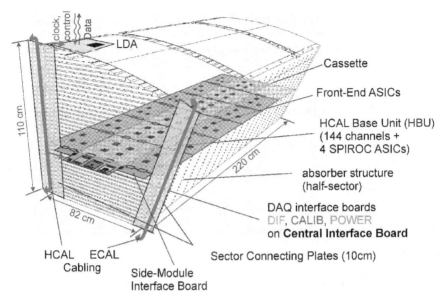

Fig. 13. View of the 1/16 of the AHCAL barrel wheel. One of 48 detection layers is shown in colour. It is made of HCAL base units HBUs. The service modules DAQ, DIF, CALIB and POWER are outside the detector volume.

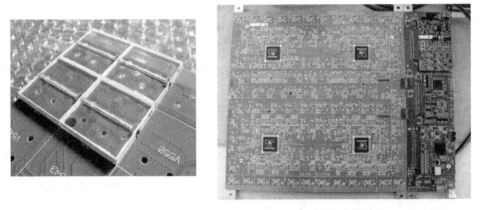

Fig. 14. (left) Four scintillation tiles assembled on the bottom of the HBU. The top side of the HBU. (right) has the front-end electronics with 4 ASICs and calibration LEDs for 144 tiles. The HBU is connected from right to DIF (bottom), CALIB (middle) and POWER (top) boards.

The new front-end ASIC chip powers, reads and amplifies the signal from 36 photodetectors (twice more than in the physics prototype). A 12-bit ADC in the chip digitizes the photodetector amplitude and provides fully digitized output. The size of the chip is reduced to 30 mm$^2$ using the 0.35 µm technology. The chip can be operated in the power pulsing mode with significantly reduced power consumption 25 µW (40 µW) per channel (including

SiPM bias voltage). After packaging the ASIC size is 28x28x1.4 mm³ (Raux, 2008). The first experience with the ASIC performance was recently reported (Terwort, 2011).

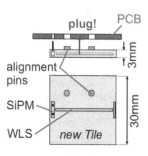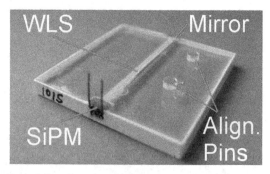

Fig. 15. The construction drawing (left) and the photo (right) of the scintillator tile with straight WLS fibre and SiPM attached to its end. The alignment pins in the tile fix the tile into the PCB.

The response of SiPMs is controlled by new calibration and monitoring system. Two concepts are currently under development:

- Each channel incorporates a SMD LED embedded on the HBU. LED illuminates the scintillator through a hole in the PCB. The control of LEDs is provided by the CALIB module on side of the HBU. This system is already implemented in the setup of Fig. 14. Each LED has its own driver – a low-component-count and effective circuit that discharges a capacitor through the LED. It was observed that pulse length depends on the LED type – blue LED generates long pulses ~40 ns and UV LEDs fast pulses ~8 ns long. This behaviour was explained by the different internal structure of LEDs which leads to different internal LED capacitance. At low intensity the driver generates single photon electron spectra for gain calibration purposes. The driver was finally tuned to deliver also high intensity light up to saturation mode of the SiPM with 796 pixels (Sauer et al., 2011).
- A luminous space LED with its driver placed on a special board. The light is distributed via notched fibres to rows of tiles (up to 72 tiles/fibre with the light output homogeneity of 20%). The fibre is positioned on the HBU on the top side. The notch flashes light through a hole in the PCB to the scintillator. The LED driver works as a heavily dumped quasi-resonant sine waves generator, where the first positive half-wave generates light and negative half-wave bias the LED negatively. Following sine waves are small and keep the LED reverse biased (non-shining mode) (Kvasnicka, 2011). It utilizes a toroidal inductor embedded in the PCB, and produces almost sinusoidal pulses which reduce the electromagnetic interference. For the toroid inductance 35 nH, the LED flash has fixed length of 3.5 ns. The LED amplitude is tuneable from low intensity light for the SiPM gain calibration to the high intensity signal equivalent to 200 MIPs in the tile. For the driver circuit scheme and the PCB with the circuit and toroidal inductor see Fig. 16.

As can be seen in Fig. 13 and in a greater detail in Fig. 14 (right), each detection plane is interfaced with the Data AcQuisition (DAQ) system via Detector InterFace (DIF) board. The calibration system is steered by the CALIB board and the power module (POWER) distributes necessary voltages and arranges the power pulsing regime. All modules are available and are currently under test to commission the communication between different systems. The detection plane or its part will undergo beam tests in the year 2012.

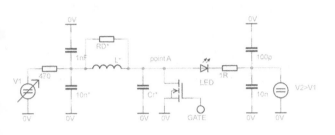 

Fig. 16. (left) Driver circuit scheme. (right) LED driver on PCB with toroidal inductor at the top left.

## 9.2 Scintillator ECAL

The development of the next generation of the scintillator electromagnetic calorimeter will be done together with the SiW ECAL (Takeshita, 2011). To decrease the price of the calorimeter determined mainly by the price of the silicon sensors, it was decided to replace approximately half of the silicon sensor planes in the ECAL by scintillator planes with orthogonal strips. As the size of the silicon sensors was decreased from 10 mm x 10 mm to 5 mm x 5 mm, the width of scintillator strips will be correspondingly reduced from 10 mm to 5 mm. The scintillator thickness will decrease from 5 mm to 2 mm to better match the width of silicon layers. It is no longer possible to insert a WLS fibre inside such a thin scintillator. Therefore a new R&D program is envisaged aimed at the direct coupling of the MPPC to the tile side. This will further simplify production of large detector.

## 10. Conclusion

Prototypes of electromagnetic and hadron calorimeters with scintillator and embedded photodetectors for a detector at a future linear collider have been successfully built. The calorimeters with unprecedented granularity were successfully tested in electron, muon and hadron beams at accelerators in CERN and Fermilab. Commissioning and operation in the test beam demonstrated that calorimeters perform according to expectations. Test data were used to set up and tune the energy calibration of calorimeters. In case of the AHCAL, two new software correction methods were developed which compensate for the calorimeter's higher response to electrons. The application of both methods to the test beam data results in improvement in the energy resolution of the hadron calorimeter by 21% in the energy range 10–80 GeV.

New semiconductor photodetectors were developed for both calorimeters which provide gain of a classical photomultiplier but in a photodetector with millimetre dimensions and bias voltage of ~ 60 V. The new photodetector is embedded directly in the scintillator tile/bar and significantly reduces the calorimeter dead space and simplifies calorimeter construction. More than 10000 of these photodetectors in total were produced by

MEPHi/PULSAR (called silicon photomultipliers) and Hamamatsu Photonic (called pixelated photon counters). The successful operation of these detectors over a period of several years opened the door for their use in applications in material research and medicine.

The experience with the physics prototypes of calorimeters will be used in the design of calorimeters for the full size detector at the linear collider. Here the total number of channels will be of the order of millions. This brings new challenges especially on electronics which must be integrated in the calorimeter volume to minimize the dead space. The electronics must have low power consumption to minimize demands on cooling. The first versions of read-out ASICs were developed. The ASIC integrates functions performed previously by the data acquisition electronics. The next generation of calorimeter prototypes has started to be built and will be ready for beam tests in 2012.

## 11. Acknowledgment

I would like to thank to all my colleagues from the CALICE collaboration for the excellent work done during the construction of calorimeters, their operation and convincing results from the beam tests. Especially I want to thank to M. Danilov, G. Eigen, E. Garutti, K. Kotera, J. Kvasnička, V. Morgunov, I. Polák, F. Sefkow, D. Ward and J. Zálešák for discussions, comments, and providing me with the material for this contribution.

This work was supported by the Ministry of Education, Youth and Sports of the Czech Republic under the projects AV0 Z3407391, AV0 Z10100502, LC527 and LA09042.

## 12. References

Abe, T., et al. (2010). The International Large Detector: Letter of Intent, In: *FERMILAB-LOI-2010-03, FERMILAB-PUB-09-682-E, DESY-2009-87, KEK-REPORT-2009-6*, ISBN 978-3-935702-42-3

Adloff, C., et. al. (CALICE Collaboration). (2010a). Construction and commissioning of the CALICE analog calorimeter prototype. *Journal of Instrumentation*, Vol.2010, No. 5, (May 2010), p. P05004, ISSN 1748-0221, e-Print: arXiv:1003.2662 [physics.ins-det]

Adloff, C., et al. (CALICE Collaboration). (2010b). Study of the interactions of pions in the CALICE silicon-tungsten calorimeter prototype. *Journal of Instrumentation*, Vol.2010, No. 5, (May 2010), pp. P05007, ISSN 1748-0221, e-Print: arXiv:1004.4996 [physics.ins-det]

Adloff, C., et. al. (CALICE Collaboration). (2011). Electromagnetic response of a highly granular hadronic calorimeter. *Journal of Instrumentation*, Vol.2011, No. 6, (June 2011), p. P04003, ISSN 1748-0221, e-Print: arXiv:1012.4343 [physics.ins-det]

Apuhn R.-D., et. al. (1997). The H1 lead / scintillating fiber calorimeter, *Nuclear Instruments & Methods in Physics Research*, Vol.A386, (1997), pp.397-408, ISSN 0168-9002

Bilki, B., et al. (2009). Measurement of Positron Showers with a Digital Hadron Calorimeter. *Journal of Instrumentation*, Vol.2009, No.4, (April 2009), p. P04006, ISSN 1748-0221, e-Print: arXiv: 0902.1699 [physics.ins-det]

Bondarenko, G., Buzhan, P., Dolgoshein, B., Golovin, V., Gushin, E., Ilyin, A., Kaplin, V., Karakash, A., Klanner, R., Pokachalov, V., Popova, E. & Smirnov, S. (2000). Limited Geiger-mode microcell silicon photodiode: New results. *Nuclear Instruments & Methods in Physics Research*, Vol.A422, No.1-3, (March 2000), pp. 187-192, ISSN 0168-9002

Brient, J.-C. & Videau, H. (2001). Calorimetry at the future e+e- collider, *Proceedings of APS/DFB/DBP summer study on the future of particle physics*, Snowmass, Colorado, USA, June30-July21, 2001

Buanes, T., Danilov, M., Eigen, G., Göttlicher, P., Markin, O., Reinecke, M. & Tarkovski, E. (2010. The CALICE hadron scintillator tile calorimeter prototype, *Nuclear Instruments & Methods in Physics Research*, Vol.A623, No.1, (November 2010), pp. 342-4, ISSN 0168-9002

Buzhan, P., Dolgoshein, B., Ilyin, A., Kantserov V., Kaplin, V., Karakash, A., Pleshko, A., Popova, E., Smirnov, S., Volkov, Yu., Filatov, L., Klemin, S. & Kayumov, F. (2001). An advanced study of silicon photomultiplier. *ICFA Instrumentation Bulletin*, Vol.Fall2001, No.3, (2001), pp. 1-14, Available from: http://www.slac.stanford.edu/pubs/icfa/fall01/paper3/paper3b.html

Calice. (n.d.). CALICE Collaboration, Available from: https://twiki.cern.ch/twiki/bin/view/CALICE/CaliceCollaboration

Chadeeva, M. (2011). Software Compensation using the CALICE calorimeters, *Proceedings of the LCWS 2011*, Granada, Spain, September 26-30, 2011

Chakraborty, D. (2005). The Tail-Catcher/Muon Tracker for the CALICE Test Beam, *Proceedings of the 2005 International Linear Collider Workshop*, PSN 0919, Stanford, California, U.S.A, March 18 – 22, 2005. Available from: http://www.slac.stanford.edu/econf/C050318/proceedings.htm

CLIC. (2011). The Compact Linear Collider Study, Available from: http://clic-study.org

Cvach, J. (2002). Calorimetry at a future e+e- collider, *Proceedings of the 31th international conference on high energy physics (ICHEP)*, pp. 922-926, ISBN 0-444-51343-4, Amsterdam, The Netherlands, July 23-29, 2002

Danilov, M. (2007a). Comparison of different multipixel Geiger photodiodes and tile-photodetector couplings, *Proceedings of the Linear Collider Workshop LCWS 2007 and ILC 2007*, pp. , ISBN 978-3-935702-27-0, Hamburg, Germany, May 30 – June 3, 2007. Available from: http://lcws07.desy.de/e14/index_eng.html

Danilov, M. (2007b). Scintillator tile hadron calorimeter with novel SiPM readout. *Nuclear Instruments & Methods in Physics Research*, Vol.A581, No.1-2, (October 2007), pp. 451-456, ISSN 0168-9002

Eigen, G. (2006). The CALICE scintillator HCAL test beam prototype, *Proceedings of the 12th International Conference on Calorimetry in High Energy Physics, AIP Conference Proceedings Series, High Energy Physics Subseries*, Vol. 867., 2006, XXIV, ISBN: 978-0-7354-0364-2, Chicago, Illinois, U.S.A., June 5-9, 2006

ILC. (2007). ILC Reference Design Report, August 2007, Available from: http://www.linearcollider.org/about/Publivations/Reference-Design_Report

Kotera, K. (2010). Study of the Granular Electromagnetic Calorimeter with PPDs and Scintillator Strips, *Proceedings of the 12th Vienna Conference on Instrumentation VCI2010*, Vienna, Austria, February 15 – 20, 2010. Available from http://indico.cern.ch/getFile.py/access?contribId=182&resId=0&materialId=paper&confId=51276

Kvasnicka, J. (2011). LED calibration systems for CALICE hadron calorimeter, *Proceedings of the 2nd conference on Technology and Instrumentation in Particle Physics TIPP 2011*, Chicago, Illinois, U.S.A., June 8-14, 2011

Laktineh, I. (2011). Construction of a technological semi-digital hadronic calorimeter using GRPC. *Journal of Physics: Conference Series*, Vol.293 (2011) p. 012077, ISSN 1742-6596

Morgunov, V.L. (2002). Calorimetry design with energy-flow concept, *Proceedings of the 10th conference on calorimetry (Calor02)*, pp. 70-84, ISBN 981-238-157-0, Pasadena, California, USA, March 25-29, 2002

Musienko, Yu. (2011). State of the art in SiPM's. *The Technology Transfer Network for Particle, Astroparticle and nuclear physics event*, CERN, Geneva, Switzerland, February 16 – 17, 2011. Available from:
http://indico.cern.ch/getFile.py/access?contribId=11&sessionId=7&resId=0&materialId=slides&confId=117424

Polak, I. (2006). Development of Calibration system for AHCAL, *International Linear Collider (ILC) Workshop*, Valencia, Spain, November 6-10, 2006. Available from: http://www-hep2.fzu.cz/calice/files/ECFA_Valencia.Ivo_CMB_Devel_nov06.pdf]

Raux L. (2008). SPIROC Measurement: Silicon Photomultiplier Integrated Readout Chip for ILC, *Proceedings of the 2008 IEEE Nuclear Science Symposium (NSS08)*, Dresden Germany, 2008

Repond, J., et al. (CALICE Collaboration). (2008). Design and Electronics Commissioning of the Physics Prototype of a Si-W Electromagnetic Calorimeter for the International Linear Collider. *Journal of Instrumentation*, Vol.2008, No. 3, (March 2008), pp. P08001, ISSN 1748-0221, e-Print: arXiv:0805.4833 [physics.ins-det]

Sauer, J., Götze, M., Weber, S. & Zeitnitz, C. (2011). Concept and status of the LED calibration system, *CALICE Collaboration meeting*, CERN Geneva, Switzerland, May 19-21, 2011. Available from:
http://indico.cern.ch/contributionDisplay.py?sessionId=10&contribId=7&confId=136864

Sakuma, T. (2010). *Performance Study of Prototype Fine-granular EM Calorimeter for ILC.* Shinshu University, Shinshu, Japan, Master thesis, 01.02.2010. Available from: http://hepl.shinshu-u.ac.jp/index.php?HE%20Lab%20Top%20Page%2FIntroduction%2FMaster%20Theses

Sefkow, F. (2006). MGPDs for calorimeter and muon systems: requirements and first experience in the CALICE test beam. *Proceedings of Science*, Vol.PD07 (2006), p. 003, ISSN 1824-8039

De La Taille, C., Martin-Chassard, G. & Raux, L. (2005). FLC–SIPM: Front-End Chip for SIPM Readout for ILC Analog HCAL, *Proceedings of the 2005 International Linear Collider Workshop*, PSN 0916, Stanford, California, U.S.A, March 18 – 22, 2005. Available from: http://www.slac.stanford.edu/econf/C050318/proceedings.htm

Thomson, M.A. (2009). Particle flow calorimetry and the PandoraPFA algorithm. *Nuclear Instruments & Methods in Physics Research*, Vol.A611, No.1, (November 2009), pp. 25-40, ISSN 0168-9002

Takeshita, T. (2011). ScECAL status report, *CALICE Collaboration meeting*, Heidelberg, Germany, September 14-16, 2011. Available from:
http://ilcagenda.linearcollider.org/conferenceTimeTable.py?confId=5213#20110916

Terwort, M. (2011). Concept and status of the CALICE analog hadron calorimeter engineering prototype, *Proceedings of the 2nd conference on Technology and Instrumentation in Particle Physics TIPP 2011*, Chicago, Illinois, U.S.A., June 8-14, 2011

Zalesak, J. (2011). Calibration issues for the CALICE 1 m³ AHCAL, *Proceedings of the LCWS 2011*, Granada, Spain, September 26-30, 2011

# Permissions

The contributors of this book come from diverse backgrounds, making this book a truly international effort. This book will bring forth new frontiers with its revolutionizing research information and detailed analysis of the nascent developments around the world.

We would like to thank Dr. Sanka Gateva, for lending her expertise to make the book truly unique. She has played a crucial role in the development of this book. Without her invaluable contribution this book wouldn't have been possible. She has made vital efforts to compile up to date information on the varied aspects of this subject to make this book a valuable addition to the collection of many professionals and students.

This book was conceptualized with the vision of imparting up-to-date information and advanced data in this field. To ensure the same, a matchless editorial board was set up. Every individual on the board went through rigorous rounds of assessment to prove their worth. After which they invested a large part of their time researching and compiling the most relevant data for our readers. Conferences and sessions were held from time to time between the editorial board and the contributing authors to present the data in the most comprehensible form. The editorial team has worked tirelessly to provide valuable and valid information to help people across the globe.

Every chapter published in this book has been scrutinized by our experts. Their significance has been extensively debated. The topics covered herein carry significant findings which will fuel the growth of the discipline. They may even be implemented as practical applications or may be referred to as a beginning point for another development. Chapters in this book were first published by InTech; hereby published with permission under the Creative Commons Attribution License or equivalent.

The editorial board has been involved in producing this book since its inception. They have spent rigorous hours researching and exploring the diverse topics which have resulted in the successful publishing of this book. They have passed on their knowledge of decades through this book. To expedite this challenging task, the publisher supported the team at every step. A small team of assistant editors was also appointed to further simplify the editing procedure and attain best results for the readers.

Our editorial team has been hand-picked from every corner of the world. Their multi-ethnicity adds dynamic inputs to the discussions which result in innovative outcomes. These outcomes are then further discussed with the researchers and contributors who give their valuable feedback and opinion regarding the same. The feedback is then collaborated with the researches and they are edited in a comprehensive manner to aid the understanding of the subject.

Apart from the editorial board, the designing team has also invested a significant amount of their time in understanding the subject and creating the most relevant covers. They scrutinized every image to scout for the most suitable representation of the subject and create an appropriate cover for the book.

The publishing team has been involved in this book since its early stages. They were actively engaged in every process, be it collecting the data, connecting with the contributors or procuring relevant information. The team has been an ardent support to the editorial, designing and production team. Their endless efforts to recruit the best for this project, has resulted in the accomplishment of this book. They are a veteran in the field of academics and their pool of knowledge is as vast as their experience in printing. Their expertise and guidance has proved useful at every step. Their uncompromising quality standards have made this book an exceptional effort. Their encouragement from time to time has been an inspiration for everyone.

The publisher and the editorial board hope that this book will prove to be a valuable piece of knowledge for researchers, students, practitioners and scholars across the globe.

# List of Contributors

**Vasily Kushpil**
Nuclear Physics Institute of Academy Science of the Czech Republic, Czech Republic

**I.I. Lee and V.G. Polovinkin**
A.V. Rzhanov Institute of Semiconductor Physics Siberian Branch of Russian Academy of Sciences, Russia

**Mikhail E. Belkin**
Moscow State Technical University of Radio-Engineering, Electronics and Automation, Faculty of Electronics, Joint Research Laboratory, "Microwave and Optoelectronic Devices", Moscow, Russian Federation

**Jonathan Bessette and Elsa Garmire**
Dartmouth College, USA

**Wilfried Uhring**
University of Strasbourg and CNRS, France

**Martin Zlatanski**
ABB Switzerland Ltd., Switzerland

**Lingze Duan and Ravi P. Gollapalli**
The University of Alabama in Huntsville, USA

**Sun-Hyun Youn**
Department of Physics, Chonnam National University, Gwangju, Korea

**Kazuya Ando and Eiji Saitoh**
Institute for Materials Research, Tohoku University, Japan

**Olga Kudryashova, Anatoly Pavlenko, Boris Vorozhtsov, Sergey Titov, Vladimir Arkhipov, Sergey Bondarchuk, Eugeny Maksimenko, Igor Akhmadeev and Eugeny Muravlev**
Institute for Problems of Chemical and Energetic Technologies SB RAS, Russia

**Sanka Gateva and Georgi Todorov**
Institute of Electronics, Bulgarian Academy of Sciences, Bulgaria

**Jaroslav Cvach and CALICE Collaboration**
Institute of Physics of the ASCR, v.v.i., Praha, Czech Republic

Printed in the USA
CPSIA information can be obtained
at www.ICGtesting.com
JSHW011442221024
72173JS00004B/910